The Major Achievements of Science

The Development
of Science
from Ancient Times
to the Present

A. E. E. McKenzie

IOWA STATE UNIVERSITY PRESS ▪ Ames

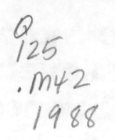

Originally published by Cambridge University Press © 1960 by Cambridge University Press
and reissued as a Touchstone Book by Simon and Schuster, Inc., in 1973.

This edition published 1988 by Iowa State University Press, Ames, Iowa 50010
Text reprinted from the original without correction by permission of Cambridge University Press
All rights reserved

Printed in the United States of America

History of Science and Technology Reprint Series
First printing, 1988
Second printing, 1989

Library of Congress Cataloging-in-Publication Data

McKenzie, A. E. E. (Arthur Edward Ellard)
    The major achievements of science.

    Originally published: Cambridge: Cambridge University Press, 1960.
    Bibliography: p.
    Includes index.
    1. Science—History. 2. Readers—Science. I. Title.
Q125.M42   1988   509   88–12778
ISBN 0–8138–0092–7

CONTENTS

PART I

v

CONTENTS

vi

CONTENTS

CONTENTS

CONTENTS

PART II

Selections from the Literature

ix

CONTENTS

CONTENTS

CONTENTS

CONTENTS

LIST OF PLATES

The limiting of scientific knowledge to a small group of men weakens the philosophical spirit of a nation and leads to its spiritual impoverishment. ALBERT EINSTEIN

...in Science it is when we take some interest in the great discoverers and their lives that it becomes endurable, and only when we begin to trace the development of ideas that it becomes fascinating. JAMES CLERK MAXWELL

It is a capital mistake to theorise before one has data. Insensibly one begins to twist facts to suit theories, instead of theories to suit facts. SHERLOCK HOLMES

PART I

CHAPTER I

SCIENCE AND TECHNOLOGY IN ANCIENT AND MEDIEVAL TIMES

MODERN SCIENCE was established in the seventeenth century and is a product of western Europe. About three hundred and fifty years ago there emerged a new experimental method of investigation and a mechanical, mathematical picture of the universe which has proved to be capable of extensive development and of remarkable control of physical processes.

The birth of modern science is one of the major events in history, comparable in its effects with the technological inventions which made possible the earliest civilisations and dwarfing such events as the fall of the Roman empire to a status of minor importance. Why science should have become established in the seventeenth century rather than earlier, and in Europe rather than, for example, in China, which was technologically more advanced until about the fourteenth century, are therefore questions of considerable interest.

Historians of science are divided on this issue into two main schools. There are those who stress ideas and the innate curiosity of man as the fundamental driving force of history. They see in modern science a continuation of a process of inquiry which has been in operation since the earliest days of mankind, accelerated and transformed in the seventeenth century by the happy occurrence of men of genius and, to some extent, by social and economic forces, but possessing an internal momentum of its own. The materialist view of history, on the other hand, maintained with varying degrees of emphasis by the followers of Karl Marx, is that all history must be accounted for in terms of the modes of economic production and exchange, that is to say in terms of man's needs rather than of his curiosity. Social changes, political revolutions and intellectual movements, according to this view, have their fundamental origin, not in thought, but in economics. The rise of science in the seventeenth century is regarded as the result of the bourgeois revolution, defined as the replacement of feudalism by a capitalist, industrial society, and the coming to power of the mercantile class or bourgeoisie. Those who support this view are concerned to minimise the debt owed by seventeenth-century science to the traditions of the science of ancient and medieval times.

The materialist conception of history, when held in an extreme form, leads to economic determinism and to a denial of free will. But it contains important elements of truth. Science, because of its intimate connections with technology, is especially sensitive to the influence of economic factors. Dr Needham explains the failure of the Chinese to create a scientific movement by the fact that their particular type of feudalism, which persisted until about 200 B.C., was replaced, not by a mercantile capitalism, but by a mandarin bureaucracy. This bureaucracy, recruited by an examination system based on literary studies, was hostile to the merchant class and effectively stifled scientific advance.

It will become apparent when we consider the scientific achievements of the seventeenth century that the internal, autonomous development of ancient and medieval science, and the external influences of economic and social conditions, both played important and possibly indispensable roles in the creation of modern science. In this chapter we shall sketch the history of technology and science prior to the modern period, in order to provide the necessary background and perspective.

*   *   *

The beginnings of technology can be traced to what is known as the palaeolithic or old stone age, when the earliest men made tools of flint, wood and bone, such as axes, knives, needles, spears and bows. The palaeolithic age was succeeded, about 5000 B.C., by the neolithic or new stone age, in which men still used mainly stone for their tools, but turned from hunting to agriculture. A wooden hoe and a wooden sickle with a flint edge were invented, and also pottery for the storage and cooking of cereals. Neolithic man invented textiles to clothe himself, instead of skins, and produced the first primitive machines for spinning thread and weaving cloth.

Some time in the millennium before 3000 B.C., the smelting and casting of metals were discovered. By heating certain types of stone with charcoal, copper was produced and later it was found that the addition of a small quantity of tin to copper gave rise to a harder metal, bronze.

The earliest civilisations arose in the river valleys of the Tigris–Euphrates, the Nile and the Indus. Here were invented the plough, the domestication and harnessing of animals, systems of irrigation, the wheeled cart and the ship. The agricultural workers were able to produce a sufficient surplus of food to maintain the ruling castes of nobles and priests and also smiths, potters and other specialist craftsmen. The virtual cessation about 2500 B.C. of the prodigious technical advances made in the previous millennium has been ascribed to the class structure of these early urban societies; the manual workers were peasants and slaves and they had no incentive or leisure to devise and apply improvements in their crafts.

The Sumerians, whose bronze-age civilisation flourished about 3000 B.C. in the valley of the Tigris–Euphrates, devised a system of writing, consisting of triangular wedges indented in soft clay tablets, known as cuneiform script, and the Egyptians developed a hieroglyphic script, written with ink on papyrus, made from the pith of reeds. Simple methods of calculation, representing the earliest form of arithmetic, were invented, and several geometrical facts useful in surveying were discovered, for example the properties of a right-angled triangle and a value for π. The Babylonians divided the circle into 360 degrees, and it is to them that we owe the fact that there are sixty minutes in an hour, and sixty seconds in a minute.

The clear and glittering skies of Babylonia and Egypt attracted men's attention to the motions of the heavenly bodies and it was in astronomy that empirical knowledge was first systematised so as to make possible the prediction of future events. It is not difficult to understand why astronomy should have been the first branch of science to be developed successfully. The data are points of light in the sky, which are simple and isolated; their periodic motion enables observations to be repeated over and over again. Astronomical observations were used to construct a calendar, necessary for the seasonal operation of seed sowing. The Babylonians achieved a remarkable accuracy and refinement; they estimated the length of the year, which is the time taken by the sun to return to the same position among the stars, to an error of only four and a half minutes, and they knew that

lunar eclipses form sequences recurring at intervals of about
eighteen years.

*   *   *

Although they took the first step in the creation of science by
systematic observation and measurement, the Babylonians and
Egyptians failed to take the second, which is abstraction and
generalisation. The latter was the splendid achievement of
the Greeks.

The Greeks were an iron-age society, the smelting of iron
having been discovered about 1400 B.C. Iron, unlike bronze, was
comparatively cheap because iron ores were abundant, so that
iron weapons and tools could be provided for all. In the bronze
age only princes and nobles could afford metal weapons, and only
a few craftsmen had metal tools. The coming of iron, therefore,
fostered democracy.

The intellectual advance made by the Greeks was partly due to
the democratic nature of their society. The conduct of government by discussion and persuasion stimulated the development
of generalised thinking and philosophy. In Babylon and Egypt
the astronomers were anonymous priests, whose observatories
were the temples, whereas Greek natural philosophers were men
who disputed publicly.

The Greeks made no revolutionary technological discoveries.
Much of the manual work in their society was performed by
slaves and this was responsible for a psychological bias which led
them to exalt the theoretical at the expense of the practical. Their
attitude was expressed by Aristotle: 'But as more arts were invented, and some were directed to the necessities of life, others
to recreation, the inventors of the latter were naturally always
regarded as wiser than the former, because their branches of
knowledge did not aim at utility.' An aristocratic attitude of this
kind towards knowledge is conducive to the development of
philosophy, but not of experimental science or technology, and
the science of the Greeks grew up as part of their philosophy.
They desired to understand the world rather than to control and
change it.

The Greeks were the first to conceive science as a body of
knowledge logically deducible from a limited number of principles.
Their most brilliant and characteristic scientific achievement was

the development of the isolated facts, discovered by the Egyptians and Babylonians in connection with land surveying, into a rigid logical system, summarised by Euclid in his *Elements of Geometry*. Euclid has exercised a profound fascination on many of the creators of modern science. Galileo and Pascal studied him delightedly in youth, despite parental obstruction, and when a friend admitted he could see no value in Euclid, Newton, who seldom smiled, broke into involuntary laughter. The geometry of the Greeks made possible their theoretical astronomy, which we shall discuss in the next chapter.

The conclusions of geometry, granted the axioms, are logically inescapable. The demonstration of what can be achieved by pure reason in geometry had a profound influence on Plato (427–347 B.C.) who, with his pupil Aristotle (384–322 B.C.), marks the summit of Greek philosophy; he is said to have placed a notice to his students at the entrance to the Academy, 'Let no one unacquainted with geometry enter here'.

We need not concern ourselves with Plato's ideas of science. His main influence occurred at the Renaissance and lay in his conviction that the design of the universe is essentially mathematical. But it is quite otherwise with Aristotle. To the men of the Middle Ages Aristotle loomed through the mists of antiquity colossal and omniscient. His works dominated scientific thought two thousand years after they had been written. They were encyclopaedic in character and embodied all the known science of his time.

Aristotle's science was a logical system in which the truths of nature were exhibited as deductions from ultimate, universal principles. These ultimate principles, such as that the heavenly bodies must move in circles because circular motion is the perfect type of motion, were perceived intuitively as correct, and, like the axioms of geometry, were fundamentally indemonstrable.

Aristotle was primarily a biologist and was constantly faced with the problem of growth. He believed that everything in nature has an aspiration or tendency to achieve an ideal state or form and that every natural process has a last stage, the form of which was present in the first. The most obvious example is a seed; in the acorn is the potential form of the oak. By analogy, he ascribed the behaviour of inanimate matter not, as we do today, to some system of mechanical relations, but to an innate

5

striving or 'occult qualities'. Thus a body falls, not because of a force, but because it is seeking its natural place. This kind of explanation, in terms of purpose or ultimate end, is called teleological and it has the disadvantage that it is incapable of further development.

Although Aristotle was one of the world's greatest biologists— Darwin said that the modern biologists, Linnaeus and Cuvier, were 'mere schoolboys to old Aristotle'—this part of his work was ignored and it was his spurious physics and astronomy that chiefly influenced his successors.

There was one school of thought among the Ionian Greeks, led by Leucippus and Democritus (*c.* 420 B.C.), which had a striking prevision of the modern scientific universe, though it was based like many Greek speculations on negligible observational foundations. Everything was considered to be made of atoms in constant motion in a void, and all natural processes were regarded as due to their chance interactions, a completely mechanistic, non-purposive system. Since Democritus went further, maintaining that thought was a physical process, and that the soul was composed of atoms, he was rejected by Plato and Aristotle, and was later assigned by Dante to a very low place in hell.

* * *

Aristotle was, for a brief time, the tutor of the youthful Alexander the Great, whose empire was divided among his generals. To the enlightened soldier Ptolemy fell the kingdom of Egypt, and Aristotle's library was transferred from Athens to Alexandria, where a new university was founded. There, during the next two centuries, the science of the ancient world reached its high-water mark and was preserved and systematised for the further five hundred years of what is known as the hellenistic period, until the destruction of the library in the sixth century.

The greatest natural philosopher educated at the university of Alexandria was Archimedes (287–212 B.C.). He discovered the law of the lever and the principles of hydrostatics; he invented the endless screw for raising water and constructed engines of war. During the hellenistic period Euclid (330–260 B.C.) wrote his textbook of geometry; Eratosthenes (*c.* 284–192 B.C.), a geographer, made an accurate determination of the diameter of the earth by measuring the elevation of the sun at midday at two places a known distance apart, and did much to improve the maps of his day; Hipparchus (2nd century B.C.) compiled a catalogue of over a thousand stars and introduced the co-ordinates of latitude and longitude for fixing a place on the earth's surface; Ptolemy (c. A.D. 90–168) wrote the *Almagest*, The Great System, summarising Greek astronomical theory; Galen (c. A.D. 130–200), physician to the emperor Marcus Aurelius, wrote books on anatomy and physiology which were to become the authoritative treatises of the Middle Ages.

It is tempting to speculate on the reasons why the modern scientific movement failed to develop in the hellenistic period. Rome gave peace and stability to the whole of the Mediterranean world, culminating in the golden age of prosperity in the second century A.D., with which Gibbon begins his history of the decline and fall. A widespread empire should have afforded the opportunity and stimulus for the development of natural resources and for technological advances. Hellenistic science, by this time, had reached a stage comparable with that in the early sixteenth century. Although men like Archimedes were deeply imbued with the aristocratic Greek attitude to knowledge, and affected to despise their own technical inventions, there was at Alexandria a school of engineering and a mild interest in technology. The modern scientific revolution came about, as we shall see, largely through the merging and mutual stimulation of the separate traditions of scholarly, speculative science and of experimental craftsmanship and technology.

It was technology which was the weakness of the ancient world. The watermill, destined to become the chief prime mover in the industrial and technological developments of the later Middle Ages, was actually invented in the Roman empire in the first century B.C. It requires, however, permanent streams with a rapid fall, and they are common in north-west Europe, but rare in the countries bordering the Mediterranean, which formed the most populous parts of the Roman empire. Moreover the use of watermills in the hellenistic world was uneconomic because slaves were cheaper. As no industrial system was built up there was no impulse for technological experiment and hence no general, momentum-gathering scientific advance.

Three other important factors deserve mention. After A.D. 100 the intellectual climate changed and cosmopolitan Alexandria became less a centre of science and technology than of religious disputations in which Christianity struggled with its pagan rivals. Secondly, printing had not been invented and paper-making was unknown, with the result that knowledge was not widely disseminated. Thirdly, the educational system of the Romans, with its emphasis on oratory, did not foster scientific interest and the Romans failed to produce a single scientific genius.

*   *   *

After the fall of the Roman empire, the heritage of Greek science was preserved and extended by the Arabs.[1] In the seventh century they overran Asia Minor, Persia, North Africa and Spain, and from the ninth to the eleventh centuries the civilisation of Islam far eclipsed that of Christendom. The streets of Cordoba were lit by lamps and paved, amenities not to be found in London and Paris for another seven centuries. Nor could any building in the west match the refinement and beauty of the Alhambra, the Moorish palace of Granada, with its cisterns and running water, its slender columns and ornamented arches.

How much the Arabs added to Alexandrian science we do not know, since a great deal of our knowledge of it has come through them. They were particularly active in the fields of alchemy and medicine. It is probable that their chief faculty lay in absorbing ideas rather than in creating them. Under the direction of cultured caliphs in Baghdad they translated the works of Greek philosophy and science into Arabic. They learnt also from Persia, India and China. Along the caravan routes through Persia from Tashkent and Samarkand there passed a steady stream of traders and of pilgrims on their way to the house of the Prophet in Mecca. Arab merchants sailed to the Indian Ocean, and through the Malacca Straits to Canton. They brought back from China the art of paper-making, and from India 'arabic' numerals. The latter, by the employment of position to denote tens, hundreds and thousands, and a symbol for zero, enormously simplified mathematical manipulation. This is obvious if one tries to multiply or divide two numbers written in Roman numerals, for example MDCCCXXIV and DCLXVII. Algebra, also learnt from the Hindus, was the subject of a book by the poet Omar Khayyam.

---

[1] The word Arabs is used to denote all Arabic-speaking, Islamic peoples and is not confined to natives of Arabia.

Perhaps the limited extent of the scientific contributions of the Arabs can be explained by the authoritarian nature of Moslem society; progress depended too much on the personalities of the caliphs.

The transmission of Arabic learning to Christendom took place mainly through Spain and, to a lesser extent, through Sicily. Many Greek works were translated from Arabic into Latin. The commentaries on Aristotle of the Islamic philosopher Averroes (1125–98) were eagerly studied by the schoolmen of medieval Europe. In the modern world the east takes its science from the west, being influenced only superficially by western religion and philosophy. There was a similar but reversed traffic in the twelfth century; the west sought the learning and technology of the Muslims while rejecting their religion.

* * *

The Christian and Muslim worlds were in sufficiently close contact for translations of works from Arabic into Latin to have been made more than two hundred years earlier than actually occurred. It was not until the end of the eleventh century that Christian scholars showed any strong interest in secular culture. By this time the Dark Ages, the five hundred years of chaos which followed the collapse of the Roman empire, were coming to an end; there was a measure of stability, trade was moving and life was becoming easier. The invention of the heavy wheeled plough and a method of harnessing the horse, whereby the pull was taken on the shoulders instead of on the windpipe, helped to raise the material standard of life of the common people to a level almost as high as that enjoyed at any time in the ancient world.

During the Dark Ages learning was preserved in the monasteries and their associated schools. Most of the works of Greek science had been lost. Thought was almost wholly theological and its tone was set by St Augustine (354–430), who believed that God created nature for the use of man and that all natural processes had an ultimate spiritual purpose. The universe was explained in terms of symbolism and allegory. This is most strikingly illustrated by the bestiaries which continued to be written throughout the Middle Ages. They were a mixture of natural history and myth based on the work of Physiologus, an anonymous writer of Alexandria in the second century. They described, for example, the albatross feeding its young on its blood as a symbol of the eucharist; the lion brought forth dead and coming to life again as a symbol of the resurrection; and the unicorn subdued by a pure maiden and leaping into her lap as a symbol of the incarnation. St Augustine said it mattered less whether an animal existed than what it symbolised.

The stars were regarded as primarily intended to foretell and to guide human events. Each sign of the zodiac governed some region of the body and each planet a bodily organ, the macrocosm of the universe reflecting the microcosm man. Astrology was not, of course, peculiar to medieval Europe. It had been practised from very early times and it survived, among the educated, until the seventeenth century. Galileo, as the most famous astronomer in Europe, was persuaded by the Grand Duchess of Tuscany to correct the horoscope of her sick husband. Galileo forecast that there were many happy years of life before the duke, who died three weeks later. Such unfortunate mistakes were ascribed to the inaccuracy of the observations rather than to the invalidity of the theory.

Between about 1100 and 1350 a succession of keen intellects, the schoolmen, were examining and disputing fundamental problems of philosophy, within the framework of orthodox Christianity. These scholars grouped themselves together to form universities, Paris in 1160 and Bologna about the same time, Oxford in 1167, Cambridge in 1209 and Padua in 1222, to name only the most outstanding. Men could move about freely from one university to another since they had a common language, Latin, and they all belonged to one unifying organisation, the Roman Catholic Church.

The effect on the schoolmen of the recovery from the Arabs of the works of Greek philosophy, particularly of Aristotle, was profound. Here was an intellectual system beyond their previous experience. Aristotle's logic, his theory and method of science were discovered in the *Analytics* in the first half of the twelfth century and his basic concepts of science in the *Physics* in the second half. The schoolmen, in their discovery of Aristotle, were like men who came upon an ancient and magnificent mansion. Some of them set about converting it, despite its classical style, into part of a gothic cathedral. But others, filled with curiosity as to how it could be supported, dug and burrowed, often using tools which they found in the mansion itself, and so undermined it during the course of several hundred years that, when the wind of the seventeenth century blew, the whole edifice collapsed. Upon the ruins were laid the foundations of modern science.

The Dominican friar, St Thomas Aquinas (1227–74) made a subtle and elaborate synthesis of Christian theology and Aristotelian science, rejecting such elements of the latter as could not be made to conform with the former. Fundamentally there was no great incompatibility between the two systems. Both assumed that everything in nature had a purpose; Aristotle's explanation was in terms of an immanent teleology and that of Christianity was in terms of the will of a personal God. The scholastic philosophy of Aquinas became orthodoxy in the later Middle Ages, and is still orthodoxy for Roman Catholics today.

There were two main critical movements which applied dialectical arguments to Aristotle's scientific conceptions and gradually discredited them. One of these movements was begun in Oxford by the Franciscan friars, John Duns Scotus (*c.* 1266–1308) and William of Ockham (*c.* 1300–49). Aristotle's ultimate principles were criticised and they were distinguished from empirical generalisations. Ockham, famous for his intellectual razor for shearing away unnecessary entities, attacked Aristotle's theory of motion and replaced it by the doctrine of impetus, which was developed by Jean Buridan in Paris.

The other main critical movement was that of the Latin Averroists, beginning in Paris in the thirteenth century. Following the Arab commentators they stressed the determinism of Aristotle's philosophy, maintaining that all action in the universe is the result of a chain of necessary causes with the implication that free will is not possessed by men or even by God. The condemnation of determinism by the Archbishop of Paris in 1277, and in the same year by the Archbishop of Canterbury, was regarded by the French historian Duhem as marking the beginning of modern science.

The school of Averroists moved from Paris to Padua. From 1404, Padua was under the protection of the republic of Venice, and became anti-clerical and anti-papal, so that Aristotle could be discussed independently of any theological framework. Aristotle's method of explaining empirical facts by deducing them from general principles, obtained intuitively and hence beyond argument, was gradually modified into what was essentially the modern scientific method: namely, to begin from empirical observations, to derive from them some hypothesis of their fundamental causes, and then to test the hypothesis by finding whether the empirical observations could be deduced from it.

The schoolmen were logicians and philosophers, with a primary interest in religion rather than in science. Because of their habitual acceptance of authority they had no expectation of any new and revolutionary discovery but only of a refinement and modification of existing knowledge. They looked back to Greece as to a golden age and were content to study and criticise the science of the ancient writers rather than to search for new truth in nature itself. The story is told that, even as late as the second decade of the seventeenth century, when Father Scheiner told his Provincial that he had seen spots on the sun, he received the reply: 'You are mistaken, my son. I have studied Aristotle and he nowhere mentions spots. Try changing your spectacles.'[1]

*   *   *

Among the most important influences in the later Middle Ages, turning men's interest to nature and providing an incentive for scientific investigation, were the remarkable technical innovations. By the thirteenth century the technology of western Europe was probably superior to that of any earlier civilisation. The sixteenth century saw the publication of two impressive treatises on mining and metallurgy, *De Re Metallica* by Agricola and *Pirotechnia* by the Italian Biringuccio.

Waterwheels were in extensive use during the Dark Ages for grinding corn; Domesday Book records that in 1086 there were five thousand mills in England. Water power was gradually extended to a variety of trades: to beating cloth to shrink it and make it more durable, to sawmills, to pumping and winding in mines, to breaking up ores, to blowing the bellows of blast furnaces, to iron rolling and wire drawing. Heavy machinery led to the growth of a primitive factory system. A new craft, that of the millwright, grew up to service the mills and furnished the skilled workmen who constructed the first mechanical clocks. New industries, for the manufacture of gunpowder, glass and paper, were introduced. Windmills were built and devices were

[1] L. T. More, *Isaac Newton.*

Fig. 1. A medieval stamp-mill for crushing ore, operated by water power. Each stamp, *D*, has an iron head, *E*, and a tappet, *G*. The tappet is engaged by a cam, *H*, on the shaft driven by the water wheel and in this way the stamp is raised. When the cam disengages from the tappet the stamp falls under its own weight to crush the ore. The illustration is taken from Agricola's *De Re Metallica*, published in Latin in 1556 and one of the finest and most superbly illustrated treatises on technology ever written. Agricola was a native of Saxony and his real name was Georg Bauer. At this time Germany led the world in mining and metallurgy. (Courtesy: *The Mining Magazine*.)

invented for turning the sails to catch the wind as the direction of the latter changed.

During the thirteenth century there was a brief heralding of experimental science. The Franciscan Roger Bacon (1214–94) experimented with lenses and gunpowder. He wrote, 'There are two methods of investigation, through argument and through experiment. Argument does not suffice, but experiment does.' At the age of 64, accused of 'suspected novelties', he was put under the surveillance of his order. About 1269 Petrus Peregrinus (the pen name of the Marquis de Maricourt) published a small but remarkable book on his researches in magnetism.

The change of interest from religious contemplation to the external world of sense is illustrated by the painting of the Italian Renaissance. Giotto (1266–1337) was the first to break away from the stylised, religious, Byzantine tradition and he was followed by a host of painters, many of whom made scientific studies of natural history, human anatomy and perspective. The most remarkable of the artist-scientists was Leonardo da Vinci (1452–1519). He dissected bodies and his anatomical knowledge was far in advance of that of the medical men of his time; he analysed the flight of birds and invented a model flying machine; he put forward suggestions for a helicopter and parachute; he was employed as a military and civil engineer and during his construction of canals made observations of fossils; he left drawings of quick-firing and breech-loading guns. Unfortunately he published nothing; his notes, often ill-phrased and incoherent, were written left-handed in a reversed script that requires a mirror for reading it.

The Renaissance spread into the rest of Europe through the invention of printing with movable type by John Gutenberg of Mainz in 1440. Caxton set up his famous press in Westminster in 1476 and it is estimated that, by the end of the century, some nine million printed books were circulated in Europe in place of a few score thousand manuscripts.

Improvements in navigation, by the invention of the magnetic compass and of the sternpost rudder, led to an expansion of sea-going trade and to the voyages of discovery. The new rudder, replacing what was virtually an oar lashed to the stern, enabled larger ships to steer closer to the wind.

The revival of trade began in Italy, being stimulated by the opening up of the Middle East by the Crusades, and the ports of Venice and Genoa became great commercial centres. In an attempt to break the Italian monopoly of trade with the east, the Portuguese, under the leadership of Prince Henry the Navigator (1415–61), began to push farther and farther down the west coast of Africa, searching for a sea route to India, and in 1486

Bartholomew Diaz rounded the Cape of Good Hope. In 1492 Christopher Columbus, the Genoese, financed by Ferdinand and Isabella of Spain, sailed across the Atlantic to the Bahamas, which he believed to be India. In 1519 Magellan set off with five ships, one of which returned after three years, having, for the first time in history, circumnavigated the globe.

The voyages of exploration had profound effects, not least on men's minds. They resulted in a shift of European commerce from the Mediterranean to the Atlantic and ruined Venice and Genoa. Italy, which in the fifteenth and sixteenth centuries was the leader of European civilisation, a place of pilgrimage for scholars, declined and sank into decadence. A further important effect was a financial one. The gold and silver of Mexico and Peru, brought across the Atlantic in the galleons of Spain, made money plentiful and gave rise to inflation. This stimulated industrial production, increased the wealth and power of the mercantile classes and impoverished the landed successors of the medieval aristocracy.

Modern science was founded by men of the late Renaissance. It began in Italy and moved to France, Holland and England. Germany, a leader in mining and metallurgy during the fifteenth and sixteenth centuries, was devastated by the Thirty Years War (1618–48), and made but a small contribution. Spain, the richest and most powerful State in the sixteenth century, inherited no strong mercantile middle class from the Middle Ages, and was cursed with the Inquisition; its scientific contribution was negligible.

CHAPTER 2

THE COPERNICAN THEORY

THE MODERN SCIENTIFIC movement first began to gather momentum in the most highly developed of the sciences, astronomy, and the man who provided the chief initial impulse was the Pole, Nicolaus Copernicus (1473–1543).

A revival of observational astronomy to meet the needs of seamen occurred in the fifteenth century. Prince Henry of Portugal set up an astronomical observatory and a navigation research institute on Cape St Vincent about 1420. The most important problem, not effectively solved until the invention of an accurate chronometer in the eighteenth century, was the determination of longitude at sea. During the voyage back from the discovery of the New World Columbus disputed with his lieutenant whether they were approaching Madeira or the Azores, a difference in longitude of 600 miles.

Longitude is found by a comparison of local time, as determined from the sun, and standard time, which is the local time at a specified place. If, for example, local time differs by two hours from standard time, the difference of longitude is one-twelfth of the way round the world, or 30°. Much astronomical observation was devoted to measuring the track of the moon among the fixed stars in the hope that the moon might be used to indicate standard time.

Another stimulus to observational astronomy in the fifteenth century was the need to reform the calendar. When the Julian calendar had been introduced in 46 B.C. it was assumed that a year was 365¼ days long, whereas it is 11 minutes 14 seconds shorter than this. The discrepancy accumulated over the centuries, and by the fifteenth century the spring equinox occurred more than a week too early. This resulted in the Church's festivals being held at improper times, and in some difficulty and confusion for astrologers because people were being born on the wrong dates.

A flourishing school of astronomy existed at Cracow, and it

PLATE I. Bust of Aristotle, by an unknown artist after a portrait in the fourth century B.C., from the Louvre.

PLATE II. Galileo, from the frontispiece to his book *Siderius Nuncius* (The Sidereal Messenger), published in 1610.

was there that Copernicus began his university studies. He was a native of Torun, a thriving port on the Vistula. After four years at Cracow he went to Bologna, where he studied law, besides mathematics and astronomy, and learned Greek, so that he was able to read the works of the Greek astronomers in the original. Later, with true Renaissance versatility, he studied medicine at Padua, and spent altogether ten years in Italy. He was a contemporary of Leonardo da Vinci, of Michelangelo and of Machiavelli.

The fundamental ideas of his theory probably germinated in his mind while he was at Cracow, and they were stimulated at Bologna, during discussions of the Pythagorean and Platonic ideas of the mathematical simplicity of the universe. He brooded over and developed these ideas during the last thirty years of his life, spent as canon of Frauenburg, a post secured for him by his powerful uncle, the bishop of Ermland. The cathedral of Frauenburg stands on a slight hill in sight of the Baltic and near to it is a tower from which Copernicus made his astronomical observations. He regretted that he had not the same excellent opportunities for observation as Ptolemy, the man whose theory he supplanted; 'the Nile does not breathe fogs as does our Vistula', he wrote. He was, however, primarily a theorist rather than an observer. He delayed the publication of his book *De Revolutionibus Orbium Coelestium* for many years until he was dying. He feared the controversy it was likely to provoke and wrote in the preface that he would be 'hissed off the stage'.

*   *   *

The astronomical theory and many of the observations which Copernicus inherited were the work of the Greeks. The Greeks conceived the idea that the sky is a revolving sphere studded with the stars like jewels—a natural and straightforward deduction from the apparent celestial motions. Stars near to the pole star, by their slow change in position as the night wears on, appear to be describing circles in the sky; others, farther away, rise on the eastern horizon, sweep through an arc, and set in the west. The axis of rotation of the sphere of the sky appears to pass through a point near to the pole star and its period of rotation is twenty-four hours.

The sun rises and sets with different stars in different seasons of the year. If the stars were visible in the daytime the sun would be seen to slip back slowly eastward among them, about twice its breadth in one day. The path of the sun among the stars, called the ecliptic, is marked by the constellations of the Zodiac, Aries, Taurus, Gemini, Cancer, Leo, Virgo, Libra, Scorpius, Sagittarius, Capricornus, Aquarius and Pisces, which are the monthly milestones in its annual circuit. The constellations are arbitrary groupings of stars, and were mapped out by Babylonian astronomers for convenience in identification about 2700 B.C.

The sun rises and sets at different points on the eastern and western horizons during the year and it is higher in the sky at midday during the summer than during the winter. About 21 March and 21 September (the equinoxes) it rises due east and sets due west; about 21 June it rises and sets at its furthest points to the north-east and north-west, while about 21 December it reaches its limits on the south-eastern and south-western horizons. These facts can be explained by assuming that the ecliptic, or path of the sun, is inclined at an angle ($23\frac{1}{2}°$) to the celestial equator, which is the equator of the apparently revolving dome of the sky.

The simple and satisfactory explanation of the motions of the stars in terms of a uniformly revolving sphere was elaborated to account for the motion of the sun. It was assumed that the sun was carried round daily by the sphere of the stars, but that it also partook of the motion of another sphere revolving once per year in the opposite direction, with its axis inclined at about $23\frac{1}{2}°$ to the axis of the first sphere.

The motion of the moon, which travels among the stars much more quickly than the sun, completing a circuit in $27\frac{1}{3}$ days, was explained in a similar manner.

The stars revolve together, without changing appreciably their relative positions, but certain heavenly bodies, called the planets, move among them, sometimes forward, sometimes backward (Fig. 2), and exhibit a considerable variation in brightness. Five planets are visible to the naked eye and were known to the ancients, Mercury, Venus, Mars, Jupiter and Saturn. Their paths are roughly confined to the ecliptic, and their speeds differ considerably, Mercury taking 80 days, Venus 9 months, Mars 2 years, Jupiter 12 years, and Saturn 30 years, to return to their original positions among the stars. Mercury and Venus are always quite near to the sun, appearing now on one side of him as morning stars and now on the other as evening stars.

The explanation of the paths of the planets in terms of uniformly revolving spheres is a more difficult problem than that of the sun and moon. Eudoxus (*c.* 408–355 B.C.), a pupil of Plato and one of the greatest mathematicians of antiquity, produced the first solution. Each planet was regarded as partaking of the motion of four spheres, being fixed to the equator of the innermost

Fig. 2. Path of Mars through the stars.

sphere, whose poles were fixed in the surface of the next sphere and so on, all four spheres being allotted suitable speeds of rotation and having their axes suitably inclined. As observations became more accurate the theory proved adequate for Jupiter and Saturn, but unsatisfactory for Mercury, Venus and Mars. Callipus therefore added a fifth sphere for each of the three last-named planets. He also required five spheres each for the sun and moon to account for their uneven speeds and other irregularities, bringing the total for the whole heavens to thirty-four spheres.

At this point Aristotle took the retrograde step of converting a mathematical device into a physical fact. The spheres became real and were thought to be crystalline because invisible. The conception of crystalline spheres ceaselessly revolving in the heavens and causing a far-off celestial music has had a special appeal for the poets:

> There's not the smallest orb which thou behold'st
> But in its motion like an angel sings.

Because he believed that the spheres were real, Aristotle thought that each set of spheres would communicate motion to each other and hence he introduced twenty-two additional spheres to allow for this effect.

At the centre of Aristotle's universe was a spherical, motionless earth, and surrounding it were fifty-five revolving crystalline spheres, carrying the sun, the moon, the planets and the stars. The outermost sphere, the primum mobile or 'first moved', was kept in motion by God: beyond it there was nothing.

The innermost sphere, that of the moon, divided the universe into two distinct parts. Inside it everything was made of the four elements, earth, air, fire and water, and was subject to change and decay. Outside it the heavenly bodies, including the moon, were made of a fifth element, changeless and incorruptible—a sublimation perhaps of the early Greek worship of the sun and moon as gods. Celestial matter moved serenely and eternally with uniform motion in a circle; terrestrial matter tended to move finitely in straight lines, seeking its natural place.

Aristotle's cosmological theory was accepted by the scholastic philosophers throughout the Middle Ages and the Renaissance, but it could not be used for accurate prediction of the positions of the planets. Between the thirteenth and fifteenth centuries it was replaced, for practical astronomers and navigators, by the system of the Alexandrian astronomer, Ptolemy (c. A.D. 140).

Ptolemy's *Almagest*, like Euclid's *Elements of Geometry*, was a text-book; it was a summary and elaboration of the work of others rather than an extension of knowledge. Its observational data were largely the work of Hipparchus (2nd century B.C.) also of Alexandria.

In the system of Ptolemy the spheres of Eudoxus and Aristotle gave place to two new devices, epicycles and eccentrics, which possessed the merit of accounting for the variation in distance of the planets from the earth, and hence their variation in brightness. In Fig. 3 (*a*) the planet P revolves uniformly in a small circle, called an epicycle, whose centre C revolves uniformly in a larger circle called a deferent, with centre the earth E. In this way P, as seen from E, may possess alternately a forward and a retrograde motion and its distance from E varies. In Fig. 3 (*b*) the planet P revolves uniformly in a circle of centre C, the earth E being situated eccentrically, while C describes a circle about E. Any eccentric system can be transformed into an epicyclic system and *vice versa*.

At the centre of Ptolemy's universe, as in Aristotle's, was the stationary globe of the earth and round it revolved the firmament. The tracks of the sun, the moon and the planets were explained by a system of some eighty epicycles. The system was the culmination of over half a millennium of Greek astronomy, its complexity being due to the fact that it accounted for every known irregularity.

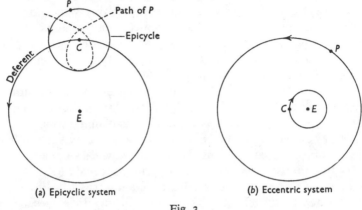

(a) Epicyclic system   (b) Eccentric system

Fig. 3

When Alphonso the Wise of Castile was introduced to the Ptolemaic system he remarked: 'If the Almighty had consulted me before the Creation I should have recommended something simpler.'

*   *   *

It had been realised long before the time of Copernicus that the revolution of the universe about the earth could equally be explained by assuming the universe to be at rest and the earth to be rotating in the opposite direction. The idea was considered and debated at Alexandria by the Arabs, and also in the fourteenth century, notably by Nicole Oresme (c. 1323–82). Oresme pointed out that, if the earth rotated, the apparent motion of the sky would be indistinguishable from a real motion. Astrology would not be falsified; eclipses, conjunctions, and oppositions would remain unaffected. Nor would the words of Scripture be contravened if apparent motion were accepted as equivalent to real. There were, however, some scriptural difficulties: for example, the sun is described as exulting like a strong man to run his course, and he could hardly do that if he were really standing still and the earth were rotating.

The chief scientific difficulties were raised by a faulty theory of dynamics. It was thought that, if the earth rotated on its axis, it would fly to pieces under the stress of centrifugal force. It was also objected that the atmosphere would be left behind and that there would be a continuous, violent wind sweeping clouds and birds away to the westward. Nevertheless this was scarcely a more difficult problem than that presented by a rotating universe. For it was realised that the stars are at a vast distance from the earth and, if the universe revolves completely every twenty-four hours, their speeds must be stupendous.

The theory that the rotating earth revolves round the sun, in company with the planets, was held by Aristarchus of Samos (born c. 310 B.C.), and it was capable of explaining much of the apparent complexity of the motion of the planets as a result of the motion of the earth. This theory was ignored or rejected for nearly two thousand years until Copernicus revived it.

Copernicus adopted a critical attitude towards the Ptolemaic theory when he found that the motions in the circles describing a planet's orbit were not uniform. Ptolemy had been compelled to resort to a device called the equant, which was a point away from the centre of a circle about which the uneven motion in the circle appeared to be uniform. Copernicus said that this was 'neither sufficiently absolute nor sufficiently pleasing to the mind'.

Copernicus came to the conclusion that, by assuming the earth to rotate on its axis and also to revolve with the planets round the sun, he could greatly simplify the Ptolemaic system. He wrote out a brief sketch of his theory in the *Commentariolus*, in which he claimed that he would be able to reduce Ptolemy's eighty circles to thirty-four.

The way in which his theory simplified the explanation of the forward and retrograde motions of the planets can be seen from Fig. 4. Suppose the circles represent the orbits of the earth and Mars about the sun *S*, and imagine

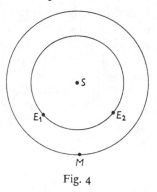

Fig. 4

Mars *M*, to be at rest. As the earth travels along the minor arc

from E_1 to E_2 Mars appears to an observer on the earth to move backwards; as the earth travels along the major arc from E_2 to E_1 Mars appears to move forwards. Suppose now that Mars is revolving also round the sun, but at a rate slower than that of the earth. Mars will appear to have a steady revolution, with a forward and retrograde motion superimposed, as it appears to have in the heavens.

Copernicus was particularly impressed with the variations in brightness of Mars, which he himself observed for many years. It can be seen from Fig. 4 that, if the earth and Mars revolve round the sun with different angular speeds, they must at times be comparatively near to each other and at other times be comparatively distant, which would explain the variations in brightness much more adequately than Ptolemy's epicycles.

The decisive objection to a heliocentric theory had always been that, as the earth moves along its orbit in space, the stars in the region from which it is moving should appear to close up slightly, and those towards which it is moving should slightly open out. If the stars were at different distances from the earth they might even be expected to change their relative positions as do objects to a traveller in a railway train. Copernicus could meet this objection only by assuming that the stars were at such vast distances from the earth that their apparent changes in position were inappreciable.

Copernicus retained uniform motions in a circle and his outlook was more medieval than modern. He obtained his system by taking the clockwork of the Ptolemaic system to pieces and putting it together again in a different way. His final system, as published in *De Revolutionibus,* was in fact more complicated than the Ptolemaic system. In the end he was obliged, partly because of his inaccurate data, to use forty-eight circles whereas the Ptolemaic system, brought up to date in the fifteenth century, required only forty. Another disturbing feature of the Copernican system was that its centre was not the sun, but a point in space near to the sun, the centre of the earth's orbit.

Copernicus spent many years of laborious and intricate calculations in producing tables to enable the positions of the sun, the moon and the planets at any instant to be calculated. His tables were scarcely more accurate than those of Ptolemy. Indeed, it required the scientific insight of men of genius like Galileo and

Kepler to realise the fundamental superiority of the Copernican system over the Ptolemaic. They sensed that the revolutions of the earth and of the planets were somehow caused by the sun.

* * *

The Copernican theory was established as physical fact by Galileo Galilei (1564–1642). Galileo devoted most of his life to persuading his contemporaries of its essential truth, and, since his chief concern was to show that it was consistent with his new concepts of motion, the main target of his attack was Aristotle, rather than Ptolemy.

The universities were strongholds of Aristotelianism. The majority of scholars accepted the authority of Aristotle as though it were holy writ and were described by Galileo, who had no gift for suffering fools gladly, as 'those silly Gulls, which too scrupulously go about to defend whatever he hath said'.

'Such people', he wrote, 'remind me of that sculptor who, having transformed a huge block of marble into the image of a Hercules or a thundering Jove, I forget which, and having with consummate art made it so lifelike and fierce that it moved everyone with terror who beheld it, he himself began to be afraid, though all its vivacity and power were the work of his own hands; and his terror was such that he no longer dared affront it with his mallet and chisel.'[1]

Galileo was not trained as an astronomer; he was a mathematician and physicist. But in 1609 fortune placed within his grasp an instrument with which he revolutionised astronomy. A report reached him that someone in Flanders had presented to Count Maurice of Nassau a device, consisting of two lenses, which made distant objects appear as though they were much nearer. He reflected on this and soon had produced a telescope of his own. Its utility was immediately appreciated when he demonstrated its powers to the Venetian Senators; from the highest bell-towers of Venice they peered through it at ships far out at sea. A telescope was presented to the Doge, and Galileo's salary, as professor of mathematics at Padua, was doubled.

It was not long before Galileo realised the telescope's potentiali-

[1] From Galileo's *Dialogue concerning the Two Chief World Systems—Ptolemaic and Copernican*, translated by Stillman Drake.

ties in the study of astronomy. On turning it to the night sky he was amazed at the multitude of new stars which it made visible. He saw at once the nature of the Milky Way: 'the Galaxy is nothing else but a mass of innumerable stars planted together in clusters.' He looked at the moon and saw that its surface was not smooth and polished but like that of the earth, 'everywhere full of vast protuberances, deep chasms and sinuosities'. He noticed spots on the sun. He discovered that Jupiter had revolving round him four moons, which he termed the Medicean stars in honour of the Grand Duke he desired as his patron; this was clearly a Copernican system in miniature.

The fame of the telescope spread throughout Europe and Galileo was pestered by princes and savants for an instrument. He made over a hundred with his own hands. There was great excitement when a telescope arrived at the French court, and a letter was dispatched to Galileo, 'Pray discover as soon as possible some heavenly body to which his Majesty's name may be fitly attached'. ·

The new astronomical discoveries placed the disciples of Aristotle in something of a quandary because their master had laid down that the heavens were complete and unchangeable. Some refused to accept the existence of the moons of Jupiter, which cut clean through Jupiter's crystalline sphere, alleging that their origin lay inside the telescope and complaining that the notion of four little planets chasing each other round another planet was, after all, a little too absurd. When one of these professors died, Galileo remarked, 'Libri did not choose to see my celestial trifles while he was on earth; perhaps he will now he is gone to heaven'. To another who suggested that the apparent valleys on the moon were filled with an invisible crystalline substance (thereby preserving a smooth, Aristotelian surface) Galileo retorted that the idea was excellent and he would like to propose that there were mountains made of this invisible substance ten times higher than those he had observed.

Such congenial controversy, however, was to become dangerous in the future because in 1610, wishing to be relieved of his teaching duties in order to devote all his time to his investigations, he left the free territory of Venice, and returned to his native Tuscany under the patronage of the Grand Duke.

Here in Florence he discovered the phases of Venus, thereby

removing one of the long-standing objections to the Copernican theory that Mercury and Venus did not show phases like the moon. Milton refers to this discovery in Paradise Lost:

> The morning planet gilds her horns.

In 1613 Galileo published his *Letters on the Solar Spots* in which he supported the Copernican theory, and two years later a *Letter Concerning the Use of Biblical Quotations in Matters of Science*. The controversy on the incompatibility of the theory of a moving earth with texts in the Psalms and Ecclesiastes, such as 'He hath made the round world so sure that it cannot be moved', was an old one and Galileo was advised by his friends to keep out of it. But he was irrepressible. When he heard that he was likely to be in trouble with the Inquisition, he hurried to Rome, hoping to convert his opponents.

The Pope, Paul V, called for an inquiry and was advised by Cardinal Bellarmine, the head of the Roman College of Jesuits and a reasonable man, that the Copernican theory was contrary to the Scriptures. Galileo was therefore admonished and, to his intense disappointment, instructed to cease holding and teaching the theory.

In 1618 three comets appeared and, as they seemed to cut through the crystalline spheres, they gave rise to considerable controversy. Galileo was inevitably drawn in, and his final reply was *Il Saggiatore* (The Assayer), published in 1623. In this book he could not resist the ironical remark that the motion of the earth, which he, as a dutiful son of the Church, must hold to be false, explains so well such a variety of astronomical observations that it might, just as deludingly, help to account for the phenomena of comets.

The accession of a new Pope (Urban VIII), who had long been a friend and protector of Galileo, prompted Galileo to seek approval for a book on the tides. He believed that he had a mechanical explanation of the tides, which the astrologers had always used as an obvious example of celestial influences; this involved the rotation and revolution of the earth, and so it would be a physical proof of the Copernican system. The Pope allowed him to proceed, but only on condition that he presented convincing arguments for the impossibility of a physical proof of any cosmological system. After many years of preparation and delay, the *Dialogue concerning the Two Chief World Systems—Ptolemaic and Copernican* appeared in 1632. The book is very different from a modern scientific work, being a dialogue between a Copernican, an Aristotelian, and an impartial commentator, similar in form to the works of Plato, whom Galileo admired, and enabling him to avoid open commitment to any particular views. It is witty polemical writing, setting forth the arguments for and against the two theories. Much of it is an attack on the physics of Aristotle and an exposition of a new mechanics which conforms with the Copernican theory.

The book was published in January. Suddenly in August an order went forth to sequestrate every copy in the booksellers' shops throughout Italy and to suspend further publication. The Pope was angry and alienated: his own favourite methodological argument had been put at the very end of the book, in the mouth of Simplicio, the somewhat obtuse Aristotelian of the *Dialogue*; and all the convincing arguments were for the Copernican system. Galileo was accused of having broken his undertaking, given on his admonition in 1616, not to support and preach the Copernican theory. He maintained, however, that in the *Dialogue* he had given the arguments impartially for both theories.

He was summoned to Rome to appear before the Inquisition. On arrival in Rome he was treated with consideration but implacable firmness and was forced into complete submission. Kneeling before the assembled cardinals of the Holy Office he agreed to abjure and abandon the Copernican doctrine. The story that he muttered under his breath 'Eppur si muove' (It does move all the same) is myth. In its original version the remark was uttered after he had left the assembly and was gazing at the sky.

The condemnation of Galileo did not prove an encouragement to Italian science, which soon declined. Milton, in his *Areopagitica*, wrote that persecution 'had dampt the glory of Italian wits that nothing had been there writt'n now these many years but flattery and fustian. There it was that I found and visited the famous Galileo, grown old, a pris'ner to the Inquisition.'

Though old and humiliated, Galileo wrote his most profound scientific work after his condemnation, *The Dialogues on the two*

New Sciences, describing his fundamental researches on motion and published in Amsterdam in 1638. Shortly before his death he became totally blind; his sight had been enfeebled long before this, possibly through looking at the sun through the telescope.

* * *

Despite the activities of the Jesuits, the Copernican theory made headway among the savants of Europe; 'Everybody in the Low Countries is for the movement of the earth', wrote Gassendi. Descartes, it is true, hastily took steps to withhold from publication his book supporting the theory, giving 'obedience to the Church as a dutiful son' as his reason.

Meanwhile a further great advance was being made by Tycho Brahe (1546–1601) and Johannes Kepler (1571–1630), whose achievements were complementary, the former being a great observer and the latter a mathematician.

Tycho Brahe was attracted to astronomy by the appearance of a new star on 11 November 1572, an amazing event in the unchangeable heavens of Aristotle. His early astronomical observations made him realise the serious discrepancies, sometimes of four or five degrees, between the observed positions of the planets and those calculated from the tables of Ptolemy and Copernicus. It was clear that no further secure advance in astronomical theory was possible until a new set of really accurate observations had been made.

In 1576 King Frederick II of Denmark offered him Hveen, an island in the sound between Copenhagen and Elsinore, on which he built an observatory which he called Uraniborg (the tower of Heaven). In this observatory, surrounded by gardens and equipped with a library, a printing shop, an alchemical laboratory and a workshop where nearly all his instruments were constructed, he made observations nightly for some twenty years with an accuracy unsurpassable without the aid of a telescope.

Tycho's fiery and arrogant temper made many enemies, and in 1597, nine years after the death of King Frederick, he was forced to leave Uraniborg, which thereupon fell into ruin. In 1599 the Emperor of Bohemia gave him a castle near Prague, where he was joined as an assistant by Kepler and, at his death two years later, his observations, so vitally important to Kepler, passed into the latter's hands.

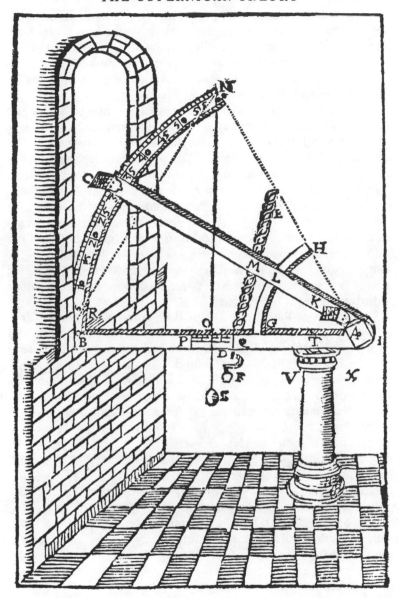

Fig. 5. Tycho Brahe's quadrant for finding the celestial altitudes of stars. The two sights on the upper arm are aligned on the star and the altitude is read on the graduated scale. (Courtesy: The Trustees of the British Museum.)

Tycho was not a supporter of the Copernican theory: he did not believe that the 'heavy, sluggish' earth could move. He held a theory, mathematically equivalent to that of Copernicus, that the planets revolved round the sun, and that the sun and the planets revolved together round the earth. Kepler, on the other hand, was an adherent of Copernicus.

Tycho's best observations had been of the planet Mars, and Kepler set about modifying the theory of Copernicus by increasing the number of epicycles to make the theory fit the facts. But, despite all efforts, there remained a discrepancy of eight minutes of arc, and this was far greater than Tycho's experimental error. As so often has happened in the history of science, a small discrepancy, which many might have ignored, was the clue to a great discovery. At length Kepler took the decisively important step of abandoning the circle and trying the ellipse. He had little hope of success because he believed that, if the orbits of the planets were ellipses, the fact would have been discovered centuries ago by Apollonius or Archimedes, who were familiar with the properties of ellipses.

It turned out that the orbits of the planets were ellipses and, although this fact was so distasteful to Kepler that he referred to it as 'a cartful of dung', it introduced a most satisfying simplicity. The complex of cycles and epicycles was swept away, leaving a single elliptical orbit for each planet. More than this, Kepler discovered the following laws:

(1) The planets move round the sun in ellipses, having the sun at one focus.

(2) Each planet revolves in such a way that an imaginary line joining it to the sun sweeps out equal areas in equal times (Fig. 6).

(3) The squares of the times of revolution of the planets bear a constant ratio to the cubes of their mean distances from the sun.

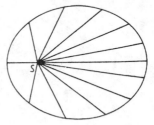

Fig. 6. Kepler's second law. The planet sweeps out equal areas in equal times. It travels faster when nearer to the Sun *S*.

The extent and laboriousness of the calculations involved in making these discoveries are quite staggering. 'If you find this work difficult and wearisome to follow', wrote Kepler, 'take pity on me, for I have repeated these calculations seventy times, nor be surprised that I have spent five years on this theory of Mars.'

Kepler was sustained by the Pythagorean mysticism of numbers, and a great deal of his work was nonsense. He believed that God created the universe according to a geometrical plan. He asked why there were only six planets, including the earth. With intense joy he realised that there are only five regular solids, with equal sides and equal angles, and that, if these are fitted inside each other, there are six spaces where the six planetary orbits can be fitted. It was, of course, long after his death that the further planets Uranus, Neptune and Pluto were discovered. In his book *The Harmony of the World* he wrote that 'The heavenly motions are nothing but a continuous song for several voices'. Saturn was credited as a basso profundo, Jupiter a bass, Mars a tenor, earth a contralto, Venus a soprano, and Mercury a falsetto.

In Kepler's work (as in that of many other scientists of the Renaissance) there is a strange mixture of wild speculation and careful respect for facts. He was always ready to try a near-mystical explanation for the phenomena, and equally ready to reject it when it did not fit.

CHAPTER 3

THE MECHANICAL UNIVERSE

THE PROGRESS of physical science from Copernicus to the end of the nineteenth century consisted of the development of the idea of a mechanical universe. It became the aim to explain all phenomena in terms of the motion of particles or atoms and the motion was thought to be under rigid law.

Before the picture of a mechanical universe could be established it was necessary to discredit Aristotle's ideas of motion and to devise new dynamical concepts. Aristotle's conception of the universe as an organism, with a sort of inherent mind, had to be replaced by that of a self-acting machine. This task was achieved primarily by Galileo. Newton used Galileo's dynamical concepts to formulate the laws of motion and showed that they applied to all motion, whether terrestrial or celestial.

*   *   *

Aristotle believed that every terrestrial body is composed of one or more of the four elements, earth, air, fire and water, each of which has a disposition to seek its natural place. Earth and water are 'heavy' elements, subject to gravity, and they tend to move down towards the centre of the universe. Air and fire are 'light' elements, subject to levity, and they tend to move upwards. Because these elements are mixed, and not sorted out into their proper layers, natural motion, without the action of a mover, can take place.

Celestial bodies, on the other hand, are in their proper place, and, being subject neither to gravity nor to levity, they move naturally in circles.

Terrestrial motion which is not up or down requires a mover, said Aristotle, and this mover must act during the whole of the motion. If the mover ceases to act the body either stops or moves vertically. The speed of the body is proportional to the force exerted by the mover and is inversely proportional to the resistance of the medium in which the body is moving. If the resistance of the medium were zero, the speed of the body would be infinite; hence Aristotle deduced that a void or vacuum cannot exist.

Since a steady application of force is necessary to preserve motion, how can the motion of a stone, when it has left the hand of the thrower, be explained? Aristotle maintained that propulsion is exerted by the air, which is compressed in front of the stone and rushes round to the back to prevent the formation of a vacuum.

Aristotle's theory has obvious weaknesses. It is unsatisfactory to regard the air as a resistant and at the same time as a propellant. A thread tied to a moving stone ought to be blown ahead and not to trail behind. Such criticisms were put forward in the fourteenth century and an alternative theory advanced. The continuous action of a mover, it was held, is unnecessary to preserve motion and a body is carried forward by its 'impetus', the impetus weakening gradually, like a hot body cooling.

The impetus theory was applied to gunnery by several writers in the sixteenth century. Tartaglia (1500–57), with its aid, discussed the path of projectiles and, although he did not develop a mathematical theory, he estimated that maximum range is achieved when the elevation of projection is 45°. The impetus theory distinguished between violent and natural motion, or between the force of projection and the force of gravity. These were thought to operate not simultaneously but consecutively, so that a projectile proceeded roughly in a straight line, until the force of projection was exhausted, and then dropped almost vertically. Tartaglia was one of the first to suggest that gravity must act, at least partially, during the whole flight and hence that the path of a projectile was curved.

The mathematical theory of the motion of a projectile was worked out by Galileo. He considered a sphere rolling on a table with uniform velocity until it came to the edge and fell in a curved path to the ground. He showed that the curved path was a semi-parabola and hence deduced that the path of a projectile is a full parabola. He was able to do this because he realised that the horizontal and vertical motions were mutually independent, that the horizontal motion continued unchanged (ignoring air resistance), and he had previously discovered the laws of vertical motion under gravity.

We must now consider in a little detail how these fundamental ideas were derived. Bodies increase in speed as they fall, and Galileo, conceiving that the motion might be subject to a simple law, speculated that the speed a body acquires may be proportional to the time of fall. Suppose a body acquires a velocity a in unit time (which is its acceleration): after a time t its velocity will be at. If it falls from rest its average velocity during the fall will be $\frac{1}{2}at$, and hence the distance fallen, which is the product of the average velocity and the time, will be $\frac{1}{2}at \times t = \frac{1}{2}at^2$. Thus the distance traversed should be proportional to the square of the time. Galileo tested and proved the truth of this deduction by 'diluting' gravity, running a ball down an incline, and timing it over different distances by means of a water clock.

His procedure was noteworthy because of the combined use of experiment and mathematics, which is the aim of every research physicist today. The schoolmen at Paris, nearly two centuries earlier, discussed 'uniform motion', 'difform motion' and 'uniform difform motion', that is, constantly accelerated motion, and they derived what was virtually the expression $s = \frac{1}{2}at^2$. They were on the verge of Galileo's discovery, but they did not make it because they did not turn to experiment.

The great achievement of Galileo in formulating and practising one of the most important of modern scientific methods has stimulated much discussion on the factors that influenced him. In developing his dynamical concepts he had the impetus theory and the writings of the gunnery engineers as a starting-point. He was professor at Padua, where the methodology of Aristotle had been debated and modified, so that the role of hypothesis was clearly understood. He was fond of visiting the arsenal at Venice and he may have learnt from the artisans there the value of experiment. He was deeply influenced by Archimedes, whose works were recovered in the middle of the sixteenth century; he spoke of 'the writings of that divine man (which moreover are extremely easy to understand, so that all other geniuses are inferior to that of Archimedes)'. He was also an admirer of Plato. Plato's advocacy of mathematics as the clue to the meaning of the universe was, however, metaphysical, whereas Galileo used mathematics, not as a preconceived structure, but as a scientific tool for the investigation of nature.

From further experiments with the inclined plane, Galileo

arrived at one of the key concepts of the mechanical universe, that of inertia, foreshadowed in the doctrine of impetus. He allowed a ball to run down one inclined plane and to run up another which was less steep. He found that the ball always rose to the same vertical height as that from which it had fallen, making a slight allowance for loss of motion due to friction. Thus if the second plane had a very gentle inclination the ball would run a considerable distance along it and Galileo asked what would happen to the ball if the second plane were horizontal. He assumed that the ball would continue to move with no tendency to stop, and this he ascribed to its 'inertia'.

Galileo, following Democritus, divided the properties of matter into primary and secondary qualities, primary qualities being mass, shape and motion, and secondary qualities colour, taste, smell and sound. He regarded the former as more fundamental than the latter. 'If therefore the organs of sense, ears, tongues and noses were removed', he wrote, 'I believe that shape, quantity and motion would remain, but there would be no more of smells, tastes and sounds.' He confined himself, as did all his successors, to the investigation of primary qualities since these alone are amenable to measurement and to mathematical treatment.

Galileo's qualitative concept of inertia, the tendency of a body to continue in uniform motion in a straight line and to resist a change of motion, was developed by Newton into the quantitative concept of mass. Suppose two bodies resting on perfectly smooth ice (Fig. 7) are pushed against the ends of a compressed

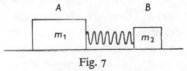

Fig. 7

spring and then released. The spring will exert equal and opposite forces on the bodies and cause them to be accelerated, the acceleration of the more massive body A being smaller than that of B. The masses of the bodies m_1 and m_2 can thus be defined in terms of the accelerations a_1 and a_2 by means of the equation $m_1/m_2 = a_2/a_1$. Some convenient body must be selected as a standard of mass with which other bodies can be compared.

The second key concept of the mechanical universe is force,

which is that which causes a body to change its motion, that is to accelerate. In the above experiment the spring exerted equal forces on the two bodies. Since $m_1a_1 = m_2a_2$ the forces may be measured by m_1a_1 and m_2a_2. Thus if a force F, acting on a mass m, causes an acceleration a, $F = ma$.

The treatment we have just given embodies the essence of Newton's laws of motion and is that devised by Ernst Mach towards the end of the nineteenth century. Newton's own treatment had several objectionable features. He defined mass in a circular manner in terms of density and volume; density is, of course, mass divided by volume. He used the concept of absolute motion, based on absolute space and absolute time, and these absolutes have been shown to be meaningless.

Even Mach's definition of mass, by means of the relative acceleration of two bodies, requires a space system of reference and does not really avoid the problem of absolute motion. In view of the logically unsatisfactory nature of Newton's laws of motion, their success in supporting the whole superstructure of classical physics is quite remarkable.

Criticisms of Newton's presentation should not be allowed to obscure his genius. The realisation of a significant concept, like mass or force, involves the rejection of a host of irrelevancies and is a major imaginative achievement.

Copernicus had made no attempt to explain why the planets circled round the sun; he assumed, with Aristotle, that circular motion is natural and requires no explanation. In the mechanics of Galileo and Newton, however, uniform motion in a straight line is natural and departure from it can be effected only by the application of a force. Descartes explained the forces on the planets as due to the action of vortices in the æther, while William Gilbert and Kepler speculated that they might be magnetic.

Newton conjectured that the force which makes the planets and the moon describe their elliptical orbits is of the same nature as that which causes an apple to fall to the ground on the earth, namely the force of gravitation. He assumed that the force is proportional to the inverse square of the distance and, knowing at what rate an apple falls, he proceeded to calculate whether the rate at which the moon should fall on this hypothesis actually corresponded with its known motion.

The moon never reaches the earth because it is too far away

and travelling too quickly in its orbit. In Fig. 8 E represents the earth and M_1 and M_2 two positions of the moon. If the earth exerted no gravitational force on the moon it would proceed with uniform velocity in the straight line M_1P; hence the moon can be regarded as falling along PM_2. Clearly M_1M_2 must be considered as very small and Newton found it necessary to invent a new branch of mathematics, which he called fluxions, similar to the differential calculus, in order to deal with this type

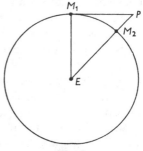

Fig. 8

of problem. If r and R are the radii of the earth and of the moon's orbit, and a and A are the accelerations of an apple and of the moon respectively, then according to the inverse square law, $a/A = R^2/r^2$. All of these four quantities were known and thus it was possible for Newton to check the truth of the equation.

Since he delayed the publication of his results for about twenty years, it has been suggested that he was dissatisfied with the discrepancy caused by the inaccurate value of the earth's radius then available, 60 miles for one degree of latitude instead of the correct value of 70, obtained by Picard some years later. It is more probable, however, that he found difficulty in proving that the earth's mass can be considered as concentrated at its centre.

*　　*　　*

Isaac Newton is commonly regarded as the greatest man of science and some have called him the greatest genius of the human race. He was born in 1642 at Woolsthorpe, eight miles south of Grantham in Lincolnshire.

As a boy he was fond of making mechanical, wooden models; 'he had got little saws, hammers and all sorts of tools, which he would use with great dexterity', said his earliest biographer, Dr Stukeley. He made a little wooden windmill and put a mouse inside it to turn the sails. He made a waterclock and sundials. This early skill with his hands foreshadowed his power as an experimenter; he had to make almost all his apparatus himself—to grind and polish his own spherical mirrors, for example.

He was sent to the King's School, Grantham, where he did not particularly distinguish himself. He was not physically strong and may have suffered from bullying, which would account for his peevish, suspicious and solitary nature.

When he had spent four years at the King's School his mother withdrew him so that he could learn to become a farmer. But he was not interested in farm work and preferred to sit under a hedge reading a book. In 1658, on the day of the great storm when Cromwell died, he performed one of his first experiments. He estimated the force of the wind by leaping into it and then in the opposite direction with the wind behind him.

Seeing how unfitted he was to be a farmer, and having some inkling of his intellectual ability, his uncle persuaded his mother to send him back to school in Grantham and then, in 1661, to Trinity College, Cambridge.

At Cambridge he studied mathematics and optics. He purchased a glass prism at Stourbridge Fair, held each autumn near what is now called Midsummer Common in Cambridge, to which merchants came from all over England and from the continent. With the prism he began his experiments on the composition of white light. In 1665 he took his degree and in the autumn of that year the University was closed because of the bubonic plague. He returned to Woolsthorpe in Lincolnshire for two years of quiet and meditation. His latent intellectual powers must have been awakened in Cambridge because in these two years, before he had reached the age of 25, his three greatest discoveries germinated: the theory of gravitation, fluxions, and the theory of the composition of white light. Einstein has said that 'a person who has not made his great contribution to science before the age of 30 will never do so', and Newton was no exception.

When he returned to Cambridge he became a Fellow of Trinity and then, at the age of 27, Lucasian Professor of Mathematics. His teaching duties as professor were not heavy and he devoted much of his time to his experiments in optics. He constructed several reflecting telescopes, the first ever made, because he believed that the ordinary refracting telescope had an incurable defect, causing coloration and blurring of the image. In 1671 he presented one of these telescopes to the Royal Society, which had received its Royal Charter nine years earlier from Charles II, and was gratified when he was elected a Fellow. This prompted him to send to the Royal Society a paper giving an account of his experiments on the dispersion of white light into colours by a prism.

The reception of this paper, the controversy it caused, the lack of understanding that his conclusions were based strictly on experiment and hence could not be subverted by mere argument, and the patronising attitude of Robert Hooke, a brilliant, vain, deformed little man who was one of the leading members of the Society, left so deep a mark on Newton's mind that he was afterwards reluctant to publish any of his discoveries. He wrote to Oldenburg, the Secretary of the Royal Society: 'I intend to be no farther solicitous about matters of philosophy and therefore I hope you will not take it ill if you find me never doing anything more in that kind.' He was by nature abnormally touchy and sensitive to criticism; as he grew older he became suspicious and watchful, quick to find disparagement and censure where none was intended.

When Oldenburg died in 1679 Hooke became a joint secretary of the Royal Society and wrote a courteous letter to Newton urging him to make further communications on scientific matters to the Society. Newton detested Hooke but, as a graceful gesture in reply, he sent a suggestion by which experimental evidence of the earth's rotation might be obtained. He said that a body falling a great distance should fall to the east of the vertical and describe a spiral line. Hooke, presenting the letter to the Royal Society, pointed out with some satisfaction that it contained an error and showed that a body would not fall in a spiral. Newton was deeply mortified by his blunder and especially that Hooke, of all people, should have been the one to discover it. He began to turn his mind once again to the studies of motion and gravitation which he had started during the years of the Great Plague.

The idea of an inverse square law of gravitation, that bodies attract each other with a force which is inversely proportional to the square of their distance apart, was in the air and three members of the Royal Society, Hooke, Halley and Wren, were convinced that it would result in the elliptical orbits of the planets, as discovered by Kepler, but they could not prove it. Hooke told Wren that he had worked out the proof 'but he would conceal it for some time, that others, trying and failing, might know how to value it when he should make it public', but Wren found this unconvincing.

In 1684 Halley decided to visit Cambridge to consult Newton and was astonished to find that Newton had worked out the proof some years earlier, together with much else besides, but had mislaid his papers. He realised that a man who could lose a proof that any other member of the Royal Society would hasten to publish might have accomplished other and even more remarkable work. He urged Newton to write a book and offered to defray the cost of publication himself.

The result was the *Philosophiae Naturalis Principia Mathematica* or Mathematical Principles of Natural Philosophy, which developed, in the most exhaustive and dazzling manner, the full implications of the inverse square law as applied to the astronomical universe.

The *Principia* took eighteen months to write, each proof being worked out afresh, a stupendous feat. It was cast in geometrical form because of Newton's admiration for the geometers of the ancient world and because it gave him satisfaction to make the proofs difficult 'to avoid being baited by little smatterers in mathematics'. This was unfortunate because geometrical methods were unsuitable for further advance and the progress of English mathematics was thereby impeded.

We have some record of the intense concentration that was needed for the writing of the *Principia* from Newton's amanuensis, Humphrey Newton (not a relative). For eighteen months Newton seldom left his rooms in Trinity, unconscious whether or not he had eaten his meals, and taking little time for sleep. He would spend whole days sitting on his bed, half-dressed, thinking. He had the power of holding a problem in his mind sufficiently long for the different pieces to fall into place and reveal the solution. Once when he was asked how he made his discoveries, he said, 'By always thinking unto them'.

While the work was being written Hooke, apparently unable to distinguish between a surmise and a rigid scientific proof, claimed that he was the true discoverer of the inverse square law and that Newton had merely applied it. Newton, irritated beyond endurance, at once decided to abandon publication of the third book of the *Principia*. He wrote sarcastically to Halley: 'Now is not this very fine? Mathematicians that find out, settle and do all the business, must content themselves with being nothing but dry calculators and drudges: and another that does nothing but pretend and grasp at all things must carry away all the invention.'

He was mollified, however, when the Royal Society took his

part against Hooke and he decided, in order to end the controversy, to acknowledge in a scholium that the law was discovered independently by Wren, Hooke and Halley.

The *Principia*, written in Latin, appeared in 1687. The first book states the laws of motion, deals with the general principles of mechanics, deduces the elliptical orbits of the planets from the inverse square law and considers mathematically a variety of astronomical phenomena. The second book is concerned mainly with the motion of fluids and demonstrates conclusively that Cartesian vortices are incompatible with Kepler's laws for the orbits of the planets. In the third book, the most spectacular part, the masses of the sun and the planets are calculated, the flattened shape of the earth is explained quantitatively, the fundamental theory of the tides is given, the main irregularities of the moon's motion due to the pull of the sun are worked out, and the slow wobble of the earth's axis, known as the precession of the equinoxes, is calculated.

There were very few people in the world who were capable of understanding the *Principia*. Demoivre, a mathematician whose fame is still alive today, tore the book into separate pages, which he carried about with him so that he could master the proofs one at a time. But it was realised by everyone that the work was one of extraordinary genius, an event unique in the intellectual history of mankind. In his later years Newton was regarded almost as a demigod. The French mathematician, de l'Hôpital, exclaimed, 'Does he eat, drink and sleep like other men?'

In 1692 and 1693 Newton suffered from a nervous breakdown. The intense mental effort of composing the *Principia*, the unhealthy life he lived at that time, with its lack of exercise and recreation, had left him in a weak and nervous state, unable to sleep properly. He had a deeply upsetting experience in February 1692. He left a candle burning in his rooms in Trinity while he went to chapel and his researches of twenty years in optics were burnt. He began to write strange letters to his friends. To John Locke he wrote: 'Being of opinion that you endeavour to embroil me with women and by other means, I was so much affected with it, as that when one told me that you were sickly and would not live, I answered, "'twere better if you were dead". I desire you to forgive me this uncharitableness.' Rumours about the state of his mind began to circulate in London and on the continent.

It was clear to Newton's friends that a complete break with Cambridge would be beneficial to him. In 1696 Charles Montague, Chancellor of the Exchequer and later Baron Halifax, appointed him Warden of the Mint. Montague was an old student of Newton's at Trinity and he was engaged in a recoinage of the currency. Three years later Newton became Master of the Mint, with a salary of £1500 a year. He was now a comparatively rich man and kept a sedan chair and three servants. He was knighted by Queen Anne.

When he rode in his sedan chair his hands would hang limply on either side. Pope said that he was bad at casting up accounts: 'though so deep in Algebra and Fluxions, [he] could not readily make up a common account; and when he was Master of the Mint, he used to get somebody to make up his accounts for him'.

Newton was elected President of the Royal Society in 1703 and re-elected annually until his death in 1727. He was a virtual dictator in all scientific affairs. The Royal Society took his side in the notorious dispute with Leibnitz over priority in the invention of the calculus. Newton invented it first but Leibnitz published his work earlier than Newton. The controversy continued for many years. Bernoulli, a Swiss mathematician, published a problem as a challenge to the mathematicians of Europe, and Newton rightly suspected that it was issued to test whether his fluxions were as powerful a method as the calculus of Leibnitz. Newton received the problem on his arrival home from the Mint at 4 o'clock and solved it before he went to bed. But his irritation was revealed in a letter to Flamsteed: 'I do not love to be... dunned and teased by foreigners about mathematical things.'

It is an extraordinary fact that science was not Newton's chief interest. In a letter to Hooke in 1679 he wrote: 'I had for some years past been endeavouring to bend myself from philosophy to other studies in so much that I have long grutched the time spent in that study unless it be perhaps at idle hours sometimes for a diversion.' It is true that he hated controversy and spoke of philosophy as 'an impertinently litigious lady', which would explain why his discoveries had to be coaxed from him. But there may have been a deeper reason for his indifference. He regarded science as the attempt to demonstrate the divine will in nature rather than to give man power to control and manipulate nature. The latter view tends to receive more emphasis today and it was also held in Newton's time, Francis Bacon being its chief exponent. Lord Keynes has suggested that Newton, having solved the riddle of the heavens by his discovery of universal gravitation and his mathematical genius, dared to conceive that he could solve the riddle of the universe if only he could find the right clues. These clues he sought in the Book of Revelation, in the prophetic books of the Old Testament, in the measurements of Solomon's Temple and in the alchemical traditions handed down from antiquity. At his death he bequeathed to his niece a collection of documents, containing over a million words in his handwriting, known as the Portsmouth collection, which reveal his vast researches into theology and alchemy.

If in theology and alchemy Newton accepted prophecy and occult tradition, in his scientific work he practised and stated explicitly for the first time the severe discipline of the scientific method. Much of his irritation in controversy was due to the inability of his opponents to understand this. 'I frame no hypotheses', runs his classic statement in the *Principia*, 'for whatever is not deduced from the phenomena is to be called an hypothesis'. By hypothesis he meant speculation or theory without basis in fact. Hypothesis in its modern sense, as a guess to be tested by fact and experiment, is a legitimate and indeed vital component of scientific method. The inverse square law itself was at first an hypothesis until, on being shown to account for the orbits of the planets, it achieved the status of law.

When he said that he framed no hypotheses Newton had in mind speculation as to the cause of gravitation. In the *Principia* he would not permit himself to speculate, but in private letters to friends he sometimes gave his thoughts free rein. He wrote to Robert Boyle that gravitation might be due to the pressure of a subtle elastic medium called the æther, filling all space and decreasing in density between two gravitating bodies. At other times he tended to ascribe gravitation to immaterial causes and the direct volition of God.

*   *   *

The *Principia* established the mechanical universe, finally replacing the medieval heavens by the vision of the stars in Meredith's poem,

> Around the ancient track march'd, rank on rank,
> The army of unalterable law.

Its impressiveness lay in its ability to predict eclipses, comets, and the exact positions of the heavenly bodies a hundred years hence or a thousand years ago, and also to forecast new phenomena. In 1845 a young English mathematician, Adams, predicted the existence of an unknown planet Neptune, from perturbations of the orbit of Uranus, and worked out its exact position. Neither Airy, the Astronomer Royal, nor Challis, the director of the Cambridge Observatory, who actually observed an unknown star with the appearance of a disc, made any attempt to verify the prediction of Adams and the planet was discovered a year later at the Berlin Observatory from similar calculations of the French mathematician Leverrier. In 1930, another planet, Pluto, was discovered by American astronomers from disturbances of the orbit of Neptune.

Newton himself did not believe that the world machine could keep running without God's constant supervision. 'This most beautiful system of the sun, planets and comets', he says in the *Principia*, 'could only proceed from the counsel and dominion of an intelligent and powerful Being', which drew the retort from Leibnitz that Newton regarded the universe as a clock which had to be regulated, every now and again, by its creator. It is said that a Cambridge lecturer in the eighteenth century was in the habit of remarking, when he had calculated the force required to keep Saturn in its orbit round the sun, that this demonstrated the enormous strength of the Deity.

Laplace, however, when asked by Napoleon why he had not mentioned God in his *Mécanique Céleste*, replied that he had 'no need of that hypothesis'. He maintained that a mathematical genie, possessed of knowledge of the present positions, masses and motions of all the bodies in the universe, would be able to calculate all past and future history, whether of stars or atoms. This view, that the mechanical system was self-sufficient and capable of a complete explanation of nature, became general in the nineteenth century because of the remarkable success of the mechanical ideas when applied to the realms of heat, light, electricity and chemistry.

CHAPTER 4

THE CIRCULATION OF THE BLOOD

THE DISCOVERY of the circulation of the blood by William
Harvey (1578–1657) marks the establishment of the experimental
method in biology and is comparable in significance with the work
of Galileo in physics. While Galileo was laying the foundations of
the mechanical universe, Harvey took the equally revolutionary
step of regarding the human body as a hydraulic machine and
the heart as a pump.

Medicine shared with astronomy the chief place in scientific
studies during the Middle Ages and in these two fields at the
Renaissance there were parallel advances. In medicine the authoritative source of knowledge, and hence the chief obstacle to progress, was not Aristotle but Galen (130–200), a great anatomist,
and physician to the emperor Marcus Aurelius.

Galen was prevented by the public opinion of his time from
making human dissections and many of his anatomical errors
were caused by his reliance on the dissection of apes, dogs and
pigs. Objection to human dissection was strange in view of the
brutality of the gladiatorial games: it was equally strange in men
of the Middle Ages, who had no scruples about torturing, burning, quartering and disembowelling. By the fourteenth century,
post-mortems were allowed in Bologna, and the bodies of criminals
and paupers were dissected; 'body-snatching' by the students was
not unknown when corpses were scarce. The anatomical errors
of Galen remained undiscovered for a further two hundred years,
however, because professors of anatomy did not perform dissections themselves. Their custom was to expound the works of
Galen while an assistant, the barber-surgeon, opened up a body
to illustrate the lecture.

Challenge to the authority of Galen came from Andreas
Vesalius (1514–64), a Fleming who was appointed Professor of
Anatomy at the University of Padua at the age of 23. The
struggle which Galileo waged against ancient authority in

45

astronomy and physics was initiated in the biological sciences by Vesalius, before Harvey's time. As a boy Vesalius dissected birds, rabbits, dogs and other animals, and when he became a professor, not merely did he perform dissections himself, but he revolutionised the technique and designed most of our modern dissecting instruments. Although always a great admirer of Galen, he discovered some two hundred anatomical errors that Galen had made. He showed, for example, that the human thigh bones are straight and not curved as in a dog. Affronted by this lack of respect for Galen, but finding it difficult to deny a matter of fact, one of his former teachers maintained that the straightness of human thigh bones was the result of the narrow trousers of the time and that these bones were curved under more natural conditions. Vesalius was attacked for denying that man has one rib fewer on the side from which Eve was formed, and because he found no trace of the body's nucleus for resurrection, the indestructible 'resurrection bone'.

His book, *De Fabrica Corporis Humani* (On the Fabric of the Human Body), was published in 1543, the same year as *De Revolutionibus Orbium Coelestium* of Copernicus. It contained a wealth of new anatomical facts, it ignored the current belief that the different organs of the body were linked with the planets and the heavenly bodies, and it was illustrated by one of the pupils of Titian.

Vesalius's brief life of research ended in 1544 when he left Italy to become physician to the delicate Emperor Charles V of Spain. He was succeeded at Padua by Fallopius and Fabricius, who maintained the reputation of its school of anatomy until the end of the century. Students flocked to Padua from all over Europe. John Caius, the second founder of Gonville and Caius College, Cambridge, lived for a time in the same house in Padua as Vesalius; William Harvey proceeded from Gonville and Caius College to study at Padua under Fabricius. Professor Butterfield has pointed out that if any single place deserves the honour of being called the home of the scientific revolution it is the University of Padua, where Vesalius and Galileo were professors and Copernicus and Harvey were students.

*   *   *

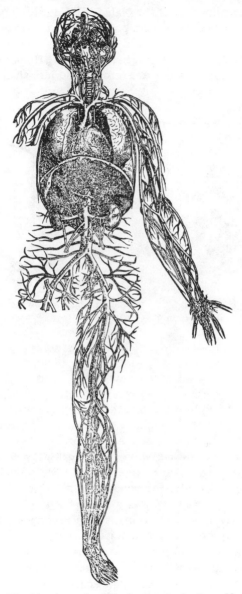

Fig. 9. The blood vessels of the human body, from Vesalius's
De Fabrica Corporis Humani.

Fig. 9 is a reproduction of the plate from Vesalius's book showing the blood vessels of the human body. There are two corresponding systems of vessels, arteries and veins, which are paired in most parts of the body. The walls of the arteries are thick and muscular, whereas those of the veins are thinner and less strong.

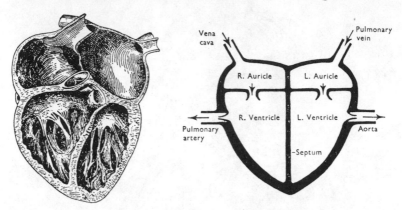

(*a*) Interior of the heart.   (*b*) Diagrammatic representation of the heart.

(From F. W. Westaway: *The Endless Quest*, by courtesy of Blackie and Son Ltd).

Fig. 10

Vesalius had an accurate knowledge of the structure of the heart. Fig. 10 (*a*) is a drawing of the interior and Fig. 10 (*b*) is a diagrammatic representation; the heart is seen to consist of four chambers, known as the right and left auricles, and the right and left ventricles. Between each auricle and ventricle there is an opening, equipped with a valve, which allows a flow only from the auricle to the ventricle. Leading into the right auricle is the trunk vein of the body, the vena cava, and into the left auricle the pulmonary vein (shown as a single vessel in Fig. 10 (*b*), but actually consisting of four veins) which branches into the lungs. Springing from the left ventricle is the aorta, the trunk artery of the body, and from the right ventricle, the pulmonary artery, which branches into the lungs. Both the aorta and the pulmonary artery are fitted with valves which enable only an outflow from the ventricles.

Accepted opinion in the time of both Vesalius and Harvey concerning movement of the blood was based upon the views of

PLATE III. Isaac Newton, painted by Sir Godfrey Kneller.

PLATE IV. William Harvey demonstrating his researches on the deer to Charles I and the boy prince, from a painting by R. Hannah (1848) in the Royal College of Physicians.

Galen. Since venous blood has a blue tinge and arterial blood is bright red, it was believed that there were two distinct blood systems, the venous system which nourished the tissues of the body, and the arterial system which gave the tissues life.

All the blood was thought to be made in the liver, from a liquid called chyle which was formed from food in the stomach and in the intestines. The liver was part of the venous system and hence it was necessary to suggest some route by which blood could be supplied to the arterial system. The reader should refer to Fig. 11, p. 52, in which the venous system is shaded and the arterial system unshaded. (The arrows represent Harvey's theory and the names of the blood vessels are post-Harvey.) Galen held that the blood entered the arterial system by exuding, drop by drop, from the right ventricle (RV) to the left ventricle (LV) of the heart through the thick, fleshy wall between them, called the septum.

In the left ventricle the blood was transformed from a venous to an arterial character. It was endowed with the basic principle of life, 'the vital spirits', from the air breathed into the lungs and transmitted to the heart through the pulmonary vein. This was the chief function of the heart, to act as a vat to brew the mixture of blood and vital spirits and hence, through the arterial system, to supply life and heat to the body.

During its passage through the tissues of the body the arterial blood was thought to become laden with impurities, which were compared to the sooty vapours from a lamp; these passed through the pulmonary vein to the lungs, there to be exhaled.

The regular pulse of the blood and the rhythm of breathing were thought to be interconnected and this was why there was ascribed to the pulmonary vein its improbable threefold task: to conduct arterial blood and waste sooty vapours from the heart to the lungs, and to conduct air in the opposite direction.

The synchronism between the beat of the heart and the pulse led to the belief that the heart and the arteries dilated together. Hence blood was thought to leave the heart when the latter dilated, because of the ebullition of its contents, and the suction of the arteries played a part in the pulsation of the blood.

Galen believed that the valves of the heart prevented the pulsation of the blood from becoming a considerable ebb and flow, which would entail a repeated withdrawal of the blood from the extremities of the body and tire the heart unnecessarily.

Without these valves, said Galen, 'the blood itself would keep travelling over this long course to no purpose...like a sea-tide patterned on Euripus, continually reversing the movement to all parts, which would in no way suit the blood'. (Euripus is the ancient name for the narrow strait between the island of Euboea and the mainland of Greece, where the tide is much more noticeable than elsewhere in the Mediterranean.)

The existence of a passage of communication between the two blood systems, necessary for the supply of blood from the venous to the arterial system, was corroborated by the fact that, by the severance of either an artery or a vein, the blood could be drained from both systems. But no one could find the pores in the septum which constituted this passage, as Galen maintained. Galen admitted that the pores were not easy to see: 'These, indeed, are seen with difficulty in the dead body, the parts being then cold, hard and rigid. Reason assures us, however, that such pores must exist.' Vesalius was unable to find the pores and, as he could suggest no alternative theory of the movement of the blood, he contented himself with the remark: 'We are driven to wonder at the handiwork of the Creator, by means of which blood sweats from the right into the left ventricle through passages which escape the human vision.'

Serveto, who worked with Vesalius in Paris before the latter went to Padua, rejected the invisible pores and accepted the impenetrability of the septum. He assumed that communication between the venous and arterial systems occurred through the lungs and, in this way, discovered the pulmonary or lesser circulation of the blood. Serveto was primarily a religious reformer and he included his discovery in an heretical theological work. He was burnt at the stake in Geneva by Calvin in 1553.

The pulmonary circulation, as expounded by Serveto, was, however, not really a circulation at all. The blood was transferred only once from the venous to the arterial system, and the theory was put forward merely to solve the problem of the impenetrability of the septum.

*   *   *

William Harvey's theory of the circulation of the blood was published in 1628 in a book entitled *Exercitatio Anatomica de*

Motu Cordis et Sanguinis in Animalibus (An Anatomical Dissertation on the Movement of the Heart and Blood in Animals).

In the introduction Harvey considered some of the difficulties and improbabilities in the prevailing theory of the action of the heart. The valve between the left ventricle and the left auricle was held to control the flow of air and of the sooty vapours from and to the lungs. But, asked Harvey, how could a valve prevent the flow of air and yet allow the flow of blood?

He asserted, as indeed Galen himself had done, that the arteries of a living animal contain only blood. If the windpipe is cut, air goes in and out, but if an artery is cut, blood comes out in one direction only, and no air goes in and out. If the windpipe of a dog is opened and air is blown in with bellows, plenty of air will be found in the lungs but none in the pulmonary vein or in the left ventricle of the heart.

Again, the right and left ventricles are similar in structure and yet they were supposed to have very different functions. The right ventricle was regarded as merely a passage-way for the venous blood on its way to nourish the lungs, whereas the left ventricle was a vat in which a mixture of blood and vital spirits was brewed.

Harvey pointed out that fishes and all creatures without lungs have only one ventricle. This suggests that, in an animal possessing lungs, one ventricle is concerned with the flow of blood to the lungs and the other ventricle with the flow to the rest of the body. Harvey stressed the advantage of studying the hearts of many different animals and complained that there was insufficient study of comparative anatomy by his contemporaries. His argument that there was too exclusive an interest in the dissection of the human body was justified but seems ironical in view of the errors promulgated by the ancients as a result of their dissection of apes, dogs and pigs instead of man.

Harvey opened up many animals so that he could observe the beating of their hearts, and for this purpose he used dogs, pigs, frogs, toads, snakes, snails, crustaceans and insects. His practice of vivisection, against which there had been a strong prejudice in classical and medieval times, indicates a revolutionary change in the climate of opinion.

Harvey noted first that when the heart contracts it becomes paler in colour, especially in serpents and frogs, and that when felt with the hand it seems harder, like the forearm muscle when the fingers are moved. He observed that it has a fibrous structure and concluded that it behaves like muscle.

He observed also that when the heart contracts, the arteries dilate. At the moment of contraction the heart rises, strikes the chest, and makes a beat. He thus exposed the error that the heart and the arteries contract and expand together. It is true that the

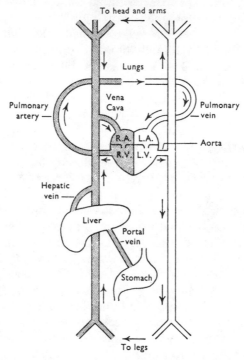

Fig. 11. The circulation of the blood as discovered by Harvey. The blood is pumped by the left ventricle of the heart through the arterial system (unshaded) whence it returns through the venous system (shaded) to the right auricle. It then passes to the right ventricle which pumps it to the lungs, where it is changed from a bluish to a bright red colour. It returns from the lungs to the left auricle and thence back to the left ventricle.

beat of the heart and the pulse occur simultaneously, but the heart contracts at its beat, whereas the arteries dilate when the pulse is felt.

Harvey maintained that the heart behaves as a pump, squeezing the blood into the arteries, as one can blow air into the fingers of a glove. He showed that if the aorta be cut or punctured, blood spurts from it when the left ventricle contracts; similarly blood spurts from the pulmonary artery when the right ventricle contracts. Thus the arteries are merely passive, stretchable tubes and do not dilate and contract under their own pulsific force, as the Galenists maintained. If a ligature is tied round a man's arm, so as to stop the flow of blood in the artery, the blood does not pulsate in the wrist; according to the Galenist theory there should still be a pulsation caused by the dilation and contraction of the arteries in the lower part of the arm and the hand.

The rhythmic movement of the heart is quick and complicated and Harvey despaired at first of being able to analyse it, but he found that when an animal was nearly dead the motion slowed down sufficiently for him to be able to follow it. He discovered that the two auricles contract together and that, almost immediately afterwards, the two ventricles contract together. He concluded that the auricles behave as priming or loading pumps, squeezing the blood through the valves into the ventricles; that the two ventricles are force pumps, each squeezing blood through the arteries round different parts of the body.

There are two circulations, the pulmonary circulation through the lungs driven by the right ventricle, and the systemic circulation through the rest of the body driven by the left ventricle (Fig. 11). In each case the blood is driven to the tissues through the arteries and returns through the veins.

Harvey asked why there is such a difference in structure between the arteries and the veins. He maintained that the arteries are thick and muscular, particularly near to the heart, because they must sustain the shock and pressure of the blood driven by the heart; in the veins the pressure has diminished, and hence they need not be so strong.

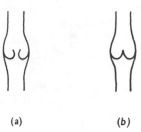

(a)   (b)

Fig. 12. Valve in a vein, whose position is indicated externally by a slight swelling. In (a) the valve is open, in (b) it is shut.

Harvey's teacher at Padua, Fabricius, made an investigation of the valves in the veins (Fig. 12) and published an account of them. The valves, which allow a flow of blood in one direction only, were thought by Fabricius merely to impede the blood and to prevent it from draining by gravity into the lower limbs of the body. But the valves in that part of the body above the heart prevent an upward flow and not a downward. When Robert Boyle asked Harvey what put into his mind the idea of the circulation of the blood he said that it was a consideration of these valves.

Harvey described some experiments on the veins and their valves (Fig. 13). He tied a ligature round a man's arm above the elbow tight enough to stop the flow of blood in the veins but not in the arteries. The veins swelled up and the valves looked like knots. He found that the blood could be streaked by the finger along a vein towards the heart, but not away from it, the flow in the latter direction being prevented by the valves. When the pressure of the ligature was increased to stop also the flow of blood through the arteries, the veins did not swell. These experiments indicated that the blood flowed from the arteries to the veins and that it flowed in the veins in one direction only, towards the heart.

Harvey's theory of the circulation of the blood implied that the blood was the vehicle which transported requirements to the tissues of the body, rather than the requirements themselves. Formerly it tended to be assumed that the blood in the arteries was absorbed by the tissues at the same rate as new blood was delivered by the heart. Harvey wrote, 'Blood is continuously passing into the arteries in greater amount than can be supplied from the food ingested'.

He supported this by a quantitative argument. He found from his dissections that a human ventricle holds rather more than two ounces of blood. He did not know how much of the blood was squeezed out of the ventricle at each contraction and suggested, for the sake of argument, that it might be taken as 'a fourth, a fifth, a sixth or even an eighth' of its contents. We now believe that the ventricle ejects nearly all of its contents, but let us assume that it ejects only one quarter, that is half an ounce. In one hour, the heart makes about 4000 beats, and hence pumps 2000 ounces or 125 lb. of blood, which is not far from the total weight of a man. Harvey concluded that the blood which the body contains is pumped continuously round and round the vascular system.

Despite his numerous first-hand observations and experiments, Harvey discovered few facts which were not known before his time. His achievement, like those of all the greatest men of science, was to conceive a comprehensive generalisation into

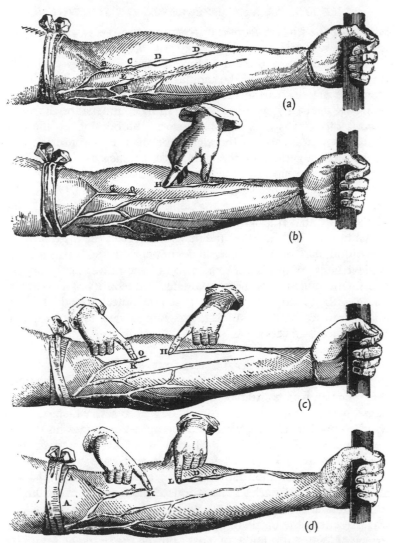

Fig. 13. Experiments on the veins in the arm, taken from Harvey's book.
(a) An arm is tied above the elbow at AA to stop the flow of blood in the veins
but not in the arteries. Nodes appear at the valves in the veins, B, C, D, E, F,
because the blood in the veins is under pressure. (b) If blood is pressed out of the
vein between O and H, by streaking a finger along it, and a finger is placed at H,
the vein remains empty showing that the blood cannot flow back from O. (c) If
a finger is pushed along the vein from K towards O, the vein between O and H
still remains empty because the valve at O prevents a flow of blood towards H.
(d) If the finger at L is removed, the vein between L and M fills up, showing that
the direction of flow of the blood is towards the heart.

55

which all the known facts fall into place, as pieces can be fitted into a jig-saw puzzle.

He did not observe the final links in the circulatory chain, the tiny blood vessels, called the capillaries, which transmit the blood from the arteries to the veins; these were discovered four years after his death by Malpighi. Nor did he understand how the venous blood was changed into arterial blood; indeed the true explanation was not forthcoming for over two centuries. He could only suggest that the change took place as a result of a straining of the blood in the fine tissues of the lungs.

General acceptance of his theory took between thirty and fifty years, being more tardy on the continent than in England. He himself said that no man over forty would accept it. The doctors of his time maintained that the new ideas about the movement of the blood had not cured a single patient. But Harvey's main achievement lay in his demonstration of the power of the experimental method, and of the permanence and security of a discovery made with its aid. He was the first great mechanist in biology and he reorientated the subject, providing a new set of problems for his successors to solve.

*   *   *

We are told by John Aubrey, the gossipy biographer of the time, that Harvey was a little man and very choleric; 'in his young days', wrote Aubrey, 'he wore a dagger (as the fashion then was...) but this Doctor would be too apt to draw-out his dagger upon every slight occasion'. Harvey was physician to Francis Bacon, whom he esteemed for his wit and style but of whose views on the scientific method he had a poor opinion. He was also physician to Charles I and espoused the Royalist cause during the Civil War, which broke out when he was 64 years of age. He was present at the battle of Edgehill and had charge of the young Prince of Wales and Duke of York; during the battle he sat under a hedge and read a book. While Harvey was away in Nottingham with the King, his house in London was broken into by the mob and all his papers, including a work on insects, the fruit of many years of research, were destroyed. 'No greife was so crucifying to him', wrote Aubrey, 'as the losse of these papers.'

CHAPTER 5

THE PRESSURE OF THE AIR

ONE OF THE MOST important scientific investigations of the seventeenth century which was responsible, perhaps more than any other, for the widespread employment of the experimental method, was that which led to the recognition of the pressure of the air. All over Europe there were men who were interested in the new experimental philosophy of science; many discussed it in groups or societies, and not a few tried to practise it.

Certain well-known phenomena, that it is difficult to open bellows when the aperture is closed, that it is difficult to separate two flat polished surfaces when they have been pressed together, and that water rushes into the barrel of a pump or syringe when the piston is drawn back, were explained by Aristotle's dictum that nature abhors a vacuum. In each of these operations a vacuum tends to be created and nature, it was thought, reacts to prevent it.

At the beginning of the seventeenth century, lift pumps were coming into extensive use for the draining of mines, and Galileo was consulted to explain why a lift pump was unable to raise water from a depth of more than 34 ft. It seemed curious that nature's abhorrence of a vacuum should extend only to this height, and Galileo gave a tentative and erroneous explanation that a column of water higher than 34 ft. might break up under its own weight.

Experiments with a column of water 34 ft. high involve difficulties, and one of Galileo's pupils, Evangelista Torricelli (1608–47) had the idea of using instead a column of quicksilver or mercury, which is nearly fourteen times heavier than water, thereby introducing a technique which has been employed by physicists to this day. A glass tube, about a yard long and sealed at one end, was filled with mercury, a finger placed over the open end, the tube upturned and the finger removed under mercury in a trough. The mercury stood to a height of about 30 in. (Fig. 14) leaving, as Torricelli believed, a vacuum above it, thereby disproving Aristotle's contention that a vacuum is impossible. But if the theory that nature abhors a vacuum must be abandoned, what is supporting the heavy column of mercury?

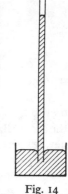

Fig. 14

The Torricellian experiment was performed in Florence in 1643 and it was known at this time that air has weight. In 1638 Galileo had published an account of an experiment to estimate its density. He found the increase in weight of a vessel when air was pumped into it with a syringe until the pressure was two or three times the normal value; he then allowed the air to escape through a tube and displace water in another vessel in order to find its volume at normal pressure.

Torricelli put forward the hypothesis that the column of mercury in his experiment was supported by the weight of the air. 'We live submerged at the bottom of an ocean of the element air', he wrote. He noted that the height of the mercury column changed slightly from day to day; 'Nature would not, as a flirtatious girl, have a different *horror vacui* on different days', he said, and he ascribed the changes in height to changes in the atmospheric pressure. His apparatus was, of course, the first barometer. His early death prevented him from performing other experiments to establish his hypothesis.

*　　*　　*

The news of Torricelli's experiment reached France in a letter to Father Mersenne, the secretary of a group of men in Paris which met every Thursday to discuss scientific matters. This group was called the *Académie Libre* and was the earliest forerunner of the present *Académie des Sciences*. Its members were adherents of the new experimental and mathematical method of acquiring knowledge, as opposed to the scholastic logic of the theologians.

One of the members of this group was Étienne Pascal, the father of Blaise Pascal (1623–62). Étienne did not send his son to school since he wished to make him a humanist; instead he undertook the task of education himself, following the principle that nothing should be taught until a child's mind is ripe to receive it. He determined to teach no mathematics until his son was 15 or 16. Blaise, knowing his father's great love and respect for mathematics, was eager to learn geometry some years before he reached this age and asked what the subject was about. He received the reply that it dealt with the relations between simple shapes. The story is told that, at the age of 12, Blaise was found by his father drawing diagrams on the floor and trying to prove that the sum of the angles of a triangle is equal to two right angles. He had invented his own terms, calling a straight line a bar and a circle a round, and had discovered for himself the greater part of the first book of Euclid.

After this the boy was allowed to attend the meetings of the *Académie Libre*, and at the age of 16 he wrote an original paper on conic sections.

In 1640 Étienne Pascal moved from Paris to Rouen to take up a post in the inland revenue. His work of assessing taxes required much calculation, and this stimulated Blaise to the invention of a calculating machine. Three years were devoted to the perfection of the machine; fifty models were made and they aroused widespread interest.

It was in October 1646 that Blaise Pascal first learnt of the Torricellian experiment. Pierre Petit, Intendant General of the Ports and Fortifications of France, on his way to Dieppe to inspect a sunken wreck, stopped at Rouen and called on his friends, the Pascals. He told how he had heard of Torricelli's experiment from Father Mersenne, and how both had attempted to repeat the experiment, without success because the glass tubes were too brittle. The Pascals said that in Rouen was one of the best glass works in France and an order for a four-foot tube was given forthwith. Fifty pounds of mercury were obtained from a pharmacy and the experiment was performed. The mercury stood to a height of about 30 in. in the tube and there was a gap between the top of the tube and the top of the mercury.

What was in the gap? Torricelli's explanation of his experiment was unknown to Petit. Could there be a vacuum in the gap despite nature's horror of the void? Blaise Pascal's interest was thoroughly aroused and during the following months he performed a number of experiments to determine whether or not it really was a vacuum and what was the force required to create it. He spent much time at the Rouen glass works supervising the making of apparatus.

He filled with water a tube 40 ft. long and sealed at one end, put a stopper in the open end, lashed the tube to a vertical ship's mast with the sealed end uppermost, and immersed the stoppered end in a vessel of water. When the stopper was removed the water sank, leaving a column 34 ft. high in the tube. Pascal measured the specific gravity of mercury, that is how many times heavier it is than water, and calculated that a height of 34 ft. of water was equivalent to a height of 30 in. of mercury. He also calculated that an equivalent height of red wine was 34·6 ft. On filling a long glass tube with wine and unstoppering the end under water he found that the wine stood to this height in the tube. Thus the force necessary to produce the apparent void in the tube had a precise value, measured by the equivalent heights of the liquid columns.

In the summer of 1647 Blaise moved to Paris. The fame of his experiments had preceded him and he received many visitors, including Descartes. Descartes had a prejudice against the Pascals and he refused to recognise the originality of the paper on conic sections which Blaise had written at the age of 16. The story of the visit is told in a letter written by Blaise's sister Jacqueline. Descartes arrived with a retinue of three friends and several small boys. Blaise, who had the support of his friend de Roberval, a difficult and disputatious man, was already an invalid. He was to be the victim of illness for the rest of his life. First Descartes examined the calculating machine and gave his congratulations to the young inventor. 'After that, they began to talk about the vacuum. Monsieur Descartes, when he had listened to an account of the experiment, said, in reply to my brother (who asked him what he thought could have got into the tube): "Why, a subtle form of matter." My brother answered as best as he could, but M. de Roberval, seeing that my brother fetched his voice with difficulty, began attacking M. Descartes (keeping just within the bounds of politeness), who replied, somewhat sourly, that he did not mind debating with my brother, who was a reasonable man, and would talk with him as long as he liked, but he had nothing to say to M. de Roberval, who, he declared, was influenced by prejudice.'[1]

Descartes was not convinced that a vacuum had been created and, later that year, he said in a letter that his young friend seemed to have a void in his head.

[1] Mary Duclaux, *Portrait of Pascal.*

News reached Pascal that in Warsaw Father Valerian Magni was also experimenting on the vacuum and so he hastily published an account of his own experiments under the title *Expériences Nouvelles Touchant le Vide*. This was written in the simple, lucid style of which he was a master. He concluded from his experiments 'that the void is not an impossible thing in nature, and that she does not flee from it with so much horror as some imagine'.

Pascal's treatise was attacked by Father Noel, an elderly Jesuit and a former teacher of Descartes. Noel suggested that the gap above the mercury in the Torricelli experiment was not a vacuum but was filled with a purified air which entered through the pores in the glass. He maintained that the space could not be a vacuum because it transmitted light and bodies took time to fall in it—for example the mercury did not fall instantaneously. As a good scholastic, he brought into his thesis the four elements, earth, air, fire and water, and the four humours, bile, phlegm, melancholy and blood. He produced an involved metaphysical argument on the nature of Nothing, but he rested his case mainly on the authority of Aristotle.

Pascal's reply was a clear exposition of the nature of experimental proof. He maintained that metaphysical arguments in a scientific matter are valueless and that, when an authority such as Aristotle is cited, one should give his demonstrations and not merely his name. An investigator of nature should not be required to reconcile his science with his religious beliefs; these should be kept separate and independent. Pascal's replies were given with politeness but delicate irony.

Pascal's most important scientific achievement was to relate the pressure exerted by the air, an invisible and at that time a mysterious and elusive substance, to the pressure exerted by liquids. 'I am inclined to import all these results to the weight and pressure of the air', he wrote, 'because I consider them only as particular cases of the general proposition respecting the equilibrium of fluids.' He performed numerous experiments on liquid pressure. For example, he lowered a tube, to which was attached a bag containing mercury (Fig. 15), into a deep vessel containing water. The mercury

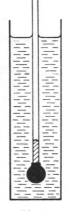

Fig. 15

rose as the tube was lowered into the water because the pressure exerted by the water became greater with increasing depth. This is analogous to the famous mountain experiment which we shall describe shortly.

Pascal calculated the weight of the earth's atmosphere, which must be the same as that of a spherical shell of water, 34 ft. thick, surrounding the earth, since the pressure of the air can support a column of water 34 ft. high. He argued that it is because the atmosphere has a finite weight that its effects must be limited. We do not feel the weight or pressure of the air because it presses from all sides, just as fishes do not feel the weight of the sea.

Air differs from the sea, however, in that it is readily compressible, and Pascal compared it to wool. 'If there were collected a great bulk of wool, say twenty or thirty fathoms high, this mass would be compressed by its own weight; the bottom layers would be far more compressed than the middle or top layers, because they are pressed by a greater quantity of wool. Similarly, the mass of the air, which is a compressible and heavy body like wool, is compressed by its own weight, and the air at the bottom, in the lowlands, is far more compressed than the higher layers on the mountain tops, because it bears a greater load of air.'

He wrote to his brother-in-law, Monsieur Perier of Clermont, asking him to take a Torricelli tube to the top of the Puy de Dôme, the highest mountain in Auvergne, to test whether the mercury fell with increasing height. The experiment was tried with complete success. The following day a tube was taken from the ground to the top of the highest tower of Notre Dame de Clermont and a small but measurable decrease in the height of the mercury column was noted. It was now clear that the tube could be used as a barometer for finding the pressure of the atmosphere, and Perier suggested that it might also be employed as an altimeter for measuring the heights of mountains.

Pascal believed with Descartes that science should be a unified system of principles and laws and that it could be presented deductively like geometry. But he laid far more stress than Descartes on the role of experiment in discovering the principles: 'Les expériences...sont les seuls principes de la physique', he wrote.

As a physicist and mathematician Pascal stands in the second rank, but as a writer of prose and as a Christian apologist he is unexcelled. On the night of 23 November 1654, he underwent a mystical religious experience and wrote of it on a piece of parchment, which he sewed in the lining of his coat and carried about with him until his death. Thereafter he forsook his work in science, devoting himself to the pursuit of truth on a different plane.

Voltaire, a century later, reproached Pascal for sewing papers in his clothes when he should have been giving France the honour of the discovery of the infinitesimal calculus ('à coudre des papiers dans ses poches, quand c'était l'heure de donner à la France la gloire du calcul de l'infini').

*   *   *

Unaware of Torricelli's experiment, Otto von Guericke (1602–86) was attempting to produce a vacuum at about the same time in Germany. His work was of a spectacular experimental nature and provided striking demonstrations of the pressure exerted by the air. It was performed between 1635 and 1645, rather earlier than that of Pascal. He lived in Magdeburg at the time of the Thirty Years War and narrowly escaped with his life when the town was sacked and burnt by Tilly in 1631. After serving as quartermaster-general to Gustavus Adolphus he returned to Magdeburg, was appointed its burgomaster, and supervised its rebuilding and refortification under Swedish protection.

His first attempts to produce a vacuum were made with the help of a water pump. He filled a wooden cask with water and set two strong men to pump out the water, reasoning that when the water had been drawn out the cask would contain a vacuum. The cask collapsed. With a second, stronger cask a sizzling noise was heard as the air forced its way in, while in the case of a third 'a noise like the twittering of a bird was heard and lasted for three days, for the wood was porous and let the air through'. The wooden cask was replaced by a copper globe and at last a vacuum was obtained.

Von Guericke then made a great technical advance by inventing an air-pump, with whose aid he drew out the air from two hollow, bronze hemispheres, the celebrated Magdeburg hemispheres, and showed that a large force was required to separate them because of the external pressure of the air. An experiment was performed in 1651 before the Emperor Ferdinand III, in

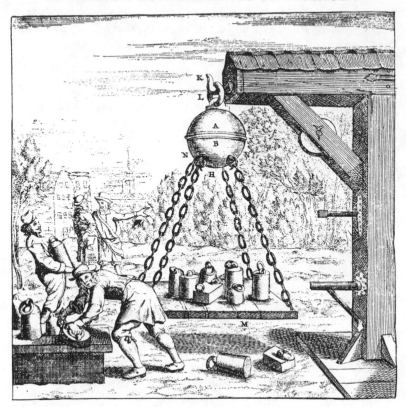

Fig. 16. An illustration from Otto von Guericke's *Experimenta Nova Magde-burgica de Vacuo Spatio* showing the great force required to separate two hollow bronze hemispheres evacuated of air by a pump.

which two teams, each of eight horses, were employed to pull the evacuated hemispheres apart.

Von Guericke used his air-pump to exhaust a metal sphere which he weighed before and after exhaustion, and was thus able to determine the density of the air. He also set up a water baro-meter at the side of his house, with a small manikin floating on the surface of the water, and made observations of the variation of the height of the water column with the weather, therewith forecasting a severe storm in 1660. He showed that light, but not sound, can be propagated through a vacuum; that in a vacuum a candle flame is extinguished and small animals die.

* * *

The chief English investigator of the phenomena associated with the pressure of the air was Robert Boyle (1627–91). He was responsible, perhaps more than any other man except Francis Bacon, for the establishment in England of the new scientific philosophy. His noble birth, his evident piety, his virtuous life in the profligate age of the Restoration, made him the most powerful spokesman against those who accused the Royal Society of a conspiracy against religion. He was a many-sided man of the Renaissance, as much a theologian as a scientist. Although he made no outstanding scientific discoveries, by his public experiments and by his prolific writings he made science known and he made it respectable.

He was the seventh son of the first Earl of Cork, a self-made man with one of the largest fortunes of his day. From his father, after a period of poverty at the close of the Civil War, he inherited £3000 a year, and this he used to equip several laboratories and to employ a staff of assistants, collectors and secretaries. Like Pascal, he experienced a religious crisis but this occurred at an early age and did not affect his scientific work. It happened in the summer of 1640, when he found himself in the midst of a terrifying thunderstorm in the Alps; he thought that the end of the world was at hand.

He was led to take up science by the influence of a group of men who began in 1645 to meet weekly in London to discuss the new experimental philosophy. They met at each other's lodgings, or in a tavern or coffee house, or sometimes in Gresham College. They were called by Boyle the Invisible College. University studies had been interrupted by the Civil War, particularly in Oxford to which Charles I transferred his court in 1642; moreover, the new ideas of the Renaissance were not taught in the scholastically dominated universities.

In 1646 Charles I escaped from Oxford and the Commonwealth forces entered the city. Nearly all the Heads of Colleges and many of the teaching officers were dismissed. Some were replaced by members of the Invisible College, among them Dr John Wilkins, who became Warden of Wadham and the centre of a group of supporters of the new philosophy. Boyle joined this group, setting up house in Oxford in 1654.

His earliest published scientific work gave an account of his experiments with an air-pump. A report of Von Guericke's air-pump had reached England and Boyle's assistant, Robert Hooke, constructed one of his own design (Fig. 17).

Boyle placed a Torricelli tube in a tall glass vessel or receiver and proceeded to evacuate the receiver. After one stroke of the pump the mercury in the tube fell from 30 to 9 or 10 in., and 'in about 3 sucks more' it fell nearly to the level in the trough. On readmission of air into the receiver the mercury rose again to 30 in. In this way Boyle proved conclusively that the column of mercury in the tube is supported by the pressure of the air on the open mercury surface in the trough. He noted the interesting fact that occasionally one stroke of the pump could be made to compress the air in the receiver, in which case the mercury rose an inch or two higher than 30 in.

Boyle was one of the founders of the Royal Society, which grew out of the Invisible College and received its Royal Charter in 1662. He was universally regarded as its most illustrious member until his fame was eclipsed by that of Newton. His air-pump, known as Boyle's Engine, was hailed as one of the greatest inventions of the age, comparable in importance with the telescope and microscope. Charles II, who was also one of the virtuosi, or amateur scientists, and had his own private laboratory in Whitehall, observed experiments made with its aid. The thrill of these early experiments, often experienced today by the young schoolboy, and the eager curiosity to try all kinds of seemingly trivial effects, is revealed in the letters and literature of the time. For example, John Evelyn wrote in his diary: 'I waited on Prince Rupert to our assembly, where we tried severall experiments in Mr Boyle's *vacuum*. A man thrusting in his arm upon exhaustion of the air had his flesh immediately swelled so as the blood was neare bursting the veins: he drawing it out, we found it all speckled.'

The Royal Society organised an expedition to repeat, on the Pic de Tenerife, Pascal's mountain experiment. Instructions were given that bladders, some half and some fully blown up, should be carried up the mountain; that living creatures should be taken up to observe the effects on their respiration, and that the experimenters should take note of any physiological changes in themselves.

The further history of the investigations into the pressure of air showed what great consequences could follow from the discovery

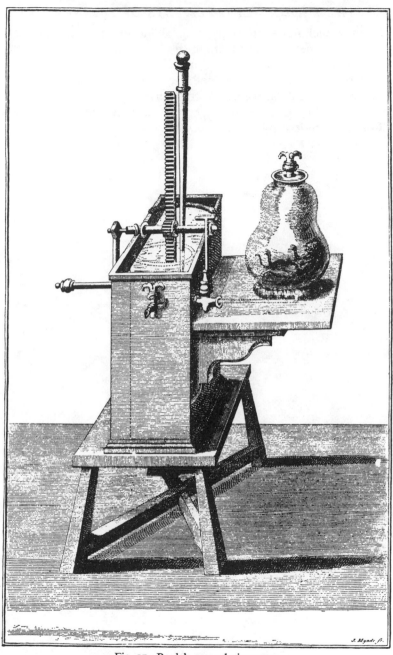

Fig. 17. Boyle's second air-pump.

of a new technique, the balancing of the pressure of gases by a column of mercury, and from the invention of a new instrument, the air-pump. Boyle measured the volumes of trapped air under different pressures and obtained the quantitative law, that the volume of a fixed mass of gas, at constant temperature, is inversely proportional to its pressure, which was the first step in the highly developed modern physical theory of gases.

CHAPTER 6

THE EARLY MICROSCOPISTS

THE SCIENTIFIC advances which sprang from the invention of the microscope were quite as remarkable as those resulting from the telescope and the air-pump.

Although the ancients used lenses as burning glasses there is no evidence that they employed them as magnifying glasses for observation. Spectacles came into use at the beginning of the thirteenth century and Roger Bacon (1214–94) wrote of the magnifying properties of lenses. But it was not until the seventeenth century that the simple microscope, consisting of a single lens, was recognised as a scientific instrument. The compound microscope, comprising two lenses and similar in principle to the telescope, was invented in 1609. Galileo published observations made with its aid on insects, remarking that it made a fly look as large as a hen. It suffered from the same defect of blurring and coloration of the image which led Newton to turn his attention to the reflecting telescope, and also from other serious defects of distortion at high magnification.

Several of the early microscopists preferred to use a single lens of very high power rather than a compound microscope because, although the latter might produce a higher magnification, this was more than offset by the distortion of the image.

One of the earliest microscopists was Robert Hooke (1635–1703), and his instrument is shown in Fig. 18. It consisted of a tube about 6 or 7 in. long, in which another tube could slide, with an objective lens at the bottom and an eye lens at the top. When he required a wide field of view, Hooke introduced a third lens in between the other two. The object was fixed on a pin attached to the base and the instrument was focused by means of a screw thread on its lower end. Hooke recognised that it was necessary 'to cast a great quantity of light' on the object since a highly magnified image is apt to be dim. 'I make choice of some Room that has only one window open to the South, and at

Fig. 18. Hooke's microscope, from his *Micrographia*.

about three or four foot distance from this Window, on a table, I place my Microscope, and then so place either a round Globe of Water, or a very deep clear plano-convex Glass (whose convex side is turned towards the Window) that there is great quantity of Rayes collected and thrown upon the Object.' At night he used the arrangement in Fig. 18, in which the light from a small lamp was focused by a globe of glass 'filled with exceeding clear Brine', the heights of the lamp and globe being adjustable.

Hooke's *Micrographia*, which appeared in 1665, described a great variety of objects as seen under the microscope and set a very high standard in the beauty of its drawings, executed by

Hooke himself. The point of a needle appeared to have 'a multitude of holes and scratches'; the edge of a sharp razor looked 'like a plow'd field with many parallels, ridges, and furrows', while the finest linen seemed like coarse matting. Crystals of snow and hoar frost, the colours of thin flakes of mica, of steel surfaces and of bubbles, moulds, moss, sponges, plants, seeds and insects, all were described. The early microscopes were often called 'flea glasses' because people were fond of observing small creatures like fleas, and the *Micrographia* had a fine plate of a flea. It also illustrated the pores of cork, which Hooke termed cells, the first use of a word which was to have a fundamental significance in biology.

The other outstanding English microscopist of the seventeenth century was Nehemiah Grew (1641–1712), whose book *The Anatomy of Plants* contained many fine plates. Grew's most remarkable discovery was the sexuality of plants; he recognised the flowers as the sexual organs, and called the stamens the male organ and the pistils the female.

*   *   *

A greater biologist than either Hooke or Grew was Marcello Malpighi (1628–94), and in his hands the microscope became a potent biological tool. Malpighi was educated, and later occupied a chair, at the university of Bologna, but much of his work was published by the Royal Society, of which he was elected an honorary Fellow in 1668. His first publication described his observation of the capillaries which carry the blood from the arteries to the veins, thereby completing Harvey's discovery of the circulation of the blood. This observation was made in the lung of a frog, water being injected into the blood vessels to wash out the blood and make them more transparent.

His famous monograph of 1669 on the structure of *Bombyx*, the silkworm, showed his skill as a microscopic anatomist. Insects have no lungs but breathe through holes at the sides of their bodies, the air being distributed by a system of tubes called *tracheae*, which Malpighi discovered and whose function he realised.

Malpighi was responsible, too, for an impressive research on the development of the heart in the embryo of a chick. His microscopic investigation of the kidney and skin are commemorated by the capsule and layer which bear his name. He shared with Grew the honour of creating the microscopic study of plant anatomy and compared plants constantly with animals.

* * *

The work of Malpighi was equalled, if not surpassed, by that of two other pioneer microscopists, both Dutchmen, Jan Swammerdam (1637–80) and Antony van Leeuwenhoek[1] (1632–1723).

Swammerdam's microscopic dissection of insects showed extraordinary skill. His work was almost unknown until, fifty-seven years after his death, his drawings and notes were published under the title *Biblia Naturae*, now recognised as one of the great classics of zoology (see Plate V).

He was an intense and intemperate worker. He spent many summer months examining the intestines and other parts of bees, working from 6 a.m. till noon and stopping only when his eyesight failed him. He would spend the rest of the day recording his observations in drawings and notes. His skill in dissection excited the admiration of all who saw it. His scalpels, knives, lancets and scissors were so fine that he had to sharpen them under the microscope. He had great dexterity in handling fine glass tubes, the thickness of a hair, which he used for inflating hollow structures and for injecting coloured wax or mercury to show up blood vessels.

His work on bees, the mayfly, the head louse, the rhinoceros beetle and other insects, undermined his health. His last six years were scientifically unproductive because he joined, with his characteristic excessive zeal, a band of religious fanatics.

* * *

Leeuwenhoek, a draper of Delft, had a passion and a flair for microscopic observation and was completely self-taught. He came into touch with the Royal Society, which was always pleased to receive scientific communications, through Oldenburg, its German-born secretary, and between 1673 and his death fifty years later at the age of 90, he sent to it several hundred letters, all in the Dutch dialect (since he knew no other language), containing a heterogeneous mass of discoveries. He had no directing hypothesis or principle guiding his observations, and their only

[1] Pronounced Laywenhook.

connecting link was the use of the microscope. He coined no new scientific terms, attempted no classification, and it took over a century to confirm, digest and classify his pioneering work. His letters are most unusual scientific documents, being written in a colloquial ingenuous style, and giving personal details with an endearing frankness.

Most of his observations were made with a single lens of high power, giving magnifications up to about × 300 and resolving objects as close as one thousandth of a millimetre. The highest power available today, by means of the electron microscope, enables objects one millionth of a millimetre apart to be resolved. Leeuwenhoek bequeathed his instruments to the Royal Society, but they were lost over a century ago, and hence we do not know whether he overcame spherical aberration, one of the most serious defects of the simple lens.

When he became famous, visitors came from far and wide to meet him and to look through his microscopes. He found these visitors a nuisance, often refused to see them, and sometimes, when he did see them, would lose patience. During the visit of the Landgrave of Hesse, 'when he had shown two or three of his microscopes, he took them away, and went to look for as many others; saying that he did this for fear lest any of them might get mislaid among the beholders, because he didn't trust people, especially Germans: and he repeated this two or three times'.

Leeuwenhoek observed with his microscope a remarkable variety of objects such as the flea, the louse, the bee, the circulation of the blood, hair, feathers, excrement, seeds, plants, and micro-organisms which he called 'little animalcules'.

His manipulative skill in micro-dissection was extremely great and he warned his contemporaries that they would not be able to repeat his dissections without practice. He knew that some persons said 'that the experiments performed by me could not possibly be performed', for example the removal of the brain from the head of a gnat, but he 'paid no attention to such insinuations'.

Perhaps his most remarkable discoveries were of his 'little animalcules'. In 1674 he observed very many of them in fresh water taken from a lake near Delft. 'And the motion of most of these animalcules in the water was so swift, and so various upwards and downwards and round about, that 'twas wonderful to see; and I judged that some of these little creatures were above a thousand times smaller than the smallest ones I have ever yet seen, upon the rind of cheese [i.e. mites], in wheaten flour mould and the like.'[1] Leeuwenhoek was observing protozoa, the simplest form of animal life, consisting of single cells.

In his eighteenth letter Leeuwenhoek describes the bell animalcule, *Vorticella*, which he said possessed a tail on the end of which was a pellet. 'These little animals were the most wretched creatures that I have ever seen; for when, with the pellet, they did but hit on any particles or little filaments (of which there are many in water, especially if it hath but stood for some days), they stuck entangled in them; and then pulled their body out into an oval, and did struggle by strongly stretching themselves, to get their tail loose.' *Vorticella*'s tail is, in fact, a stalk by which it fastens itself to the stems of freshwater plants, and it is said to be sessile.

Leeuwenhoek examined an infusion of macerated peppercorns under the microscope. Infusions, the results of soaking the parts of plants in water, are now a recognised way of producing vast numbers of minute organisms which are called by the general term, infusorians. There is usually a distinct succession of events; bacteria, which may have been present on the surface of the plants, or in the water, or have settled from the air, multiply and cause the rotting of the plants; then minute flagellate infusorians, which feed on the results of the rotting process, multiply prodigiously; then slipper animalcules, that feed mainly on the bacteria, appear and finally the vast population dies away.

Leeuwenhoek described his observations of spirilla, the 'little eels', in pepper water, as follows:

The same day, about 3 o'clock in the afternoon, I saw still more animalcules, both the round ones and those that were twice as long as broad; and besides these a sort which were still smaller; and also incredibly many of the very little animalcules whose shape, this morning, I could not make out. I now saw very plainly that these were little eels, or worms, lying all huddled up together and wriggling; just as if you saw, with the naked eye,

---

[1] This extract and those that follow are taken from *Antony van Leeuwenhoek and his 'Little Animals'*, by Clifford Dobell, F.R.S. Although Leeuwenhoek's letters were translated from the Dutch by Oldenburg, they were often abbreviated and sometimes mistranslated. Dr Dobell has retranslated them, having not only to learn the Dutch dialect, but also to master the almost indecipherable script in which the letters were written.

a whole tubful of very little eels and water, with the eels a-squirming among one another; and the whole water seemed to be alive with these multifarious animalcules. This was for me, among all the marvels that I have discovered in nature, the most marvellous of all; and I must say, for my part, that no more pleasant sight has ever yet come before my eye than these many thousands of living creatures, seen all alive in a little drop of water, moving among one another, each several creature having its own proper motion: and even if I said that there were a hundred thousand animalcules in one small drop of water which I took from the surface, I should not err. Others, seeing this, would reckon the number at quite ten times as many, thereof I have instances; but I say the least. My method for seeing the smallest animalcules, and the little eels, I do not impart to others; nor yet that for seeing very many animalcules all at once; but I keep that for myself alone.

Leeuwenhoek went on to observe the infusorians in ginger water, clove water, and nutmeg water. Many people, including Huygens, expressed doubts about the possibility of such vast numbers of organisms inhabiting so small a quantity of water. 'My counting', said Leeuwenhoek, 'is always as uncertain as that of folks who, when they see a big flock of sheep being driven, say, by merely casting their eyes upon them, how many sheep there be in the whole flock.' He decided to call in reliable independent witnesses, such as notaries and ministers of religion, to confirm his statements.

I said also that when I should again have a great number of living creatures in water, I would send you testimonials thereof, for the satisfaction of yourself and the other Philosophers: these I now send you herewith, from eight several Gentlemen, some of whom say that they have seen 10,000 living creatures in a parcel of water the bigness of a millet-seed, while others say 30,000 and also 45,000. I have generally counselled these Gentlemen, when giving their testimony, to put down but half the number that they judged they had seen; for the reason that the number of animalcules in so small a quantity of water would else be so big, that 'twould not be credited: and when I stated in my letter of 9 October 1676, that there were upwards of 1,000,000 living creatures in one drop of pepper water, I might with truth have put the number of eight times as many.

Later Huygens made observations himself and withdrew his sceptical remarks. Robert Hooke prepared infusions of pepper water so that members of the Royal Society, including Charles II, could see the animalcules. Hooke spoke of 'gygantick monsters in comparison of a lesser sort', the former being protozoa and the latter bacteria.

Leeuwenhoek discovered the animalcules which inhabit the human mouth and saw many of them in his spittle. He examined the white stuff lodged between his teeth.

I didn't clean my teeth (on purpose) for three days running, and then took the stuff that had lodged in very small quantity on the gums above my front teeth; and I mixed it both with spit and with fair rain-water; and I found a few living animalcules in it too.

While I was talking to an old man (who leads a sober life, and never drinks brandy or tobacco, and very seldom any wine), my eye fell upon his teeth, which were all coated over; so I asked him when he had last

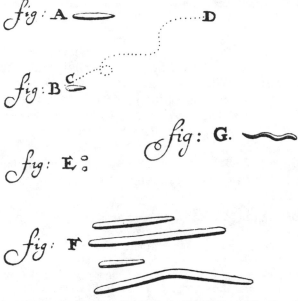

Fig. 19. Leeuwenhoek's figures of bacteria from the human mouth. *A*, a motile *Bacillus*; *B*, *Selenomonas sputigena*. The dotted line shows its motion; *E*, Micrococci; *F*, *Leptothrix buccalis*; *G*, A spirochaete. (From C. Dobell: *Antony van Leeuwenhoek and his 'Little Animals'*, by courtesy of Mrs Dobell.)

cleaned his mouth? And I got for an answer that he'd never washed his mouth in all his life, so I took some spittle out of his mouth and examined it; but I could find in it nought but what I had found in my own and other people's. I also took some of the matter that was lodged between and against his teeth, and mixing it with his own spit, and also with fair water (in which there were no animalcules), I found an unbelievably great company of living animalcules, a-swimming more nimbly than any I had ever seen up to this time. The biggest sort (whereof there were a great plenty) bent their body into curves in going forwards [as in Fig. 19 *G*]. Moreover, the other animalcules were in such enormous numbers that all the water (notwithstanding only a very little of the matter taken from between the teeth mingled with it) seemed to be alive. The long particles too, as before described, were also in great plenty.

I have also taken the spittle, and the white matter that was lodged upon and betwixt the teeth, from an old man who makes a practice of drinking brandy every morning, and wine and tobacco in the afternoon; wondering whether the animalcules, with such continual boozing, could e'en remain alive. I judged that this man, because his teeth were so uncommon foul, never washed his mouth. So I asked him, and got for an answer: 'Never in all my life with water, but it gets a good swill with wine or brandy every day.' Yet I couldn't find anything beyond the ordinary in his spittle.

When he showed the 'eels in vinegar' (*Anguilla aceti*), observed earlier by others, to several gentlewomen in his house they were so disgusted at the sight that they vowed they would never use vinegar again. 'But what if one should tell such people in future', commented Leeuwenhoek, 'that there are more animals living in the scum of the teeth than there are men in a whole kingdom.'

Summing up the letters of Leeuwenhoek, Dr Dobell writes: 'on reading these passages, the modern protozoologist, knowing the patience and perseverance needed to make such observations —even with adequate instruments and with the accumulated information of the next two hundred years to help him—can only regard this extraordinary old man, as he regarded his "little animals", with dumbfounded admiration.'

*   *   *

The first exciting period of microscopic investigation came to a close just after the end of the seventeenth century, and was followed by a necessary era of classification. The microscope introduced a severe discipline into observation and produced a profound impression by its revelation of the intricate complexity of plant and animal structure, and by the discovery of the teeming world of micro-organisms. The next great age of microscopic investigation commenced about the middle of the nineteenth century and has lasted to the present day. Achromatic microscopes, in which colour distortion was eliminated, appeared about 1830, and the modern microscope, first manufactured by Zeiss of Jena, appeared in 1878.

The reader may like to know something of the classification of Leeuwenhoek's 'little animals'. They have been divided into

(i) one-celled (or non-cellular) animals called protozoa, (ii) one-celled plants called protophyta, (iii) bacteria, best regarded neither as plants nor animals, but as a kingdom of their own. They vary in size from roughly 1 to 30 thousandths of a millimetre.

The protozoa may be subdivided as follows:

(*a*) Infusorians, many of which move about very rapidly by means of living lashes, called cilia or flagella, including *Paramecium* and *Vorticella*, well known to students of biology;

(*b*) Rhizopoda, having bodies without a fixed shape, which flow, the best known being the Amoebae;

(*c*) Sporozoa, sluggish parasitic organisms which multiply by spores, for example the malarial parasite, *Plasmodium*.

The protophyta include algae and fungi. One of the best known of the unicellular algae is *Chlamydomonas*, largely responsible for the green colour of water in a raintub or ditch, particularly in the spring.

Bacteria are classified according to their shape: a bacillus is a rod, a coccus is a sphere, and the spirillum is a spiral. Under favourable conditions bacteria may divide every half hour, each forming two individuals. Hence in 24 hours a single bacterium could give rise to 2^{48} or 280 million million bacteria. The increase stops because of the limitation of the food supply and poisonous waste products. A pound of soil may contain about 2000 million bacteria, about 300 million protozoa and 100 million algae.

Bacteria play an essential part in the economy of nature. They are responsible for the rotting and putrefaction of vegetable and animal refuse. The healthy human mouth has a large population of many different species of bacteria, and the human bowel contains countless numbers of them. Some bacteria are responsible for diseases, as we shall see in chapter 17.

CHAPTER 7

THE SEVENTEENTH CENTURY

THE CHANGE in the outlook towards the universe of western man in the seventeenth century, a change from a medieval, religious and teleological outlook to a modern, scientific and mechanistic one, has largely fashioned our present culture. What were the fundamental causes of this far-reaching change?

We mentioned in the first chapter that there are two extreme points of view among historians of science. There are those who look upon seventeenth-century science as merely a sudden step forward in a process of inquiry which has continued sporadically since the earliest days of mankind. We must allow a great deal of truth to this view when we consider how the work of Copernicus grew out of that of Ptolemy, how Galileo reacted to Aristotle and was influenced by Archimedes, and how much Vesalius and Harvey owed to Galen.

The other point of view is that of the Marxist historical materialists, who maintain that the creation of modern science was the result of economic changes. We will examine, in a little detail, the arguments in favour of this view and then consider the objections to it.

The growth of trade and the beginnings of capitalism can be traced throughout the later Middle Ages; there was, however, a pronounced economic and industrial surge forward in the sixteenth and seventeenth centuries, coinciding with the modern scientific movement. The revival of science began in Italy when the Italian cities were pre-eminent for their traders, bankers and craftsmen. As European trade and industry moved to the Atlantic seaboard, the centre of scientific advance moved to Holland, France and England.

In England a minor industrial revolution occurred in the reigns of Elizabeth I and Charles II. The production of the coal industry, one of the earliest fields of capitalist enterprise, increased fourteen-fold between 1550 and 1650. Blast furnaces, sawmills, sugar refineries, breweries, tanneries, soap-boiling and candle-making works sprang up, and new looms were invented for the manufacture of textiles.

Early in the sixteenth century there began a steady rise in prices which affected the whole of western Europe, and favoured the new mercantile classes. By the middle of the seventeenth century prices had ceased to rise and begun to fall, causing keen international trading and competition. It was at this precise moment that the English and French scientific societies were formed. Merchants saw in these societies a means of stimulating inventions for reducing labour costs and multiplying production.

The interaction between technology and science may be abundantly illustrated. The medieval clock was the source of the researches on the pendulum of Galileo and Huygens, and was thereby enormously improved. The growing needs of navigation led to the interest in astronomy which gave rise to the work of Copernicus, Tycho Brahe, Kepler, Galileo and Newton. The manufacture of spectacles, notably in the Low Countries and Venice, made possible the invention of the telescope and microscope. The needs of the mining industry in Germany, Hungary and the Low Countries, for the draining of mines by pumping, led to experiments on the vacuum, to the invention of the airpump by Otto von Guericke, and to the researches on gases of Robert Boyle. The key idea of a pump lay behind Harvey's researches on the circulation of the blood, and also behind Newcomen's invention in 1712 of the steam-engine.

Familiarity with watermills, windmills, pumps, cranes, mechanical clocks and other machines must have been largely responsible for the congenial nature of the hypothesis of the mechanical universe, which was so decisive a break with the medieval concept of the world as a moral order.

Another novel concept of the new scientific thought, differentiating it sharply from medieval philosophy, was its quantitative nature. Attempts were made to measure physical quantities and, to this end, scientific instruments were invented or improved. Galileo invented the thermometer, Torricelli the barometer, Gunter the slide rule, Gascoigne the micrometer and Huygens perfected the pendulum clock. The habit of measurement may have developed from the book-keeping of early capitalism.

The most revolutionary innovation of all was the habitual use

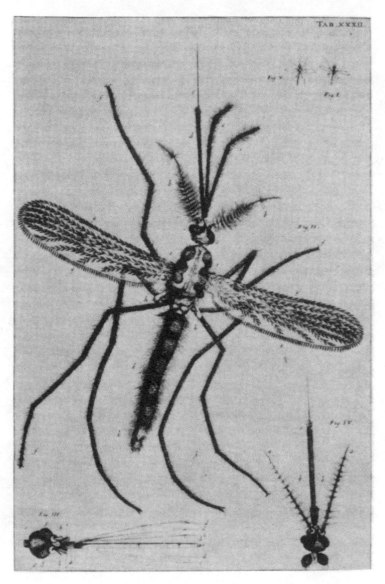

PLATE V. A Male Gnat from Swammerdam's *Biblia Naturae*. Fig. I shows the appearance of the gnat to the naked eye and Fig. II its appearance under the microscope. Fig. III illustrates the sting and Fig. IV the head and antennae.

PLATE VI. Antony van Leeuwenhoek, a portrait by Verkolje.

of experiment. It would be untrue to say that the philosophers of Greece or the schoolmen of the Middle Ages never performed an experiment, but they made no general practice of experimental investigation. They worked almost entirely with their heads and scarcely at all with their hands, building up a tradition of logical thinking and abstract generalisation. The 'new philosophy' of science emerged from the fusion of this scholarly tradition with that of the manual worker. By the end of the sixteenth century there was a comparatively large class of skilled artisans—miners, smiths, foundrymen, glass workers, surveyors, navigators and makers of instruments. These men were mainly uneducated and largely illiterate but, often under the spur of economic competition, they experimented with the materials of their craft and made improvements and minor discoveries. The rise of a capitalist society brought the upper classes into contact with industrial production and gave them an interest in crafts, and in the methods of the artisan.

A clear and early example of the combination of the two traditions is provided by the work of William Gilbert, at one time a Fellow of St John's College, Cambridge, and physician to Queen Elizabeth I. His *De Magnete*, published in 1600, was the first learned work on experimental physics to appear in England. He deliberately cultivated the acquaintance of mariners, miners and foundrymen and learnt from smiths how to forge. He was indebted to Robert Norman, a retired mariner and compass-maker of London, for the discovery that a magnetic needle, suitably suspended, dips at an angle to the horizontal. He put forward the theory that the earth is a giant magnet and constructed spherical lodestones with which to demonstrate how the angle of dip should vary over the earth's surface. He was scornful of the scholastic philosophy of the universities and dedicated his book to those 'who look for knowledge not in books but in things themselves'. While his experiments in magnetism were inspired by the needs of navigators and by a growing iron industry, his researches in static electricity had no economic incentive since there was then no obvious practical application.

Here we glimpse the Achilles heel of the Marxist explanation of the establishment of modern science. Despite the strong evidence of the influence of economic factors, it is not possible to sustain the argument that the chief incentive of most of the pioneers of science was an economic one. Galileo and Newton were both university professors, and universities are places where the disinterested quest for knowledge can be followed without economic motives. Newton wrote to Bentley, the Master of Trinity, that in the *Principia* he 'had an eye upon such principles as might work with considering men, for the ⸱ ⸱lief of a Deity'. Boyle spent much of his time and energy in an attempt to reconcile science and religion. Harvey and the early microscopists, many of them medical men, had no economic incentives. Indeed the economic background of the seventeenth century provided the conditions under which science could emerge and flourish, but it did not furnish the underlying driving force.[1]

*   *   *

Two of the most influential thinkers in the seventeenth century were Francis Bacon (1561–1626) and René Descartes (1596–1650). The former was responsible for disseminating the ideas of the inductive, experimental method, and of the control and exploitation of nature, while the latter was the chief promulgator of the philosophy of a mathematical and mechanical universe.

Bacon's lifetime coincided with the period of rapid industrial development in England. At the age of fifteen he conceived the idea, which occupied him until his death, that science could win such power over nature as to transform man's conditions of life. He wrote: 'The difference between civilized man and savages is almost that between gods and men. And this difference comes not from soil, not from climate, not from race, but from the arts.' He was referring to the mechanical arts. He believed that an equal further step forward could be taken by means of the development of applied science.

To achieve this he advocated a new method, entirely different from the philosophical methods of the ancients and of the medieval schoolmen, which he criticised as barren of effect on the improvement of human conditions. These methods were like the statues of the gods, he said, which were worshipped but could not move.

Bacon's method was essentially experimental. With his never-failing gift of metaphor, he wrote: 'Nature, like a witness, reveals

[1] See *Science and Social Welfare in the Age of Newton* (Oxford, 1937) in which Sir G. N. Clark replies to the Marxist thesis of Professor B. Hessen, 'The Social and Economic Roots of Newton's *Principia*', in *Science at the Crossroads* (London, 1931).

her secrets when put to torture.' He advocated the collection of masses of facts by observation and experiment from which generalisations could be induced.

The type of observation and experiment he envisaged are given in his fable *The New Atlantis*, published in 1626 and running through ten editions by 1670. In the mythical island of New Atlantis, in the South Seas, was a vast institute of research known as Salomon's House, where many types of applied science were practised. The conduct of research showed that Bacon had no working knowledge of the scientific method, and it is a curious and significant fact that he did not recognise the importance of the work of some of his greatest scientific contemporaries—of his own physician Harvey, of William Gilbert and of Galileo.

Some of the Fellows of Salomon's House collected information from foreign countries, some did experiments which others recorded and classified, some tried to make generalisations from the experiments, some devised new experiments, and finally the 'Interpreters of Nature' drew grand generalisations from all the discoveries made. This is a travesty of scientific method, and it inspired Swift's professor of the Academy of Lagado, who employed a machine which shook up all the words of the language into fortuitous combinations: 'the professor showed me several volumes in large folio already collected of broken sentences, which he intended to piece together, and out of these rich materials to give the world a complete body of all arts and sciences'.

Despite Swift's ridicule, Bacon was right to stress the inductive method, as opposed to the deductive method of Aristotle and the scholastics. He pointed out, with justice, that Aristotle obtained his major premises, on which he based his science, without properly consulting experience.

In the *Novum Organum*, so called because Aristotle's work on logic was known as the *Organum*, Bacon put forward a method of induction whereby satisfactory and, as he thought, conclusive and necessary generalisations about nature could be made. The simplest form of induction is induction by simple enumeration; it is based purely on a large number of instances. This Bacon rejected as uncertain. For example, that all swans are white is a generalisation to which many of us might subscribe, but it was shown to be invalid by the discovery of black swans in Australia.

To avoid the uncertainty of induction by simple enumeration

Bacon devised a method of induction by elimination. Instead of collecting only positive instances he advocated the collecting of negative instances, and then eliminating them. He took as an example, to find the cause of heat. He first drew up a long table of examples of heat—flames, sparks from the percussion of flint and steel, quicklime sprinkled with water, fresh horse dung, and so on. He then prepared a table of cases like those in the first table which were not instances of heat, and proceeded to eliminate as causes of heat whatever were common in the two tables. In a third table he collected instances of heats varying in degree and ruled out any circumstances not varying with the heat. He concluded that the cause of heat is motion.

Although his conclusion was the same as that derived two centuries later by experimental scientists, it had little influence since his method is quite different from that of normal scientific advance. He did not recognise the part, often calling for the exercise of the highest scientific ability, that conjecture or hypothesis must play in deciding which facts to seek, and which experiments to perform.

Bacon made no important scientific discovery himself. 'I shall content myself', he wrote to Lord Salisbury, 'to awake better spirits, like a bell-ringer, which is first up to call others to church.' He was the inspiration of the founders of the Royal Society and of the French encyclopaedists in the eighteenth century, so that, as Macaulay said, he moved the intellects who have moved the world. James I, on the other hand, compared Bacon's philosophy to the peace of God which passeth all understanding, and Harvey said, slightingly, that Bacon wrote philosophy like a Lord Chancellor.

At the height of his career, with the titles of Lord Verulam and Viscount St Albans, and holding the office of Lord Chancellor of England, Bacon was accused of accepting bribes. He could do no other than admit the charge and pleaded the corruption of the age. He was removed from office, imprisoned for a few days in the Tower, and sentenced to a fine of £40,000, which, however, was remitted by the king.

*   *   *

Whereas Bacon's method was experimental and inductive that of Descartes was mathematical and deductive. From simple self-evident principles Descartes deduced the outlines of a mechanical universe which fired the imagination of his contemporaries. 'Give me motion and extension', he said, 'and I will construct the world.' He believed that his scientific system could be worked out in full detail within a few generations. It provided an intense stimulus to scientific progress at the moment when the bondage of Aristotle was being broken, and it swept irresistibly through the universities of Europe.

Descartes was educated at La Flèche, one of the most celebrated of Jesuit colleges. As a young man he served as a soldier in Germany and on 10 November 1619, as he himself related, in a locked room heated by a stove, he conceived an idea for a complete reform of philosophy. The idea was the mathematical method. Since mathematics excels all the branches of thought in clarity and certainty, its method should be applied universally. Descartes was himself a distinguished mathematician, being the founder of co-ordinate geometry, a combination of algebra and geometry which is more suited to handling the problems of dynamics than pure geometry.

The science and philosophy of Descartes were presented like geometry. His starting-point was the self-evident axiom: 'I think, therefore I exist.' From this he deduced the theorem that God exists and another theorem that matter exists. He believed that there are in our minds innate ideas, fogged and confused by the mistakes in what we have read, and he was resolved to accept only those which were absolutely clear and convincing. He reached many scientifically false conclusions from his mathematico-philosophical reasoning; for example, that the smallest particle is always further divisible, that the variety of matter depends on motion, that a vacuum cannot exist. His style was beautifully limpid and precise and his works were widely read, with the result that the ladies of Molière found that they could not 'suffer a vacuum'.

Like Bacon he did not appreciate or understand the work of Galileo. He did not realise the part that experiment must play in establishing the fundamental principles of science. Indeed he criticised Galileo for not laying down intuitive, fundamental principles: 'without considering the first causes of nature he sought only for the causes of a few particular effects and thus built without a foundation.'

Although not in the first rank as a scientist, Descartes was the first great modern philosopher, making a decisive break with medieval scholasticism. He was a dualist, believing that there are two independent realities, matter and mind. The world of matter, he held, was self-operating, like a machine, and in this world he placed animals, which he regarded as automata. Mind or soul was purely spiritual, and its essence was to be conscious and to think. It was outside the mechanistic world of matter, its point of connection with matter being the pineal gland. Hence, religion and science dealt with independent spheres, and there could be no fundamental clash between them.

Descartes was a man who did much of his thinking as he lay in bed in the morning, or when he was sitting in a really warm room. At the age of 53 he was invited by the imperious Queen Christina to Sweden, as her tutor, and she sent a warship to fetch him. She liked to begin her studies at 5 a.m. and, at one of these early morning sessions, Descartes caught a chill. In a few weeks he was dead.

* * *

The pursuit of scientific studies largely outside the universities led to the formation of scientific societies or academies for discussion, experiment, and the communication of ideas. One of the earliest of these was the *Accademia dei Lincei* (Academy of the Lynx-eyed), founded in Rome in 1601 by Federigo Cesi, Marquis of Montebelli. Galileo was the sixth member of this academy and he greatly valued the contacts which it afforded. The academy came to an end with the death of Cesi in 1630. Another famous Italian society was that which flourished between 1657 and 1667 in Florence, under the patronage of the Medici family, known as the *Accademia del Cimento* (Academy of Experiment). The main achievement of this academy was the perfecting of scientific instruments and the publication of a book which became a laboratory manual in the eighteenth century.

In France a small group, including Fermat, Gassendi, Roberval and Pascal centred round Father Mersenne, who was in communication with Galileo, Descartes and Hobbes and acted as a clearing-house for scientific information. Another group met at the house of the wealthy Habert de Montmor. In 1666 the Paris *Académie des Sciences* was founded by Colbert and consisted of twenty members, each paid a pension by the king. It owed some of its inspiration to the Royal Society of London; Moray wrote to Oldenburg, 'Colbert intends to sett up a Society lyke ours'. Its meetings were held in the king's library every Wednesday and Saturday from 3 to 5 in the afternoon. In 1699 the academy was reorganised and its rules were signed by Louis XIV at Versailles. It was now a more numerous body, including all the most famous persons interested in science, and met in a more spacious apartment in the Louvre. The varied subjects which engaged the attention of the academy can be illustrated by a sample of the titles of the non-mathematical memoirs submitted by its members: *Upon the Feathers of Birds*; *Upon the Kind of Madness which is called Hydrophoby*; *Upon the circulation of the Blood in Fishes*; *Upon the Effects of the Spring of Air in Gunpowder and in Thunder*; *Upon Yellow Amber*; *Upon the Matter and Substance of Fire*; and *Upon the Trees Killed by the Frost in 1709*.

In England the Royal Society was founded by Charles II in 1662. Its object, as defined in its charter, was 'the improving of Natural Knowledge by experiment'. Its origin was a club, formed in London in 1645 by men who had been influenced by Bacon's advocacy of the 'new philosophy'.

The headquarters of the Society was Gresham College, founded in 1597 under the will of Sir Thomas Gresham for the teaching of such subjects as Geometry, Astronomy, Geography and Navigation. It was governed not by churchmen, like most educational foundations, but by the Company of Mercers, and by the Lord Mayor and Aldermen of London. It became, far more than the universities, the focus of scientific activity. Joseph Glanvill, in his Ballad of Gresham College, wrote:

> Thy Colledg, Gresham, shall hereafter
> Be the whole world's universitie,
> Oxford and Cambridge are our laughter;
> Their learning is but Pedantry.

Merchants played a considerable part in the foundation of the Royal Society. Committees were set up to investigate the histories and technical requirements of various trades such as shipping, mining, brewing, the manufacture of wool, varnish-making, engraving on copper and the making of bread.

Among the Fellows of the Society were most of the eminent men of the age. The Society could be divided, however, into two main groups; one consisting of genuine investigators who were advancing science and a much larger group of the nobility and wealthy, the 'virtuosi', many of whom collected scientific curios and oddities, following the teaching of Bacon, in the expectation that scientific laws would emerge. The virtuosi discussed such things as the miraculous properties of the salamander and experimented on the ability of a spider to escape from a circle of powdered horn of 'unicorn'.

Charles II gave the Royal Society no money and it was always in financial difficulties. The weekly contributions of one shilling, required to defray the expenses of the Secretary and Curator, of performing experiments and of the publication of the *Philosophical Transactions*, were often in arrears.

One of the most valuable activities of the scientific societies was the promotion of contacts between men of science in different countries. Oldenburg, the Secretary of the Royal Society, was forced to use the pseudonym, Mr Grubendol, because his heavy foreign correspondence aroused suspicion; even so he spent two months in the Tower in 1667.

The spirit of the Royal Society at its foundation was essentially Baconian; by the end of the century it had become Newtonian. Bacon advocated the amassing of observations and stressed the application of science to trade and industry. Newton, an unknown young man of twenty in 1662, revealed in his work the true method of scientific research and dominated the Society from his appointment as President in 1703 until his death in 1727. In the eighteenth century the early enthusiasm of the virtuosi flagged and died, and the Royal Society fell on bad times; its fine collection of instruments became covered with dust and many of them were broken or lost.

*   *   *

Most men of science in seventeenth-century England were able to retain their religious beliefs and some of them, like Robert Boyle, were at great pains to demonstrate that there was no essential conflict between science and religion. Several Fellows of the Royal Society were appointed bishops; John Wilkins became bishop of Chester, Seth Ward bishop of Salisbury, and Thomas Sprat bishop of Rochester. Nevertheless, the Society was violently attacked as subversive of religion and as a conspiracy against the church. Joseph Glanvill records that Robert Crosse, the vicar of Chew Magna, Somerset,

travelled up and down to tell his stories of the Royal Society, and to vent his spite against the honorable assembly. He took care to inform every tapster of the danger of their designs; and would scarce take his horse out a hostler's hands, till he had first let him know how he had confuted the Virtuosi. He set his everlasting tongue at work in every coffee-house, and drew the apron-men about him, as Ballad-singers do the rout in Fairs and Markets: they admired the man and wondered what the strange thing called the Royal Society could be.

There was a strong predominance of Protestants over Catholics among the members of the scientific societies. Thomas Sprat compared the origin of the Royal Society to the break with Rome of the Anglican church: 'They both may lay claim to the word Reformation, the one having compassed it in Religion, the other purposing it in Philosophy.'

At the end of the century a number of English divines, stigmatised by their High Church brethren as latitudinarians, preached the reasonableness of Christianity. Assured by Newton and Boyle that there was no fundamental conflict between science and religion and that reason was the ally rather than the enemy of religion, they held that nothing contrary to reason should be believed, although some religious truths might be above reason, and that the validity of revelation should be tested by reason. An argument which appealed to many Protestants was that, if reason could not be trusted, there was no defence against Roman Catholicism.

But Newtonian physics was insidiously undermining religious faith. The God which science revealed was remote and impersonal; he was the great Designer and, for Newton, a constant Regulator of the motion of the planets, but the conviction grew that his only intervention in the universe was at its creation. The deists believed in God only as a First Cause; they felt that Christian revelation was a tissue of superstition, and they based their religion solely on reason. Deism was never very respectable in England; in the eighteenth century, however, it swept over France in a great wave of religious scepticism.

The most profound effect of science was the destruction of the belief in the universe as a moral order and of values as absolute. The effect was incipient in the seventeenth century and it was clearly foreshadowed in the work of the philosopher Thomas Hobbes (1588–1679). As a young man Hobbes was a friend of Bacon, he lived for a time in Paris, where he met Descartes and was a member of Mersenne's group of natural philosophers, and he travelled to Italy to meet Galileo. He took over the concepts of the mechanical universe and applied them to things mental and spiritual. He was a complete materialist, rejecting the belief that mind is non-material; he regarded the human body as a machine and held that all our thoughts and aspirations are caused by the movement of atoms. He asserted that there is no such thing as absolute good; good is merely that which pleases us. 'But whatsoever is the object of any mans Appetite or Desire; that is it for which he for his part calleth Good:... There being nothing simply and absolutely so.'

These ideas of Hobbes were abhorrent to the Royal Society and, to his disappointment, he was never elected a Fellow. But they are widely held, in a modified form, today.

Landmarks of science and philosophy in the
sixteenth and seventeenth centuries

1543   Nicolaus Copernicus, *De Revolutionibus Orbium Coelestium*.
1543   Andreas Vesalius, *Fabrica Humani Corporis*.
1600   William Gilbert, *De Magnete*.
1609   Johannes Kepler, *Astronomia Nova* (1st and 2nd laws).
1620   Francis Bacon, *Novum Organum*.
1628   William Harvey, *De Motu Cordis et Sanguinis*.
1632   Galileo Galilei, *Dialogo... Tolemaico e Copernicano*.
1637   René Descartes, *Discours de la Méthode*.
1644   Evangelista Torricelli, Letter on his experiment to Michelangelo Ricci.
1647   Blaise Pascal, *Expériences Nouvelles touchant le Vide*.
1651   Thomas Hobbes, *Leviathan*.
1660   Robert Boyle, *Touching the Spring and Weight of Air*.
1665   Robert Hooke, *Micrographia*.
1672   Otto von Guericke, *Experimenta nova Magdeburgica de vacuo spatio*.
1675 and 1679   Marcello Malpighi, *Anatome Plantarum*.
1682   Nehemiah Grew, *The Anatomy of Plants*.
1682   Jan Swammerdam, *Histoire générale des Insectes*.
1687   Isaac Newton, *Principia*.
1690   Christiaan Huygens, *Traité de la Lumière*.
1690   John Locke, *Essay concerning Human Understanding*.
1695   Antony van Leeuwenhoek, *Arcana Naturae Detecta*.

(Copernicus and Vesalius were one-book men; most of the others wrote several important works of which only'one is given above. Some of the titles are slightly abbreviated.)

CHAPTER 8

THE CREATION OF MODERN CHEMISTRY

LAVOISIER'S *Traité Élémentaire de Chimie*, published in 1789, is comparable in importance with Newton's *Principia* and with Darwin's *Origin of Species* because, like them, it marks the beginning, in its modern form, of one of the major sciences. The remark of Wurtz, 'La chimie est une science française, elle fut constituée par Lavoisier' is, however, an overstatement—the exaggeration of patriotic pride. Lavoisier's long series of experiments resulted in the discovery of no new chemical substance and of few new chemical facts. His achievement was to interpret the facts discovered by others in terms of a new theory of combustion, to make clear the fundamental simplicity of chemical combination, and to devise a new chemical nomenclature. He laid the foundations on which the framework of modern chemistry, the atomic theory, was soon afterwards to be erected.

At the time he began his researches in chemistry the ancient Greek theory, that matter consists of four elements, earth, air, fire and water, still lingered in men's minds, but the generalisation which was used to correlate and interpret chemical facts was the phlogiston theory, first put forward by Johann Joachim Becher (1635–82), and developed by his disciple Georg Ernst Stahl (1660–1734). Stahl coined the word phlogiston from the Greek (*phlox* = flame) and the theory was primarily intended to account for combustion; but it was gradually extended and adapted, in a loose, uncoordinated manner, to explain almost all chemical observations. Lavoisier said of it: 'It is a veritable Proteus that changes its form every instant.'

When a substance, such as a piece of wood, burns, it gives off flame and smoke and is reduced to ashes. During combustion, asserted Stahl, it gives off fire-stuff or phlogiston. A substance like charcoal, which leaves very little ash, was regarded as almost pure phlogiston.

Charcoal, used from time immemorial for the extraction of

metals by heating it with their ores, is used up as the ore changes to metal:

$$Ore + charcoal = metal.$$

A metal was, therefore, considered to contain phlogiston, which it absorbed from the charcoal.

Some metals, such as lead, when heated until they are molten, leave a dross on the surface, called a calx, the process being known as calcination. This was explained as the giving up by the metal of its phlogiston, under the action of heat, to form the calx:

$$Metal - phlogiston = calx.$$

It was known, however, that the calx is heavier than the metal from which it is formed. An explanation of this fact, given by some supporters of the phlogiston theory, was that, just as flames rush upward, so phlogiston has levity rather than gravity—in other words, that it has negative weight. Robert Boyle (1627–91) attributed the increase in weight of the calx of tin, when tin was calcined in a closed glass vessel, to particles from the fire which passed through the pores of the glass vessel to the tin.

The view that a calx is simpler than the metal had some justification because most metals do not occur free in nature and have to be manufactured. Moreover, different metals have similar properties and it was thought that phlogiston provided the metallising principle.

Another fact about combustion (and calcination), which was established beyond doubt by the experiments of men like Boyle in the seventeenth century was that air is essential to it. A candle will not burn in a vacuum; in a limited supply of air, it will burn only for a limited time. This was explained by assuming that air is essential for the absorption of the phlogiston given off by a burning substance, and that the amount of phlogiston which air can take up is limited.

* * *

Combustion could not be understood until more was known about gases than in the seventeenth century, and the scientific career of Antoine Laurent Lavoisier (1743–94) commenced at a time when the study of gases was a dominating interest. Until the middle of the eighteenth century only one kind of gas, the air, was recognised, although it was believed that the air could take up, or be 'loaded with' various impurities in addition to phlogiston. In 1756 Joseph Black, a Scottish chemist, took a decisive step forward by establishing the existence of another gas which he called fixed air, now known as carbon dioxide (CO_2).

In September 1772 Lavoisier bought some phosphorus, which gives off white fumes[1] in the air and glows with a ghostly greenish light in the dark. He put it in a bottle and brought it near to a fire; it started to crackle. On further heating it burnt with a flame and emitted a dense white smoke. Lavoisier attached a bladder of air to the neck of the bottle and demonstrated that a large volume of air was absorbed by the phosphorus in burning. He found that in a vessel of limited air capacity only part of the phosphorus would burn and he showed that the weight of the phosphorus fumes was greater than that of the original phosphorus.

He performed similar experiments with sulphur, which takes fire when sufficiently heated and burns with a blue flame.

What impressed Lavoisier was the prodigious quantity of air absorbed by phosphorus and by sulphur when burning. This seemed to him to be the chief clue to the solution of the problem of combustion. According to the phlogiston theory, phosphorus and sulphur on burning, so far from absorbing anything, emit phlogiston.

On 1 November 1772 Lavoisier deposited a sealed note at the *Académie*, describing his experiments and his theory that combustion involves the absorption of large quantities of air. He directed that the envelope should remain sealed until the time that he made his experiments known. His intention was to secure priority for his discoveries.

Lavoisier, then aged 29, was a man of immense energy and ambition. He felt that he had found the clue to a problem of fundamental importance and one which was worthy of his powers. Chemistry, to the alchemists, had been a craft for transmuting metals and for obtaining the elixir of life; to Lavoisier, it was an intellectual exercise, a quest for the understanding and codifying of chemical processes. He wrote a memorandum, for his own future guidance, which breathes a sense of purpose and ardent dedication to his task. He realised that he had to master the researches of Black, Priestley and other workers in the field of gases and combustion. He foresaw a long series of experiments to confirm their results and to further his own theory. He

---

[1] The white fumes are phosphorus pentoxide, P_2O_5.

conceived that his mission was to accomplish nothing less than the establishment of chemistry as a coherent system in which the unrelated facts discovered by his predecessors would find their logical place. He wrote: 'The importance of the end in view prompted me to undertake all this work, which seemed to me destined to bring about a revolution in physics and in chemistry.... The results of the other authors, whom I have named, considered from this point of view, appeared to me like separate pieces of a great chain; these authors have linked only some pieces of the chain. But an immense series of experiments remains to be made in order to lead to a continuous whole.'

*　　*　　*

Lavoisier's scientific researches occupied only a small part of his time. He was elected to the *Académie des Sciences* in 1768 at the early age of 25 and, during the next twenty-five years, took part in about two hundred commissions set up by the *Académie*, which issued reports on such diverse subjects as fire and hydrogen balloons, cosmetics, ink, the art of the upholsterer, mesmerism, hospitals and prisons.

In 1768 he bought a share, from the fortune inherited from his mother, in the Ferme Générale. This was a private company of financiers who, for a fixed annual sum paid to the French government, were allowed to collect taxes on salt, tobacco, certain beverages and the customs duties at the ports and at the gates of Paris. The popular opinion of the farmers-general was epitomised in a quip of Voltaire. At a dinner party the guests began to exchange stories of robbers at whose hands they had suffered, and when it came to Voltaire's turn he said: 'Jadis, il y avait un fermier-général—ma foi, Messieurs, j'ai oublié le reste.' Lavoisier's decision to join the Ferme resulted in his death at the guillotine twenty-six years later.

In 1771 Lavoisier married Marie Paulze, the daughter of a fellow farmer-general. Although he was 28 and she was only 14, the marriage was a complete success. She was quick and intelligent and soon was acting as her husband's secretary and laboratory assistant, as well as hostess to the French and foreign savants whom Lavoisier, as a man of wealth, entertained in style. She learnt English and translated some of the scientific papers of Priestley and Cavendish into French. She took drawing lessons from the painter David and drew the sketches of her husband's apparatus for his books. After Lavoisier's untimely death she fiercely defended his honour. With many years of life still before her, she sought to recapture the happiness of participating in lively scientific research by marrying another man of genius, Count Rumford. The marriage soon ended in his locking out her guests, and in her pouring boiling water over his flowers.

In 1775 Lavoisier was appointed a Commissioner for Gunpowder and went to live at the Paris Arsenal, where he installed a magnificent laboratory. Sparing no expense, he procured the finest instruments that the craftsmen of Paris could provide. The hours of 6–9 a.m. and 7–10 p.m. were devoted to his scientific work and, in addition, one whole day a week. On this day some of his scientific friends from the *Académie* together with admiring young men would meet to discuss the work in progress and later have lunch in the laboratory.

In January 1774, little more than a year after his dedication to his task of chemical investigation, Lavoisier published *Opuscules Physiques et Chimiques*. The first fifteen chapters are an historical introduction, and it is perhaps not surprising, in view of the shortness of time in which researches in several foreign languages had to be assimilated, that the account of the work of Joseph Priestley, the outstanding experimenter on gases in Europe, was full of mistakes. Lavoisier praised Priestley's work, but made the true and devastating comment that it was 'a web of experiments almost uninterrupted by any reasoning'.

The second half of the book is devoted to an account of his experiments. In one of these he discovered the new and significant fact that, when the calx of a metal is heated with charcoal, the gas which is driven off is Black's fixed air (CO_2).

In April 1774 Lavoisier submitted to the *Académie* a memoir on his work concerning the calcination of metals in sealed vessels. Boyle had heated tin and lead in sealed vessels, unsealing the vessels after cooling, and found an increase in weight, which he ascribed to absorption of igneous particles from the furnace. Lavoisier realised that, if the increase in weight of the calx was derived, as he believed, from absorption from the air in the vessel, the total weight of the vessel after heating, but before unsealing, should be the same as before heating. Boyle had not weighed the vessel before unsealing. Lavoisier therefore repeated Boyle's experiment, heating lead in a sealed retort on a charcoal furnace, until no further calcination took place. He cooled the retort and found that its weight was unchanged. When he broke the seal he heard a whistling noise as air rushed into the retort. The increase in the total weight of the unsealed retort and its contents was equal to the increase in weight of the calx. During these experiments he wore goggles and an iron mask because of the great danger of the bursting of the sealed glass retorts when they were heated, due to the expansion of the air they contained.

No sooner had he published his results than he received a letter from Turin informing him that Father Beccaria had performed similar experiments fifteen years earlier. He had the further mortification of discovering that Jean Rey, as long ago as 1630, had anticipated the essential idea in his theory, and attributed increase in weight on calcination to absorption of air.

Lavoisier's experiments showed that the calcination of a metal is limited by the amount of air available and that only a part of the air (actually about one-fifth) is absorbed. The problem that baffled him was why only part of the air was absorbed and what was the nature of that part. He could not see his way clear for a fresh assault on the problem until the clue for which he was searching was provided by Priestley.

* * *

Joseph Priestley (1733–1804), a Unitarian minister, was a skilful experimenter with a flair and resource that resulted in a stream of chemical discoveries. Lavoisier's quantitative measurements, despite his magnificent apparatus, seldom equalled in accuracy those of Priestley. It has been said that Priestley's work was haphazard and that he had no plan of attack. But his primary interest was the investigation of gases, in which he was a pioneer, and his discoveries appear random because they were published as he made them. He discovered about ten new gases, employing the simple but effective device of collecting them over mercury (Fig. 20).

As a theoretical chemist he was vastly inferior to Lavoisier. He lacked that single-minded, ruthless determination of Lavoisier to arrange the facts of chemistry in a coherent and logical pattern. Scientific investigators can be roughly classified into two types; those whose gifts and interests lie in experiment and observation and those of a more speculative turn of mind whose main contribution is to theoretical structure. Priestley was an admirable example of the former type; Lavoisier, like Newton, was a blend of the two types—primarily a theorist but also a skilled experimenter.

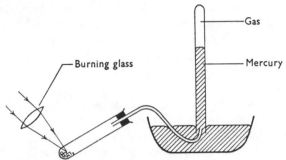

Fig. 20. Priestley's method of collecting a gas over mercury.

From 1773 to 1780 Priestley was librarian and literary companion to Lord Shelburne; it was at Lord Shelburne's seat at Calne that many of his finest experiments were performed. His fearless advocacy of civil and religious liberty (as a Dissenter the universities had been closed to him) proved an embarrassment to Lord Shelburne, who released him from his post and gave him an annuity of £150 per annum. In 1780 he took up a ministry at Birmingham and here he became a member of the Lunar Society, so called because it held monthly dinners at the time of the full moon. It consisted of a group of scientists and inventors, including James Watt and his partner in the manufacture of steam-engines, Matthew Boulton.

While in Birmingham Priestley published thirty pamphlets, and became notorious throughout the whole country. In one of these he allowed his eloquence full rein in likening the force of the truth, held by himself and his fellow Dissenters, to the force of an explosion: 'We are, as it were, laying gunpowder, grain by grain, under the old building of error and superstition, which a single spark may hereafter inflame, so as to produce an instantaneous explosion.' For this he was nicknamed by his opponent pamphleteers, 'Gunpowder Priestley', and a rumour was spread of a plot to blow up churches.

Priestley was rashly outspoken in his support of the French

97

Revolution. On 14 July 1791, a dinner was held in Birmingham by the Friends of the Revolution, to commemorate the second anniversary of the Fall of the Bastille. Priestley was not present at the dinner and was playing backgammon at his home when a friend arrived breathless to inform him that the mob was setting fire to his chapel and would soon be on its way to his house. He and his family escaped to the home of a friend, half a mile away, and from there they heard, in the moonlit summer night, the shouts of the mob as they wrecked his house. All his apparatus and his furniture were smashed, and his books and pamphlets were scattered.

Many people in England, from George III downwards, derived a malicious pleasure from Priestley's misfortune. His radical views had alienated even his scientific colleagues of the Royal Society, of which he had been a Fellow since 1766. He was offered the citizenship of France—and this did not warm the feelings of his fellow-countrymen. He emigrated to America in 1794, the year that Lavoisier was guillotined.

* * *

On 1 August 1774, Priestley, recently the possessor of a new and powerful burning glass of 12 in. diameter and 20 in. focal length, which provided a more intense heat than a flame, heated red calx of mercury (mercuric oxide, HgO) and discovered that a gas was driven off and the calx changed to mercury.[1] This was one of a series of experiments he had planned to find the effect of heating various substances by the burning glass.

What astonished him about the new gas was that a candle burned in it with a remarkably vigorous flame. He had already discovered another gas, nitrous air (nitric oxide, NO),[2] in which combustion was more vigorous than in common air. He could see no possible connection between the new gas and nitrous air because mercury calx cannot contain 'nitre'.

[1] If mercury is heated in air just above its boiling point for some days a red powder of mercuric oxide is formed. Heated above 500° C. the oxide is decomposed.

$$2Hg \ + \ O_2 \rightleftharpoons \ 2HgO$$
mercury   oxygen   mercuric oxide.

[2] Nitric oxide decomposes readily at 600° C. and supports the combustion of burning substances; it gives out considerable heat in decomposition and hence the substance burns as vigorously as it would in oxygen.

Later he mixed the new gas with nitrous air and found a reduction in volume and the formation of a red gas.[1] He found that exactly the same thing happened when nitrous air was mixed with common air but there was a smaller diminution in volume and a less red colour.

He realised that the new gas might, like common air, be respirable and on 8 March 1775 he put a mouse in it. The mouse would be expected to live a quarter of an hour in the same volume of common air, but this mouse lived for half an hour. It was taken out apparently dead but revived in front of a fire. He began to realise that the new gas was 'better' than common air and that the most accurate test of the 'goodness' of an air was not to use mice, but to mix the air with nitrous air and measure the diminution in volume.

He decided to try inhaling the new gas himself. 'The feeling of it to my lungs was not sensibly different from that of common air; but I fancied that my breast felt peculiarly light and easy for some time afterwards. Who can tell but that, in time, this pure air may become a fashionable article in luxury? Hitherto only two mice and myself have had the privilege of breathing it.'

Priestley decided to call the new gas 'dephlogisticated air' because, being such an excellent supporter of combustion, it so readily took up phlogiston. It is now known as oxygen.

In the latter half of 1774 Priestley accompanied his patron, Lord Shelburne, on a continental tour and in October he met Lavoisier in Paris. He was ten years senior to Lavoisier and, at that time, more celebrated. He enjoyed being lionised by the French savants, but his nonconformist conscience was outraged by their lack of religious belief. He informed Lavoisier of his experiments which were then in progress.

Having made the discovery of dephlogisticated air some time before I was in Paris, in the year 1774, I mentioned it at the table of Mr Lavoisier, when most of the philosophical people of the city were present, saying that it was a kind of air in which a candle burnt much better than in common air, but I had not then given it any name. At this all the company, and

---

[1] Nitric oxide combines spontaneously with oxygen and gives red fumes of nitrogen peroxide:

$$2NO + O_2 = 2NO_2$$

nitric   oxygen   nitrogen
oxide   peroxide.

Mr and Mrs Lavoisier as much as any, expressed great surprise. I told them I had gotten it from *precipitate per se* (calx of mercury) and also from red lead. Speaking French very imperfectly, and being little acquainted with the terms of chemistry, I said *plombe rouge*, which was not understood till Mr Macquer said I must mean *minium*.

Lavoisier repeated Priestley's experiments with the burning glass and mercury calx and, on 26 April 1775, he read to the *Académie* his most famous memoir *On the Nature of the Principle that combines with the Metals on Calcination and increases their Weight*. In this memoir he pointed out that to convert most calces to a metal they must be heated with charcoal. But mercury calx can be converted to mercury by heating it alone without charcoal. This is of the utmost significance.

He first heated mercury calx with charcoal and collected the gas given off. The gas precipitated lime water, it extinguished a candle flame and killed animals that were placed in it in a few seconds. It was obviously Black's fixed air.

He then heated mercury calx alone and collected the gas given off. This gas did not precipitate lime water, a candle flame burnt in it with unusual brilliance, and it supported the respiration of animals. It seemed to be especially pure air.

Lavoisier drew the conclusion that what combines with substances in combustion and calcination is highly pure air. He had not yet achieved the novel conception that the air is not an element, but a mixture of two elements. His theory was correct in principle, but still inaccurate in detail.

Although Priestley had drawn his attention to the fact that mercury calx, when heated, gives off highly pure air, Priestley was not mentioned in the memoir. When a friend remonstrated with him about this, Lavoisier laughed and said: 'My friend, you know that those who start the hare do not always catch it.'

Lavoisier's besetting fault was a reluctance to give due credit to other men working in the same field as himself. Priestley happened to be preparing the second volume of his *Observations* for the press and he took the opportunity of inserting a rebuke to Lavoisier: 'I am not conscious to myself of having concealed the least hint that was suggested to me by any person whatsoever.'

Lavoisier was always loth to admit, what history has since confirmed, that Priestley must be allowed priority in the discovery of oxygen. He had been searching for it and he alone recognised its fundamental importance; he felt almost cheated that someone else had discovered it before him.

Lavoisier reached a complete solution of the problem of calcination in 1777–8. In a classic experiment he showed that air consists of two gases: one gas (oxygen) which supports combustion and respiration, and is responsible for the increase in weight of metals on calcination; another gas (nitrogen) which supports neither combustion nor respiration.

He heated four ounces of mercury in a retort which communicated with air over mercury in a bell jar (Fig. 21). Red specks

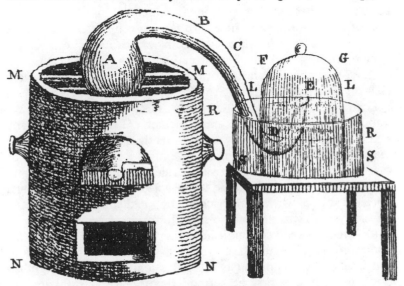

Fig. 21. Lavoisier's apparatus for investigating the composition of air, from his *Traité Élémentaire de Chimie*.

of mercury calx appeared on the surface of the mercury and he continued the heating until no more calx was formed. The volume of air in the retort and bell jar diminished from 50 cubic inches to 42 cubic inches and the 42 cubic inches which remained behind would not support combustion or respiration. He then collected the red mercury calx and heated it to a higher temperature than before in a smaller retort; the calx changed back to mercury and gave off 8 cubic inches of gas which supported combustion brilliantly and had all the properties of Priestley's dephlogisticated air. Finally, on mixing the 42 cubic inches of spent air and the 8 cubic inches of 'highly respirable air', he obtained the 50 cubic inches of common air with which he started.

The explanation on the phlogiston theory of the formation of mercury calx, by heating mercury in oxgyen, was that the mercury releases phlogiston into the oxygen. But Lavoisier demonstrated that the volume of the oxygen is reduced and that the oxygen can, in fact, be absorbed by the mercury 'to the last bubble'. This was a *reductio ad absurdum* of the phlogiston explanation.

* * *

Lavoisier's theory, that combustion consists of the combination of the burning substance with the oxygen of the air, did not gain immediate acceptance. Its truth was brought home to his contemporaries, and the phlogiston theory was finally overthrown, by Lavoisier's interpretation of the experiments of Henry Cavendish on the synthesis of water.

Henry Cavendish (1731–1810) was an eccentric aristocrat, a millionaire, and a superb experimenter. He was morbidly shy and is said to have spoken fewer words than any other man who reached the age of 80. Almost his only social contacts were at the weekly dinners of the Royal Society Club at the Crown and Anchor in London. He is said to have fled in terror from a Royal Society soirée when approached by a distinguished foreign physician who was anxious to meet him.

In 1766 Cavendish published his first chemical paper *On Factitious Airs*. The term factitious air was first used by Boyle and was intended to denote a gas other than common air. Cavendish obtained a gas, which he termed inflammable air (hydrogen) because it burns in air, from the action of acids on metals. He interpreted his experiment in terms of the phlogiston theory, regarding the metal as a combination of a calx and phlogiston:

$$\underbrace{calx + phlogiston}_{metal} + acid = \underbrace{calx + acid}_{salt} + \underset{inflammable\ air.}{phlogiston}$$

This led him to regard inflammable air as almost pure phlogiston. He wrote of the metals used in his experiments: 'their phlogiston flies off, without having its nature changed by the acid, and forms inflammable air'.

In 1781 Priestley exploded, by means of an electric spark, a mixture of inflammable air (hydrogen) and dephlogisticated air (oxygen) and noticed a deposition of dew. In the summer of that year Cavendish undertook a detailed investigation of the phenomenon and his results were published three years later in a memoir entitled *Experiments on Air*. He exploded inflammable air and common air and found that almost all the inflammable air and one-fifth of the common air condensed to dew. To find the nature of the dew he collected 135 grains of it by burning a large volume of inflammable air in common air. He found that the liquid had no taste or smell and that it left no sediment when evaporated to dryness. It seemed to be pure water. He then sparked inflammable air and dephlogisticated air and found that the volumes of the two gases which disappeared were in the ratio 2·02 : 1. He decided that inflammable air was more likely to be a mixture of phlogiston and water than pure phlogiston, because it does not react with dephlogisticated air until a flame is applied. He regarded dephlogisticated air as water deprived of its phlogiston. His explanation of his experiments was therefore as follows:

Inflammable air + dephlogisticated air = water.

water + phlogiston   water − phlogiston

In June 1783 Blagden, Cavendish's assistant, visited Paris and told Lavoisier of the experiments. Lavoisier was at first incredulous because he had long been baffled by the problem of what is produced when inflammable air burns; according to his ideas the product should have been acid. On 24 June 1783 Lavoisier repeated Cavendish's unpublished experiments in a rough, qualitative manner and the next day he announced to the *Académie* that he had burnt inflammable air in dephlogisticated air and that the result was very pure water. He further asserted, and this could only have been derived from Blagden's report of Cavendish's experiments, that the weight of the water was equal to the sum of the weights of the two gases. Cavendish was barely mentioned in the announcement. In 1786 Blagden charged Lavoisier with plagiarism and to this charge Lavoisier never made a reply.

Although Lavoisier's claim to the discovery of the composition of water is untenable, it was he who gave the correct explanation of Cavendish's experiments. He perceived, with the clarity of genius, that inflammable air and dephlogisticated air are elements; that water is no element but is a compound of the two; and that chemical combination, in its ultimate simplicity, is here revealed. Lavoisier's perspicacity is apparent by contrasting his views with those of Cavendish, who suggested that the two gases are water with an excess or deficiency of phlogiston. Nor could Cavendish recognise the merits of the new ideas when they were presented to him: 'The commonly received principle of phlogiston', he wrote, 'explains all phenomena, at least as well as Mr Lavoisier's.'

Cavendish had synthesised water; Lavoisier proceeded to analyse it. He passed water drop by drop through a red-hot, slightly inclined, iron gun-barrel. The oxygen of the water combined with the iron of the barrel to form an oxide of iron, and the hydrogen of the water was released.

Lavoisier's theory now began to gain acceptance. Berthollet publicly announced his adherence to it at a meeting of the *Académie* in April 1785; de Morveau followed suit in 1786 and de Fourcroy in 1787. Black had been quietly teaching the new theory in his lectures at Edinburgh since before 1784 and his adherence, when it became known, was particularly influential, since he was regarded as the pre-eminent chemist of his time. Most British chemists capitulated after 1789. But Priestley and Cavendish, on whose work much of the new theory was based, clung to the phlogiston theory to the end of their lives. They provide a vivid example of the conservatism of the human mind, and of the painful difficulty of abandoning ingrained ideas and of assimilating new ones. It is true that there were chemical phenomena whose nature was not fully understood, which could not be explained by the new theory, and of these Priestley made full play. But to the historian, looking back, the case for Lavoisier's theory seems to have been overwhelming.

The historian too would endorse the claim of Lavoisier, savouring though it does of too obvious egoism: 'This theory is not, as I have heard it called, the theory of the French chemists in general, it is *mine*, and it is a possession to which I lay claim before my contemporaries and before posterity.'

*   *   *

Lavoisier now sought to complete the revolution in chemistry which he had envisaged in his memorandum of 1772, by modernising chemical nomenclature, and in this task he obtained the collaboration of de Morveau, Berthollet and de Fourcroy. He maintained that the names of substances should convey their chemical composition and should form part of a rational, coherent system. Many of the current names were devised by the alchemists and he cited 'powder of Algaroth, sal alembroth, pompholix, phagedenic water, turbith mineral, aethiops and colcothar' as confusing, arbitrary and difficult to remember. He also mentioned butter of arsenic, which has an appetising sound, but is in reality a deadly poison.

The new words were derived from Latin and Greek. Dephlogisticated air or vital air was given the name oxygen (*oxys*, acid and *gennao*, I beget), and the calces of the metals were called oxides. Inflammable air was called hydrogen (*hydor*, water and *gennao*, I beget). The pure principle of charcoal became carbon, fixed air carbonic acid, and its salts carbonates. Vitriolic acid, the acid formed by the combination of sulphur and oxygen, was called sulphuric acid and its salts were called sulphates; Vitriol of Venus became copper sulphate.

The new nomenclature was received without enthusiasm by the French Academy and with open hostility abroad. Its critics maintained that it would render the writings of earlier chemists unintelligible, and that it was based upon a new theory of Lavoisier which had not stood the test of time. Even Black was bitter and sarcastic about 'the junto of French chemists'.

Lavoisier decided that the new system could be established only by a text-book and this led him to write his *Traité Élémentaire de Chimie*, published in 1789. In the book a list of known elements was given for the first time. An element was defined by Lavoisier as a substance which it had been found impossible to decompose into anything simpler. The list of thirty-three elements contained twenty-three genuine elements, in the modern sense, and it included caloric and light—a reminder that the complete modern view of combustion was not to be known for another half century, when the kinetic theory of heat (that heat is the energy of movement of the molecules of a body) was formulated.

In Lavoisier's book was the first clear statement of the law of the Conservation of Mass, which had been tacitly assumed in the researches of Black and Cavendish, namely that in a chemical change no mass is lost or gained, the mass of the final products being the same as the mass of the original substances.

The new nomenclature has survived almost unchanged to the

present day. It was perfectly adapted to the atomic theory, founded twenty years later by John Dalton.

*   *   *

The fall of the Bastille took place in the year that Lavoisier's *Traité Élémentaire de Chimie* was published, and as the Revolution took its course he found himself in greater and greater peril. He was openly accused of stealing other men's ideas by Marat, the ugly, malevolent little doctor who was a prominent member of the Paris Commune, and who was later murdered in his bath by Charlotte Corday. Marat had never forgiven Lavoisier for the rejection by the *Académie* of his researches on flame in 1780.

Lavoisier was engaged at this time in experiments on respiration, animal heat and the combustion of foodstuffs—experiments which still excite the admiration of workers in the field of nutrition. He was almost reconciled to the confiscation of his wealth and he told a friend that if the worst came to the worst he could always get a job in a chemist's shop. On 28 November 1793, at the height of the Terror, he was arrested and lodged in the Port Libre prison. On 8 May 1794 he was tried with thirty other farmers-general, and sentenced to the guillotine. When it was represented to the Court 'that the opinion of most of the scientists of Europe assigns to Citizen Lavoisier a distinguished place among those who have brought honour to France', the President of the Court, Coffinhal, uttered his celebrated *bon mot*: 'The Republic has no need of men of science.'

On the way to the guillotine, from the Conciergerie to the Place de la Révolution, the tumbrils were held up, and one of the farmers-general, regarding the unwashed mob, remarked: 'What hurts me most is to have such unpleasant heirs.' With Lavoisier perished his father-in-law; so that Madame Lavoisier, in one day, lost both husband and father.

In revolutions dog eats dog; Coffinhal, the President of the Court, and Fouquier-Tinville, the Public Prosecutor, both met their ends at the same guillotine to which they had consigned one of the greatest of all Frenchmen.

CHAPTER 9

THE HEROIC AGE OF GEOLOGY

THE FUNDAMENTAL principles of geology, which is the study of rocks and landforms, were discovered in the three decades 1790–1820. These decades have been termed the heroic age of geology. Geology developed into a science during the Industrial Revolution, being stimulated by the prospecting for coal and mineral ores, and its rise coincided also with the Romantic Movement, when mountains and the wild recesses of nature became suffused with a new interest.

The bulk of the earth's crust is formed of rocks consisting of layers or strata (Fig. 22) which, by the eighteenth century, were

Fig. 22. Strata.

recognised as formed from sediment deposited on the floor of an ocean; hence the term sedimentary rocks. Further evidence that the land had once been covered by the sea was known to the ancients. Inside the sedimentary rocks are numerous casts or moulds of plant and animal life, called fossils,[1] and these were

---

[1] See Plate XV(*b*). The term fossils includes also bones, shells, and other remains of ancient life.

considered by the Greeks and Romans to be the remains of marine animals—their presence on dry land and even on the summits of mountains being attributed to vast inundations of the past.

The true nature of fossils was not universally accepted, however, until the latter half of the seventeenth century. The Arabian, Avicenna (980–1037), thought that fossils were the unfinished work of a *vis plastica* or creative force, capable of changing the inorganic into the organic, which had given to the fossils form, but not life. Writers of the Renaissance considered them to be sports or jokes of Nature, and thought that they were reproduced in the rocks rather as the human form is imitated by the mandrake or the human voice by the parrot. Other writers held that fossils were created to be the ornaments of the secret parts of the earth, as flowers adorn its surface; that they grew from seeds or seed-bearing vapours; that they were influenced by the stars. Leonardo da Vinci (1452–1519), who as a young man was engaged in the construction of canals, and saw many marine fossils in the rocks, poured scorn on these fantastic notions and deduced that the mountains of North Italy had been covered at one time by the sea.

Writers of the seventeenth century endeavoured to explain fossils in terms of Noah's flood, but Robert Hooke pointed out that the fossils could not have been laid down during the 150 days that the waters covered the earth since the thickness of the beds is too great. Even he had no suspicion that the time required was of the order of millions of years. Many writers indulged in wild speculation far beyond any sober interpretation of the facts, in an endeavour to reconcile observation with the biblical account of the creation and the flood. Buffon, the French naturalist, said caustically that they were like the Roman augurs, who imposed such fantasies upon the credulous that they could not meet without laughing. In 1749 he denounced the idea of a universal deluge, but his book was withdrawn owing to ecclesiastical opposition.

*   *   *

One of the founders of geology was Abraham Gottlob Werner (1749–1817), professor of Mineralogy at the Mining Academy of Freiberg. His authority during his lifetime was unrivalled, and his eloquence and charm drew students from all over Europe to his crowded classes, from which they went forth with the ardour of evangelists. Saxony had been a mining centre for a thousand years and it was here that field geology began. Werner's great contribution was the concept of the regular succession of the rocks. He showed that the earth's crust was not a chaotic jumble of rocks, but an orderly succession of layers, and he established the elementary but fundamental principle that, where the strata were undisturbed, the order of superposition must be the order of formation: thus the older rocks lie in general at the bottom and the younger rocks at the top. These ideas were not new, but Werner placed them beyond dispute by his detailed and accurate studies.

Werner prided himself upon basing his deductions on first-hand observations in the field, but as these were limited to Saxony, he became the chief exponent of the untenable Neptunist theory of the origin of the earth. He held that in the beginning the earth, with naked mountains and depressions very much as they are today, was completely submerged beneath a thick and turbid primeval ocean which held in solution and suspension the materials of the present crust. The first rocks to be deposited were those which he found beneath the stratified rocks of Saxony, such as granite, and these, because of their crystalline nature, he regarded as chemical precipitates. Slates and schists were next laid down, then suspended sediments forming, for example, sandstones and limestones, and finally alluvial deposits such as sand, clay, gravel and shingle, which, being confined chiefly to the lowlands, showed that the ocean had subsided. Steeply inclined strata were laid down on mountain sides while the folding, contortion, and gashing of strata were caused by the powerful currents, the furious winds and storms which raged in and above the primeval ocean. The mineral ores, to be found in veins and layers in the rocks, were chemical precipitates.

Werner never explained satisfactorily the disappearance of all the water, which had been sufficient to cover the highest mountains. Some Neptunists maintained that it had vanished into the interior of the earth, while Werner himself speculated on the possibility of a passing heavenly body attracting it away into space. Nor could he account for the great variation in the thickness of the deposits. But where his theory foundered was in the role it assigned to volcanoes. In 1777 he visited the Mittelgebirge of Bohemia, where the coal seams had once caught fire and baked the rocks, giving them a volcanic appearance. From this he deduced that volcanoes were of subordinate importance, being caused by burning coal, and he denied that there was a vast store of subterranean heat.

The opponents of the Neptunists were the Vulcanists, who ascribed an igneous origin to certain rocks, and a long controversy focused itself on the origin of basalt, a heavy, black, lava-like rock which is found in a striking formation of polygonal columns. During his study of the Ertz mountains Werner discovered layers of basalt lying above strata of sandstone and clay, and he deduced that basalt was an aqueous deposit.

The true origin of basalt was discovered in that strange region of central France, now a place of pilgrimage to students of geology, the extinct volcanoes of Auvergne (Plate IX(a)). Jean Étienne Guettard (1715–86) was the first to discover that these mountains were ancient volcanoes. While travelling in Auvergne he noticed that the milestones were made of a black lava-like rock and, on reaching Riom, found that the houses were built of it. Although he had never seen a live volcano he recognised that the rock was quarried from a stream of solid lava. From the top of the Puy de Dôme he saw, with enlightened eyes, an array of cones and craters, now mantled on their lower slopes with vegetation and serried with streams and waterfalls.

The work of Guettard was continued by Nicholas Desmarest (1725–1815), who traced the lava flows from the craters of the Puys and discovered amongst them the unmistakable, columnar formation of basalt. Where the surface layer of volcanic ashes, slag and glassy lava had been eroded, the basalt, formed from more slowly cooling lava, was exposed. The evidence that basalt was an igneous, rather than an aqueous or sedimentary rock, was irrefutable, and although Desmarest took no part in the dispute between the Neptunists and Vulcanists, whenever he was consulted he replied 'Go and see'.

The students of Werner left Freiberg ardent Neptunists, in loyalty to their teacher, but one by one they faltered and recanted. The most distinguished was Leopold von Buch. In 1798 he visited Rome and studied the volcanic rocks of the Alban Hills. The evidence for the volcanic origin of basalt was so strong that he said he could hardly believe his eyes. In the following year he proceeded to Naples where Vesuvius was active. The basalt among the volcanic rocks was plain to see; he was impressed with the terrific power of the volcano and could find no trace of burning beds of coal. He was obliged to admit that basalt could be an igneous rock although he still supported his master's contention that the beds of basalt in Saxony were aqueous in origin.

On Werner's retirement the number of students at Freiberg rapidly declined and a clever young geologist, Ferdinand Reich, was dispatched to Auvergne to see if he could explain the rocks there on the Neptunist theory, to revive the reputation of the Academy. His report, which was never published, was a complete capitulation to the Vulcanist position.

* * *

While the attention of the geologists in Germany and France was mainly absorbed by the Neptunist–Vulcanist controversy, James Hutton (1726–97) was quietly making observations in the wild glens and coasts of his native Scotland to prove and illustrate a theory which was to become the basis of modern geology. Scotland, and particularly Edinburgh where Hutton resided, was the centre of a remarkable outburst of intellectual activity in the second half of the eighteenth century. Hutton numbered among his personal friends, Joseph Black the physicist and chemist, James Watt the engineer, Adam Smith the economist, and Henry Raeburn who painted his portrait; he must have been acquainted too with David Hume the philosopher.

Geological facts were beginning to outrun theory, and the time was ripe for some comprehensive principle to co-ordinate them. Hutton provided the principle, whose importance in geology is comparable with that of evolution in biology: 'the present is the key to the past'. He refused to postulate the universal deluges, the frightful convulsions and catastrophes which were the stock-in-trade of contemporary geologists; he observed the processes which were occurring in his own day and assumed that their operation in the past was responsible for the present features of the earth's crust. 'No powers are to be employed', he said, 'that are not natural to the globe, no action to be admitted to except those of which we know the principle, and no extraordinary events to be alleged in order to explain a common appearance.'

The first process, of which he saw signs on every hand, was that the rocks are decaying and 'the lofty mountains are going into ruin'. The chemical effects of the atmosphere cause the decomposition of the rocks; minerals become oxidised and some constituents of the rocks are dissolved by rain water which has become acid by the solution of carbon dioxide from the air. Pure rock is quickly tarnished on exposure to the atmosphere and the weathering of old buildings is a small example of what is happening on a universal scale. The strains set up in the rocks by expansion under the heat of the sun cause cracking and flaking; the expansion on freezing of water which has collected in crevices splits off fragments.

Water carries the fragments and soluble materials into the mountain streams which, clear and sparkling in fine weather, become turbid after rain. Streams and rivers behave like sinuous files wearing away their beds by the bumping, rubbing and grinding of the sand, gravel and small boulders which they carry.[1]

Hutton acknowledged his indebtedness to Desmarest, who had given proofs that valleys are made by rivers and that a landscape passes through clearly defined stages of development; also to the evidence of the erosion by running water in the Alpine valleys furnished by de Saussure, a pioneer of mountain climbing and the first to make the ascent of Mont Blanc. Hutton wrote of the Alps: 'From the top of those decaying pyramids to the sea, we have a chain of facts which clearly demonstrate this proposition, that the materials of the wasted mountains have travelled through the rivers.'

Hutton maintained that valleys are excavated by rivers, sometimes being cut through the solid rock, and supported his contention by the gentle and uniform inclination of the tributary valleys as well as of the main trunk valley. But his view was not accepted even by Lyell and Darwin who believed deep valleys to be fractures in the earth's crust caused by earthquakes or convulsions. 'We do not believe that one-thousandth part of our present valleys was excavated by the power of existing streams', wrote Buckland in 1823. In the latter half of the nineteenth century, however, when American geologists showed that even

---

[1] It is calculated that the Thames sweeps annually between one and two million tons of rock waste into the sea, while the Mississippi discharges into the Gulf of Mexico five hundred million tons of dissolved and suspended matter each year. At the present rate, the rivers of the world would reduce all land to sea level in ten million years.

PLATE VII. Lavoisier, shortly before his death. The engraving is said to have been made during his detention in 1793.

PLATE VIII. James Hutton, an engraving from the portrait by Raeburn.

(a)

(b)

PLATE IX. (a) The extinct volcanoes of Auvergne—a view from the Puy de Pariou. (b) Perched block of dark grit resting on limestone at Norber in the West Riding of Yorkshire. The block was carried on a glacier, which melted and deposited it here.

PLATE X. John Dalton collecting marsh gas, from the painting by Ford Madox Brown in the Manchester Art Gallery.

the great cañons of the Colorado are wholly due to water erosion, Hutton's view was completely vindicated.

The coarser detritus, such as gravel and sand, discharged by the rivers is deposited in shallow water, being smoothed and rounded by the scouring of the tides, whereas fine silt and mud are swept out into deeper water. The ceaseless agitation of the sea further sifts and grades the beds of sediment which, as centuries pass, slowly become thicker and thicker.

Although the rivers carry in solution vastly more lime than common salt, sea-water contains less of the former than of the latter because many marine animals take up lime to build their shells, the oyster being a large and familiar example. Beds of limestone are formed from the shells of innumerable tiny creatures and massive limestone reefs are built by coral polyps.

The logical consequence of the ceaseless wearing away of the rocks and their deposition under the sea is the ultimate disappearance of dry land unless there is the compensatory process of the raising of the sea floor. That such a process occurs is suggested by the existence of sedimentary rocks on dry land and even on the tops of the highest mountains, and by the tilting, bending, folding and breakage of the strata. Hutton suggested that the force necessary for such gigantic upheavals is to be found in the subterranean heat of the earth, having in mind the force exerted by heat in the steam-engines of his friend James Watt. In effect he regarded the earth as a gigantic heat-engine. He spoke of 'an internal fire or power of heat, and a force of irresistible expansion, in the body of the earth'. The Neptunists thought of the centre of the earth as cold, and regarded volcanoes as local phenomena. Hutton saw volcanoes as universal safety valves, the vents through which the molten rock, now known as magma, could well up from the earth's interior when the pressure became uncontainable. The modern view is that the raising of the land is due partly to the great weight of sinking sediment, thousands of feet deep, on the plastic interior of the earth; acting like a wedge, it gradually buckles and wrinkles the crust on each side. Under this lateral pressure the strata may fold like a breaking sea wave, and override each other. Rocks like the clays yield, but sandstones and limestones are liable to snap, causing slipping of the strata, known as faults, and giving rise to earthquakes.

Hutton looked for and discovered in the rocks evidence to

support his view that there have been in the past long periods of quiet deposition interspersed with periods of violent upheavals. This evidence was provided by the famous unconformities, consisting of layers of sedimentary rocks resting on the worn surfaces of other sedimentary rocks dipping at a different angle. Fig. 23 shows the striking unconformity which he found along the river Jed, near Jedburgh, where the river had hollowed out a deep gully. Vertical strata of a hard rock known as schist are overlaid with horizontal strata of softer sandstone and marl, with an intervening horizontal layer of puddingstone.

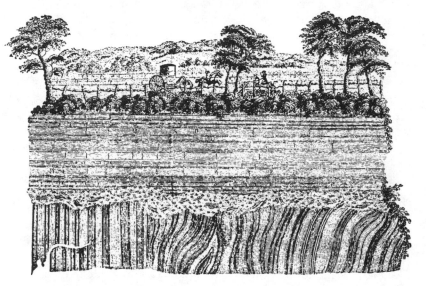

Fig. 23. Unconformity along the river Jed, from James Hutton's
Theory of the Earth.

Hutton asked how this could have occurred. The lowest strata of schist must have been laid down under the sea, and brought into their vertical orientation by some tremendous buckling of the earth's surface. The fact that their rounded tops are worn away indicates erosion by rivers, winds or tides, so that they must have been raised more or less above the surface of the sea. The layer of puddingstone immediately above them is composed of stones and gravel, which appear to be products of this erosion of the schist. The top strata of sandstone and marl could have been laid down only after another subsidence beneath the sea, and the dry land of today by a final elevation. Hutton rightly argued that an unconformity revealed 'a succession of former worlds', and there is now abundant evidence that Britain has several times been long beneath the sea.

Hutton deduced from his theory of the internal heat of the earth that the frequent occurrence of veins and fissures of foreign rocks containing no fossils, in broken and intersected strata, was due to the intrusion of molten material from below. He ascribed such an igneous origin to granite, which, being a crystalline rock, was regarded by the Neptunists as a chemical precipitate from water. It was plausible to regard large sheets or beds of granite as precipitated from the sea, but, during his visit to Glen Tilt in the north of Perthshire, Hutton saw veins of red granite in the black schist which must have been intruded. 'The sight of objects which verified at once so many important conclusions in his system', wrote Playfair, 'filled him with delight, and as his feelings on such occasions were always strongly expressed, the guides who accompanied him were convinced that it must be nothing less than the discovery of a vein of silver or gold that could call forth such strong marks of joy and exultation.'

The Vulcanists maintained that basalt was an igneous rock, but Hutton went a step further in recognising granite as igneous also. He and his followers, as an extreme sect of Vulcanism, were labelled Plutonists. Charles Darwin, who attended the lectures on geology at Edinburgh University by Professor Jamieson, an ardent disciple of Werner, records that he was taken with a party of students to Salisbury Crags to study a dyke 'which, said the professor, was a fissure filled with sediment from above, adding with a sneer that there were men who maintained that it had been injected from beneath in a molten condition'.

Sir James Hall, after many conversations and arguments with Hutton in his home at Dunglass, became a strong supporter of the latter's views. An accident at the Leith glass works prompted him to test them by experiment. A large mass of green glass, allowed to cool slowly, became white, opaque and crystalline; when remelted and cooled rapidly it regained its glassy appearance. This suggested the reason for the difference between the appearances of glassy lava poured forth on the surface of the earth's crust and the crystalline igneous rocks, formed from substantially the same material, which were intruded undergound. Hall took specimens of the whinstones from the dykes around Edinburgh, heated them in the reverberatory furnace of an iron foundry, and found that they became glassy on rapid cooling, but crystalline on slow cooling. He obtained similar results with the lavas he had brought years earlier from Vesuvius and Etna. The large crystals in granite, visible to the unaided eye, indicate that it cooled very slowly under great pressure inside the strata into which it had been intruded. The granite of Dartmoor, for example, cooled deep under the surface and owes its exposure to erosion.

Hutton had some conception of the nature of metamorphic rocks, which are sedimentary rocks transformed by the heat of intruded igneous rocks. Sir James Hall, by heating carbonate of lime in firmly closed gun barrels, proved that limestone, under the action of great heat and pressure, becomes marble.

The consolidation of sediment into rock—for example, the compacting of loose sand into sandstones, of mud into shales and slates, of marine ooze and shells into limestone, of loose pebbles into conglomerate—was ascribed by Hutton to the action of subterranean heat. In this he was mistaken, for it is due mainly to alternate wetting and drying, whereby dissolved matter, especially lime, left as a film between the grains, acts as a kind of cement.

Hutton's theory of the decay of the land by erosion and its replacement by the upheaval of the sea bed is best summarised in his own words: 'This earth, like the body of an animal, is wasted at the same time that it is repaired. It has a state of growth and augmentation; it has another state, which is that of diminution and decay. This world is thus destroyed in one part, but it is renewed in another; and the operations by which this world is thus constantly renewed, are as evident to the scientific eye, as are those in which it is necessarily destroyed.'

The theory demanded a vast expanse of time; Hutton's dictum, 'we find no vestige of a beginning—no prospect of an end', caused him to be denounced as a freethinker and an atheist. His original paper, read before the Royal Society of Edinburgh in 1785, attracted little attention. Stimulated by an attack by a member of the Royal Irish Academy he composed his larger work, *Theory of the Earth, with Proofs and Illustrations*, which was published in 1795. His friend, John Playfair, popularised his ideas in *Illustrations to the Huttonian Theory*, 1802.

*   *   *

The discovery of a further fundamental principle, providing a sure method of identifying and ascribing a geological age to the strata, hitherto classified according to their chemical composition, mineral content and position, completed the heroic age of geology.

William Smith (1769–1838) was an English engineer and surveyor who, as a young man, supervised the construction of the Somerset Coal Canal, and later in the course of his professional duties travelled to all parts of England, sometimes covering as much as 10,000 miles in a year. Wherever he went he took notes

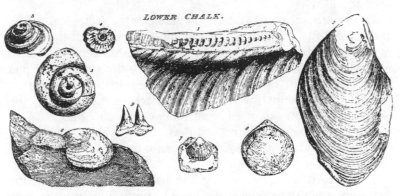

Fig. 24. Fossils in the Lower Chalk from William Smith's *Strata Identified by Organized Fossils*. The drawings were made from Smith's collection, deposited in the British Museum, of about 700 species of fossil shells, zoophytes and other fossils found in England and Wales. 1, *Inoceramus cuvieri*; 2, *Inoceramus*; 3, cast of the inside of a *Trochus*; 4, ammonite; 5, *Cirrus depressus*; 6, Terebratula; 7, Terebratula; 8, Terebratula subundata; 9, shark's teeth.

of the geological structure. He was a self-taught geologist, plain, matter of fact, with a range of interest limited to the strata, and became known to his friends as 'Strata Smith'. His great discovery was that particular fossils are peculiar to, and are found only in, particular strata. This enabled him to distinguish the different limestones, the different clays and so on in the west of England and to identify the same beds when they reappeared in other parts of the country after a gap, perhaps caused by erosion, faults, or eruptive rocks. Many fossils are not restricted to one particular stratum, but become gradually less abundant in the contiguous strata, and then disappear altogether. Thus a stratum is identified in general rather by its assemblage of fossils than by the presence of a particular fossil.

Smith discovered that there is a constancy in the order of superposition of the strata and hence he was able to forecast the fossils likely to be found in a particular district. He recorded of a geological excursion in January 1802: 'As we advanced near to the foot of the Dunstable Chalk-hills, I ventured on a prediction which in former times might have stamped me for a wizard. I said "If there be any broken ground about the foot of these hills, we may find sharks' teeth"; when Farey, presently pointing to the white bank of a new fence-ditch, we left our horses, and soon found six.'

Smith spent most of his salary in pursuing his geological hobby and spoke of 'the interminable labour of working out the truths of science'. His great ambition was to publish a geological map of England, and this he accomplished in 1815, on a scale of five miles to an inch, beautifully hand-painted in twenty different colours. He did not invent new terms for the strata from the Greek and Latin, but used local names such as London Clay, Kentish Rag, Purbeck Stone and Lias. In 1831 he received from the Geological Society the first Wollaston medal and was named by the President 'The Father of English Geology', a title which was well merited since, during the half century following his discovery, English geologists developed a time scale for measuring the age of all the sedimentary and igneous rocks.

While Smith was making his observations in England a similar investigation was being made of the Paris basin by Cuvier and Brongniart, who drew up a systematic table of the strata and of the characteristic fossils in each. Cuvier records that he made a geological excursion nearly every week for four years into the country round Paris to study the geological structure.

Cuvier was a great anatomist, as well as a geologist, and he established that the fossils are the remains of animals now extinct. He created a tremendous impression by reconstructing from a few bones some of the great saurians. He adhered firmly to the view that species are fixed and unalterable, and accounted for the extinct forms of life revealed in the fossils by a number of great catastrophes, after each of which new species were created. 'Life upon the earth in those times was often overtaken by these frightful occurrences', he wrote. 'Living things without number were swept out of existence by catastrophes. Those inhabiting the dry lands were engulfed by deluges, others whose home was in the waters perished when the sea bottom suddenly became dry land; whole races were extinguished leaving mere traces of their existence, which are now difficult of recognition, even by the naturalist. The evidences of those great and terrible events are everywhere to be clearly seen by anyone who knows how to read the record of the rocks.'

Cuvier's fellow-countryman, Lamarck, put forward an alternative view, that the fossils are evidence of a gradual evolutionary process, but this gained no acceptance until the publication, in 1859, of Darwin's *Origin of Species*.

* * *

The ideas of the heroic age of geology, particularly those of Hutton, were not immediately accepted; the man who gained for them universal credence was Charles Lyell (1797–1875), born in the year that Hutton died. Lyell was not a particularly original thinker but his *Principles of Geology*, published in three volumes from 1830 to 1833, was the most influential book on geology which has ever appeared.

At that time almost all geologists were catastrophists. Speaking of two of his eminent contemporaries, Lyell complained that, to account for the geological facts, they required three universal deluges besides that of Noah, and 'God knows how many catastrophes'. He maintained, with Hutton, that changes in the earth's surface in the past could be explained by causes operating in the present. His theory of slow and gradual geological change came to be known as uniformitarianism.

While Hutton believed that there were periods of activity when mountains were raised, followed by long periods of repose, Lyell held that the raising of mountains was occurring as rapidly now as in the past. He confirmed by his own observations that the coast of Sweden was rising at the rate of several feet a century, and it was well known that sudden changes in the level of the ground could occur through earthquakes. By means of such changes, acting over a long period of time, the highest mountains could have been formed without any catastrophic convulsion.

The curious phenomenon of erratic boulders, some of them weighing many tons, on the heights of the Jura, formed of rock similar to that in the Alps but quite different from that in the Jura, was explained by von Buch as the result of a frightful convulsion in the Alps, which hurled the boulders across the valley of Geneva. Erratic boulders were found in other parts of the world (Plate IX(b)), and they presented a puzzling problem. One explanation was that they were swept from the place of their origin to their present position by one of the great deluges.

In 1815 a chamois hunter of the Alps, Perraudin by name, having noticed large boulders being slowly transported by glaciers, realised that if the Alpine glaciers once extended as far as the Jura they could be responsible for the position of the erratic boulders. He consulted Jean de Charpentier, a skilled geologist and the director of the mines at Bex, who merely laughed at the idea and forgot about it. Eight years later Perraudin told his theory to M. Venetz, a civil engineer, and he was sufficiently impressed to read a paper on it to a local society. The paper stimulated Charpentier to make a survey and he found further evidence of glacial action in the Jura, for example scratched and polished outcropping rocks, and moraine deposits. He brought the theory to the notice of Louis Agassiz who was spending the summer in the Alps. Agassiz looked for and found unmistakable evidence of glacial action almost everywhere and put forward the theory of a universal ice age. It is now believed that an ice sheet once covered the British Isles as far south as a line joining the estuaries of the Thames and the Severn.

* * *

Geology had a great influence in the nineteenth century. There were repeated controversies between geologists and the upholders of the literal truth of Genesis, in which the whole future of Christianity was held by the latter to be at stake. The geologists eventually emerged victorious; the observation of nature triumphed over the revelation of scripture. It was established that the earth had not been created about 5000 years ago, but that it had been in existence for millions of years; that Noah's flood was merely an episode in a series of great local inundations. Tennyson wrote:

> There rolls the deep where grew the tree.
> O Earth, what changes hast thou seen.
> There, where the long street roars, hath been
> The stillness of the central sea.

The widespread interest in the controversies was shown by the large sale of books on geology in the first half of the nineteenth century, and the result was a growth of scepticism in religion. Man's view of his place in the scheme of things underwent a profound change.

CHAPTER 10

THE EIGHTEENTH CENTURY

THERE WAS a pause in the advance of science in the first half of the eighteenth century. In England the general enthusiasm for scientific experiment died away and the activities of the Royal Society declined.

The most far-reaching scientific achievement of the century was the establishment of modern chemistry as a result of the work of Black, Priestley, Cavendish and Lavoisier. The next most fundamental advance was the formulation of the basic principle of geology by Hutton.

Two other scientific developments were remarkable as reconnaissances preparatory to major advances in the nineteenth century. The experimental investigation of electricity generated by rubbing led to the work of Galvani (1737–98) and Volta (1745–1827) and to the discovery in 1800 of a continuous source of electric current, the voltaic cell. The classificatory and descriptive work in biology of Linnaeus (1707–78) and Buffon (1707–85) presaged evolution.

* * *

The eighteenth century, although comparatively undistinguished in its science, was outstanding in its philosophy. During the hundred years following 1690 a succession of philosophers, Locke, Berkeley, Hume and Kant, made a sustained attempt to assimilate, as it were, seventeenth-century science into philosophy. The work of Locke, Berkeley and Hume represents the highest achievement of British philosophy, while Kant is commonly regarded as the greatest of modern philosophers.

John Locke (1632–1704) was a friend of Boyle and Newton; he became a Fellow of the Royal Society and performed experiments in the field of medicine. His *Essay Concerning Human Understanding*, published in 1690, was the opening stage in a long philosophical argument. It profoundly influenced Voltaire and the other writers of the Enlightenment, who regarded it as initiating a science of the mind comparable with Newton's science of matter.

Locke's philosophical investigations originated in a discussion with half a dozen friends at which they decided that, to reach any secure solutions to fundamental problems, 'it was necessary to examine our own abilities, and see what objects our understandings were, or were not, fitted to deal with'. Locke set to work to examine his own mind and came to the conclusion that the sole source of knowledge is experience. The school of philosophy which he founded is known as empiricism, in contrast to the rationalism of Descartes, who believed that we acquire knowledge through reason, by deducing propositions, rather in the manner of geometry, from intuitive ideas innate in the mind. Locke denied that the mind is equipped with innate ideas, and held that all ideas come to us from experience.

Locke compared the mind at birth to a blank sheet of paper. By means of the senses knowledge is acquired, and this is remembered, ordered and organised. Locke was thus led to distinguish two types of ideas, ideas of sensation and ideas of reflection. By means of reflection the mind organises the simple ideas of sensation into more complex ideas. For example, sensations of hardness, smoothness, brownness and roundness may be organised into the idea of a table.

In effect, Locke carried over the atomic theory from physical science into psychology. His ideas of sensation were atomic entities, introduced by the external senses into the mind, where they were combined into compound entities by the 'internal sense' or the faculty of reflection.

Locke did not question the existence of external material objects, such as tables, since there must be something which is responsible for our ideas of sensation. But he took over from physics the distinction between primary and secondary qualities. He held that the primary qualities of size, shape, motion and number correspond to something objective in the external world. Secondary qualities, such as colour, taste and smell, on the other hand, are entirely subjective; they are merely ideas in the mind, which differ from person to person, and under different conditions. Thus we can have objective knowledge of the shape, mass and state of motion of a table, but not of its colour or smell.

Since all we can know from our ideas of sensation is appearance, it follows that we cannot know the real nature of matter. If we ask ourselves what is the nature of pure substance we can reply only that matter is the unknowable cause of our sensations. On the other hand, matter is somehow associated with the primary qualities of science.

Locke was not a consistent thinker, but a man of great common sense. His philosophy was widely accepted because it conformed with the science of his age.

The inconsistencies of Locke were criticised by George Berkeley (1685–1753), a Fellow of Trinity College, Dublin, and later bishop of Cloyne. Berkeley was concerned to refute the materialism and scepticism about religion inspired by the success of science. He believed that knowledge of reality must be acquired by religious and moral experience rather than by the senses and the scientific method.

He attacked Locke's distinction between primary and secondary qualities. He used the same arguments as Locke had done in the case of secondary qualities to show that primary qualities are subjective, and merely ideas in the mind. A table from one position may appear round and from another oval. Its apparent size depends on its distance away. He pointed out that it is impossible to separate primary and secondary qualities in our perceptions: 'For my own part I see evidently that it is not in my power to frame an idea of a body extended and moving, but I must withal give it some colour or other sensible quality which is acknowledged to exist only in the mind. In short, extension, figure and motion, abstracted from all other qualities, are inconceivable. Where therefore the other sensible qualities are, these must be also, to wit, in the mind and nowhere else.'

Berkeley concluded that there is no need to posit material substances external to the mind. Locke had admitted that the only things we can know are our own sensations and ideas. His external world of material objects, inferred as responsible for our sensations, cannot be known and can therefore be ignored. It is an unnecessary duplication of our ideas.

The external world is merely a collection of sensations and a material object is nothing more than the sum of its sensible qualities. It follows therefore, since sensations exist only in a mind, that an object cannot exist unless it is being perceived. Berkeley laid down the principle that being is to be perceived (*esse* is *percipi*) and hence that nothing can exist without a mind. To account for the continued existence of objects when they cease to be perceived by a human mind, Berkeley contended that they are always perceived by the mind of God. That which is responsible for the sensations and ideas which come to us without any volition on our part is something mental or spiritual.

Berkeley's argument that matter does not exist except in a mind did not appeal very much to his contemporaries. Dr Johnson's method of refuting it was to give a great kick to a stone. Dean Swift kept Berkeley waiting at his front door remarking that, if Berkeley's theory were true, he should be able to pass without difficulty through a closed door.

David Hume (1711–76) carried the argument a step farther than Berkeley. Just as Berkeley rejected Locke's inference of the existence of material substance to explain our perception of primary qualities, so Hume rejected Berkeley's inference of the mind of God, or of spiritual substance, to explain our ideas. He came to the conclusion that all that we really know to exist are our own ideas.

Hume pointed out that the idea of substance, whether material or spiritual, is inferred because certain perceptions are associated together. We believe that a table exists because we associate perceptions of hardness, smoothness, brownness and roundness. But there is no necessary connection between these perceptions. Hume asserted that the association of ideas, without which our experience would be chaos, must be accepted as a fact without understanding its mechanism, just as the principle of universal gravitation is accepted in physics to account for the association or attraction of material particles.

If all we can know is our own ideas, how can we distinguish between our sense impressions and our imaginings? Hume's answer was that our sense impressions have more force and a greater 'degree of liveliness' than our imaginary ideas. This is not very satisfying, since a nightmare may be distinctly livelier than our normal observation of the external world.

Hume realised that the inference of substance rests ultimately on the principle of causation, that for every effect there must be a cause. His examination of this principle is his most famous achievement and it has compelled attention from every subsequent philosopher.

Hume denied that there are such things as causes, in the sense of one event making another event happen. When we see one event follow another we cannot infer that the second event was caused by the first. After we have seen the two events in conjunction a number of times we may, by association of ideas, assume that the second event is a necessary consequence of the first, but there is no conclusive, logical reason for this. When we say that the sun will rise tomorrow or that sugar dissolves in water we can conceive the contradictory, and hence our statements are not made on necessitation.

If Hume is correct it follows that there is no logical basis for inference from the past to the future and hence for scientific methods, which are essentially the making of generalisations from past events to enable prediction about future events.

Hume did not commence his investigations as a sceptic. He merely carried the empiricism of Locke and Berkeley to its logical conclusion and thereby made it incredible. He said that he was quite confounded with the scepticism to which his inquiries had led him and, after dining and playing backgammon with his friends, his findings appeared 'so cold, strain'd and ridiculous' that he could scarcely bring himself to enter into them any farther.

The deadlock reached by Hume's scepticism was resolved by Immanuel Kant (1724-1804). Kant rejected the assumption of the empiricists, that the mind is purely passive like a blank sheet of paper, and assumed, instead, that by its activity it adds a contribution of its own to what is acquired from experience. His book, *The Critique of Pure Reason*, which first appeared in 1781, examined the part played by the mind in the gaining of knowledge and was, in a sense, a reconciliation of empiricism and rationalism. Kant did not believe that knowledge could transcend experience; he held, however, that it was partly *a priori*, that is to say, prior to experience or innate in the mind. He believed that this was a great philosophical discovery and compared it to the Copernican revolution in science. Knowledge, according to Kant, has two components: that which is given by experience or the *a posteriori* element, and that which is provided by the mind or the *a priori* element.

Kant did not accept Berkeley's argument that what is in a mind is any more real than what is not. Perception is possible only if an external world exists; our means of perception order our knowledge of this external world.

Kant believed that causality is an *a priori* concept. It is something which our minds put into the phenomena; it is a principle necessary to make a unity of our discontinuous perceptions. Similarly space and time are not things in themselves but are essential parts of our means of perception. Experience is organised in our minds into a spatial and temporal order. Kant drew up a list of twelve *a priori* concepts, which he called categories, and which he regarded as the elements in the working of our minds.

Kant was himself a talented physicist and it was one of the major aims of his philosophy to explain how mathematics and science are possible. The certain and necessary deductions of mathematics are the result of our *a priori* perception of space and time. The physical world is amenable to mathematical treatment because we ourselves, as it were, put mathematics into the phenomena as a result of our means of perception. In science we are dealing with independently existing external objects, or things in themselves, but it is possible, so Kant maintained, to investigate only their appearance, and not their true nature.

Of all the great philosophers Kant is the most congenial and illuminating to the modern physicist, although, as we now believe, he postulated too much rigidity in the human mind. For Kant all geometry was Euclidean and all physical science was Newtonian, both representing final truth. He was concerned to show how the mind could apprehend Euclidean space and construct deterministic science, that is to say science in which the law of causality is strictly true. The modern physicist, however, is faced with non-Euclidean spaces and indeterministic quantum theory. For him the problem is not how a unique Newtonian science is possible but how freely invented and alternative scientific theories are possible. We shall discuss this more fully in the last chapter.

*   *   *

The eighteenth century was pre-eminently the Age of the Enlightenment which sparkled in France, then the centre of European culture, in the dying years of the *ancien régime*. The French intellectuals or *philosophes* had a deep conviction that reason and science, if only allowed freedom of expression, would inevitably produce a new and enlightened outlook. They believed in freedom of thought, in tolerance, in reason, in science and in progress. The salons of Paris were clearing-houses of ideas, exchanged with brilliance, wit and urbanity.

The *philosophes* were the intellectual heirs of Descartes, Bacon, Newton and Locke; from Descartes they learnt the method of systematic doubt, from Bacon the idea of science as an instrument of social progress and the policy of amassing encyclopaedic knowledge, from Newton the method of physical science, and from Locke the science of the mind.

Natural laws, similar to the law of gravitation in physics, were sought in psychology, political science, economics, history and religion. The idea of nature became a substitute for ethics and theology. 'Nothing that exists can be against or outside nature', wrote Diderot; Montesquieu proclaimed that slavery was against nature; and Rousseau popularised the idea of the noble, natural savage. Science, it was believed, would teach man to behave naturally and to model his institutions rationally. A natural and rational environment would enable man's innate goodness to develop towards happiness and perfection. In this the *philosophes* differed fundamentally from the Christian view that man is originally sinful and that the earthly environment is of secondary importance.

The theory of scientific mechanism in the material world was carried over to the realm of mind and spirit. Descartes' metaphysical dualism of mind and matter had the weakness that the interaction of mind and matter was obscure and it seemed more natural that matter could think than that the mind existed independently. Locke's theory that the mind derives all knowledge through sensation and experience gave rise to a psychology of sensationalism and to a morality based on self-interest rather than an innate moral sense.

The abbé de Condillac centred his book *Traité des Sensations* (1754) upon the idea of a marble statue which acquires one by one all the human senses, beginning with smell, and in this way he analysed the development of ideas. La Mettrie, in his *L'Homme Machine* (1748), praised Descartes for saying that animals are machines, and went on to describe men as 'only animals and perpendicularly crawling machines'.

The chief intellectual monument of the *philosophes* was the

Encyclopédie, published in twenty-one volumes between 1751 and 1765. Its main editor was Diderot and among its contributors were most of the eminent French thinkers and writers of the age— Voltaire, Rousseau, Montesquieu, d'Alembert, Buffon and Turgot. The contributors did not all hold the same views, but most of them were progressive thinkers, fired with the idea of enlightenment in every field of knowledge. Much of what they wrote was subversive of the Christian religion; they had to pay lip service to established institutions, however, and their criticisms were expressed obliquely in the form of a suggestive parallel or equivocal praise. Even so the *Encyclopédie* was banned and publicly burnt. Public opinion did not support the action of the authorities and it is said that dummies were burnt while the volumes of the *Encyclopédie* remained on the shelves of the police. After the publication of the first seven volumes, as their further sale was prohibited, it was decided to bring out the remaining volumes secretly in a single issue. Diderot was driven to frenzy when he found that the printer had expurgated the edition, excising everything which he thought would give offence. Voltaire tells a story of Madame Pompadour's efforts to remove the ban on the *Encyclopédie*. During an argument at the court on the composition of gunpowder and of rouge for the cheeks, she remarked to Louis XV what a pity it was that their encyclo- paedias had been confiscated, since reference to them could have solved the problems they had been discussing.

Many of the *philosophes* were deists, that is to say they believed in the existence of a God and in natural religion without accepting revelation; others were atheists. David Hume was dining one day at the house of the Baron d'Holbach, a German baron who settled in Paris and wrote many of the scientific articles for the *Encyclopédie*. Hume said, 'Atheists! I don't believe there are any; I have never met one.' 'Then you have been unlucky', replied the Baron, 'here you are at table with seventeen.'

Voltaire remained all his life a deist. His God was the First Cause and Designer of Newton's universe. 'Nobody can doubt that a painted landscape or drawn animals are the works of a skilled artist', he wrote: 'Could copies possibly spring from the intelligence and the originals not?' He was constantly attacking revealed religion, however. He said that he was tired of hearing that twelve men had established Christianity, and he wanted to prove that it needed only one man to destroy it. Most of his letters ended with the phrase 'Écrasons l'infâme'—'We must crush the vile thing'. The vile thing was superstition, whether ecclesiastical or secular, and its crushing was the principal aim of all the *philosophes*. Voltaire spent a great deal of his time and energy attempting to undo judicial errors. He fought for the vindication of Jean Calas, broken on the wheel and burnt at Toulouse on the wrongful conviction of killing his son because the latter had become a Catholic; and of the young Chevalier de la Barre, accused of defacing a crucifix and of singing improper songs, for which he was tortured, beheaded and burnt at Abbeville.

The deterministic materialism which emerged from the writings of the *philosophes* required another century for its full development. The spirit of the Enlightenment, optimistic, humane and superficial, is still a living element in European thought.

* * *

Towards the end of the eighteenth century there arose in Germany a school of thought which abominated the materialism of the Enlightenment. This was the school of *Naturphilosophie*, of which the most prominent figures were the poet Goethe (1749–1832), the philosopher Schelling (1775–1854) and the biologist Oken (1779–1851).

Goethe looked upon nature not as a machine but as a goddess, fecund and beautiful. He believed that the way to understand her was that of the lover or the poet: sympathetic observation and, above all, intuition. In his biological studies he searched for the 'inner necessity and truth' governing the development and organisation of living things. He discovered what he thought to be the primal or basic plant form from which he could deduce all varieties of plants. He refused to accept the criticism of Schiller that his primal plant was not a fact but merely an idea. On the Lido in Venice he picked up a sheep's skull, and this prompted the theory that the animal skull is a modification of its spinal column.

The scientific investigations of which he was most proud were those on colour. When one wishes to understand a woman, there are often revealing traits which provide the clue to her whole character; so, Goethe thought, there are certain phenomena in nature of basic and illuminating importance. He selected as the basic phenomena of colour those produced by scattering—the blue colour of distant hills in the haze of a summer day and the red colour of the sun seen through mist. These are perhaps the most difficult of colour phenomena to understand and did not receive a satisfactory scientific explanation until the end of the nineteenth century. Goethe's abuse of Newton, whose experiments he could not bring himself to repeat and with whose approach to Nature he was completely out of sympathy, merely accentuated the unscientific nature of his own work.

Schelling, like Goethe, had a low opinion of the work of Newton. He believed that science should seek for spiritual forces in nature, for fundamental causes rather than for functional relationships between physical events. The science which he advocated had no practical value and was a kind of mysticism. Many secret brotherhoods were formed at this time by students at German universities to share such esoteric knowledge.

Oken has earned an honourable place in the history of science by being the first to organise meetings of scientists; the British Association for the Advancement of Science owes its origin to his example. But as a biologist he suffered from an unbridled imagination. Those of his ideas which were scientifically valuable were engulfed in an ocean of mysticism. He maintained, for example, that there must be five classes of animals to correspond with the five senses of man and that there were four parts of a plant related to the four elements, earth, air, fire and water.

Naturphilosophie was part of a wider cultural movement, the Romantic Movement, which began in Germany as a revolt against rationalism and stressed sentiment rather than reason. In many of its manifestations it involved a mystical communion with nature, familiar in the poems of Wordsworth.

Naturphilosophie had little influence on the science of England and France; some of its ideas bore fruit, however, in the almost exclusively German contributions to biology of the cell theory and embryology. Even in Germany there came sharp reaction against its indisciplined speculation. Papers without a basis of experimental or observational fact were rejected by learned scientific periodicals. Mayer's theory of the conservation of energy was at first ignored because it seemed to be tinged with the speculative spirit of the Nature philosophers.

Landmarks of science and philosophy in the eighteenth century

Science
1704   Isaac Newton, *Opticks*.
1735   Carl Linnæus, *Systema Naturae*.
1749–88   Georges Buffon, *Histoire Naturelle*.
1756   Joseph Black, *Experiments upon Magnesia Alba, etc.*
1774–86   Joseph Priestley, *Experiments and Observations*.
1775–81   William Herschel, *Investigation of the System of the Stars*.
1784–5   Henry Cavendish, *Experiments on the Composition of Water*.
1788   Joseph Lagrange, *Mécanique Analytique*.
1789   Antoine Lavoisier, *Traité Élémentaire de Chimie*.
1791   Aloisio Galvani's experiments on 'animal electricity'.
1795   James Hutton, *Theory of the Earth*.
1800   Alessandro Volta's discovery of the electric cell.
1799–1825   Pierre de Laplace, *Mécanique Céleste*.

Philosophy
1690   John Locke, *Essay concerning Human Understanding*.
1710   George Berkeley, *Treatise concerning the Principles of Human Knowledge*.
1748   David Hume, *An Enquiry concerning Human Understanding*.
1781   Immanuel Kant, *The Critique of Pure Reason*.

CHAPTER II

THE ATOMIC THEORY

CHEMISTRY has been shaped by the atomic theory as a modern building is constructed round its girder frame. Without the atomic theory it might well have remained an incoherent mass of chemical observations, like a jumbled pile of bricks incapable of soaring into the air.

The idea that matter is not continuous, but made up of very small, indestructible particles, known as atoms, dates back to the Ionian Greeks, Leucippus (500 B.C.) and Democritus (c. 420 B.C.). None of their works has survived except a few fragments, but an account of their theory was given by later writers.

It was held that in solids the atoms are interlocked by a kind of hook and eye or by the entanglement of antlers; that in liquids they are smooth and slippery—the drinking of water was compared to the sucking of poppy seeds out of the hand; that in gases they are far apart in ceaseless motion. Lucretius, one of the great poets of antiquity and an ardent disciple of Epicurus, who made atomism the basis of his philosophy, has left us a description of the theory in his long poem *De Rerum Natura* (On the Nature of Things). He held that the continuous appearance of a body is the result of the very small size of the atoms; a flock of sheep on a distant hillside, for example, appears like a stationary white patch although the sheep are moving in a random fashion.

Aristotle did not accept atomism. He did not believe in the existence of the vacuum between the atoms, nor did he find credible the existence of indivisible atoms with parts, such as hooks and eyes. Plato disliked the theory so much, and particularly the idea that the soul is made of atoms, that he wished to see all the works of Democritus burnt. Galen also was an opponent of the theory; he objected that an atom was too much like a brick to be a constituent of a seed.

After a long period of disfavour in later antiquity and the Middle Ages, the atomic theory was revived in the seventeenth century. One of those chiefly responsible for its revival was Pierre Gassendi (1592–1655), who wrote a life of Epicurus and freed the atomic theory from its association with godlessness. A French writer has said that he made the atoms Christians. Even more important than the revival of Greek atomism was the general notion of explaining phenomena in terms of particles, put forward by Descartes and developed by Boyle and Newton.

Newton, in his *Principia*, proved mathematically that Boyle's law (the pressure of a gas is inversely proportional to its volume) follows if a gas is made up of mutually repulsive particles, the forces between which are inversely proportional to the distances between their centres. 'Whether elastic fluids do really consist of particles so repelling one another', wrote Newton, 'is a physical question. We have here demonstrated mathematically the properties of fluids consisting of this kind and hence philosophers may take occasion to discuss that question.' The particles of which a gas is composed are not, in fact, mutually repulsive; the pressure of a gas is caused by the ceaseless motion of the particles and their bombardment of the walls of the vessel in which they are contained. But the starting-point of the speculations of John Dalton, with whose name the modern atomic theory is chiefly associated, was Newton's mutually repulsive particles.

*　　*　　*

Before discussing his scientific ideas, we will look briefly at the life and character of the stubborn, self-educated man whose name figures so prominently in the history of science. John Dalton (1766–1844) was the son of a humble Quaker hand-weaver living in the Cumberland village of Eaglesfield. His strong mind received its scientific bent from the influence of two men, Elihu Robinson and John Gough. Elihu Robinson, a Quaker gentleman of scientific attainments and a keen meteorologist, took an interest in the poor weaver's son and taught him science. The blind John Gough, also a man of scientific interests, encouraged Dalton to record the meteorological observations which he continued to make for fifty-seven years until the day before he died.

Dalton became a teacher of mathematics in the New College of Manchester, founded by the Presbyterians and the forerunner of Manchester College, Oxford. He was elected a member of the Literary and Philosophical Society of Manchester and later became its Secretary and President. His earliest paper read to this society was on the subject of colour blindness, a defect he discovered in himself. It is said that he went to a Manchester tailor for a suit of 'some good strong drab cloth' and, to the amazement of the tailor, selected a piece of scarlet material used for hunting coats.

He loved to spend his holidays walking on the Cumberland fells and he carried with him his home-made barometer and thermometer. He would measure the drop in the atmospheric pressure at the summit of Helvellyn, a favourite fell which he ascended forty times, find the temperature of boiling water on the top of Skiddaw, and collect marsh gas from the floating island in Derwentwater.

Dalton lectured several times at the Royal Institution, then the resort of intellect and fashion. He was not a success. He had no facility in devising impressive experimental demonstrations nor was he very competent in making work those which he did attempt.

The truth is that Dalton's manners and speech were rough and sometimes crude. When he was presented at court, William IV said to him, 'Well, Dr Dalton, how are you getting on in Manchester—all quiet, I suppose?' 'Well, I doan't know', replied Dalton, 'just middlin', I think.'

*   *   *

Dalton's atomic theory sprang from his meditation on the mutually repulsive particles which Newton had suggested might constitute a gas, and his application of this conception to the constitution of the atmosphere.

According to the law of gravitation, particles of matter should attract each other and, to account for the repulsion of the particles of a gas, Dalton imagined each particle to consist of an atom of matter embedded in a globe of heat. Heat was then thought to be a weightless, self-repulsive fluid known as caloric, so that the large globes of heat surrounding the atoms repelled each other.

When a gas is heated the pressure increases; this was explained by an increase in the size of the globes of heat. On condensation to water, steam gives out much heat and its volume is greatly diminished; this was explained by a great reduction in the size of the globes of heat.

In one of his lectures, Dalton revealed how his mind had been working with regard to the structure of the atmosphere:

Having long been accustomed to make meteorological observations, and to speculate upon the nature and constitution of the atmosphere, it often struck me with wonder how a *compound* atmosphere, or a mixture of two or more elastic fluids, should constitute apparently a homogeneous mass, or one in all mechanical relations agreeing with a simple atmosphere.... The same difficulty occurred to Dr Priestley, who discovered this compound nature of the atmosphere. He could not conceive why the oxygen gas, being specifically heaviest, should not form a distinct stratum of air at the bottom of the atmosphere, and the azotic gas [nitrogen], one at the top of the atmosphere.

To form a model of the atmosphere, 'I set to work', wrote Dalton, 'to combine my atoms on paper'. The prevailing scientific opinion was that the homogeneity of the atmosphere could be accounted for only by some kind of chemical combination of the gases constituting it. Dalton soon found that he had insufficient water and oxygen atoms, and far too many nitrogen atoms, for a simple chemical combination.

As he considered how physical equilibrium between his different types of atom could be obtained, he hit upon an idea in the autumn of 1801 which seemed to solve his problem.

The distinguishing feature of the new theory was, that the particles of one gas are not elastic or repulsive in regard to the particles of another gas, but only to particles of their own kind. Consequently when a vessel contains a mixture of two such elastic fluids, each acts independently upon the vessel, with its proper elasticity, just as if the other were absent, whilst no mutual action among the fluids themselves is observed.

Thus the gases of the atmosphere diffuse readily through each other and give rise to a homogeneous mixture. The conclusion is, of course, correct, but Dalton's explanation was erroneous.

Dalton pictured a gas as rather like a pile of jostling balloons with a pellet of shot at the centre of each. Each pellet was an atom and the balloon was weightless caloric.

*   *   *

It was in the year 1803 that Dalton began to apply his picture of atoms to chemistry. 'It occurred to me that I had never contemplated the effect of *difference in size* in the particles of elastic fluids', he wrote. The atoms of different elements might be of different size and hence of different weight. How could the relative weights of the atoms of the elements be determined?

Water is a compound of oxygen and hydrogen; an analysis of water by weight shows that eight parts of oxygen combine with one part of hydrogen (these are the approximate modern figures). If one atom of oxygen combines with one atom of hydrogen to form a compound atom (now called a molecule) of water, it follows that an atom of oxygen is eight times as heavy as an atom of hydrogen. By collecting the analyses by weight of a number of compounds, Dalton drew up the first table of atomic weights, taking hydrogen as unity.

The difficulty that confronted him, to which he never found a satisfactory answer, was to decide how many atoms of each element were contained in a compound atom. He thought it most probable that one atom of an element would combine with one atom of another element because two atoms of the same element would be mutually repulsive and one would be driven away. He was wrong in assuming that one atom of hydrogen combines with one atom of oxygen to form water. The reason for our belief that a molecule of water consists of two atoms of hydrogen and one of oxygen will appear later in this chapter. Dalton's value for the atomic weight of oxygen was, of course, half the correct value. Similarly, he took ammonia to consist of one atom of nitrogen and one of hydrogen, NH, instead of NH_3, with the result that the atomic weight he ascribed to nitrogen was only one-third of its correct value.

Some elements combine in more than one proportion, to form more than one compound. Dalton examined the analyses, made by Sir Humphry Davy, of nitrous oxide and of nitric oxide. He decided, correctly, that a compound atom of the former consists of two atoms of nitrogen and one of oxygen (N_2O), while a compound atom of the latter consists of one atom of nitrogen and one of oxygen (NO). He himself made analyses of carburetted hydrogen gas (now known as marsh gas or methane), and of olefiant gas (ethylene). He considered carburetted hydrogen gas to be CH_2, whereas it is now known to be CH_4, and olefiant gas to be CH, now known to be C_2H_4.

The conception of one atom of an element combining both with one and with two, or perhaps more, atoms of another element led Dalton to his discovery of the law of multiple proportions, namely that if two elements combine together to form more than one compound, then the weights of one element which combine with a fixed weight of the other element are in a simple multiple ratio to one another.

It was once thought that Dalton, from his analyses of carburetted hydrogen gas and of olefiant gas, discovered the law of multiple proportions first, and that from the law he was led to the atomic theory. But science does not always grow in an orderly manner. It should proceed, so it was believed in the nineteenth century, from the discovery of facts to the induction of laws and from thence to the elaboration of an explanatory theory. Dalton reversed this proper order. He started with an obstinate, preconceived idea of an atomic theory; from the theory he deduced laws; and if the facts did not fit the laws he refused to accept the accuracy of the facts.

* * *

Dalton was a crude experimenter. He provided the primitive theoretical structure of atomic theory, but the accurate analyses on which it ultimately rested were the work of other men.

Three laws form the foundations of the atomic theory. The law of constant composition, which states that all specimens of any particular compound possess the same composition by weight; the law of multiple proportions (p. 137); and the law of reciprocal proportions, which may be expressed as follows: if x parts of A combine with y parts of B, and x parts of A combine with z parts of C, then y parts of B combine with z parts of C. Hence x, y and z are called equivalent weights (when referred to hydrogen as unity).

The law of constant composition confirms Dalton's view that all compound atoms of a particular substance are made up of the same number and kind of atoms, and that all atoms of the same element have the same weight. Proust, the discoverer of the law, was involved in controversy with Berthollet, then the recognised leader of science in France. Berthollet thought that the force of chemical affinity, like gravity, must be proportional to the masses of the acting substances and maintained that the composition of a compound can vary considerably. He found that metals such as copper, tin and lead, on being heated in air, can take up almost any quantity of oxygen up to a fixed limit. Proust showed that he was dealing with mixtures of two or more different oxides of each metal and, for the first time, a clear distinction was drawn between compounds and mixtures.

The man who was chiefly responsible for placing the atomic theory on a firm experimental basis was the Swedish chemist, Berzelius. He always maintained that Dalton had taken the greatest step ever achieved in chemistry by the publication of the atomic theory, and most of his experimental work was inspired by it. The immense number and accuracy of the experimental analyses of Berzelius made him the arbiter of chemistry in Europe. In the decade 1810–20 he purified and analysed over two thousand inorganic compounds.

The influence of Berzelius as a teacher was far-reaching and men came to him for the instruction which could not be obtained in the universities. His laboratory consisted of a couple of living rooms and a kitchen in his house, equipped in the most spare and primitive manner. Yet Wöhler, one of his most distinguished pupils, wrote: 'As he led me into his laboratory I was, as it were, in a dream, doubting whether it was really true that I was in this famous place.'

Berzelius established the law of multiple proportions beyond any doubt and wrote to Dalton: 'You are right in this, that the theory of multiple proportions is a mystery without the atomic hypothesis; and so far as I have been able to see, all the results gained hitherto contribute to justify this hypothesis.'

Berzelius was responsible for our modern chemical symbols. The alchemists had an ancient sign language, including the symbols of the planets for the corresponding metals; thus a circle represented gold, a crescent, silver, and a scythe, lead. Dalton introduced his own symbolism and its unsuitability is evident from his symbol for potash alum, shown in Fig. 25.

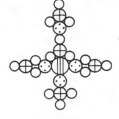

Potash alum
Fig. 25

Berzelius selected the initial letter of the Latin name of each element, two letters sometimes being necessary, for example:

Gold (aurum)	Au	Potassium (kalium)	K
Silver (argentum)	Ag	Sulphur	S
Aluminium	Al	Silicon	Si
Carbon (carbo)	C	Antimony (stibium)	Sb
Copper (cuprum)	Cu	Tin (stannum)	Sn
Oxygen	O		

Each symbol stood for one atom and hence for its atomic weight. The symbol for potash alum was $KAl(SO^4)^2$. The superscripts were later changed to subscripts: $KAl(SO_4)_2$.

Dalton reacted violently against the symbols of Berzelius; 'they cloud the beauty and simplicity of the atomic theory', he said. He felt that the great advantage of his own symbols was that they indicated the actual positions of the individual atoms in a compound atom; similar atoms were supposed to repel each other into positions of equilibrium.

* * *

It is only fair, in an account of the establishment of the atomic theory, to say a brief word about the unfortunate William Higgins. Higgins published a book in 1789, fourteen years before Dalton's first draft of his chemical atomic theory, which was mainly concerned with support for Lavoisier's views in opposition to the phlogiston theory, but which also adumbrated an atomic theory. He wrote, for example, 'water is composed of molecules, formed by the union of a single particle of oxygen to an ultimate particle of hydrogen'. But he did not draw up a table of atomic weights, nor did he deduce the law of multiple proportions. He was a pioneer, but he did not go far enough. The chemists of his time were too much concerned with the controversy over the phlogiston theory to be receptive to a new theory about atoms. His words fell upon deaf ears and his work was unheeded.

In 1814, when Dalton's work had been published and acclaimed, Higgins wrote a book to endeavour to establish his own claim to priority. He made a veiled accusation against Dalton: 'I cannot with propriety or delicacy say that Mr Dalton is a plagiarist, although appearances are against him. Probably he never read my book; yet it appears extraordinary that a man of Mr Dalton's industry and learning should neglect one of the few works that were expressly written on the subject of the theory. At the time it was published, there were one thousand copies of it sold.'

Dalton's friends maintained that he had not known of Higgins's book. Nor did Dalton ever claim to have originated the atomic conception. Whatever the rights of priority, it was Dalton, and not Higgins, who caught the ear of his contemporaries, and introduced the atomic theory into chemistry.

* * *

There were few chemists in England during Dalton's lifetime who really believed in the existence of atoms. Wollaston, by his researches, did much to confirm the law of multiple proportions, but he preferred to speak of equivalents rather than of atomic weights. His views were shared by Sir Humphry Davy. 'It is impossible not to admire the ingenuity and talent with which Mr Dalton has arranged, combined, weighed, measured and figured his atoms', said Davy, with, perhaps, some slight condescension, 'but it is not, I conceive, on any speculations upon the ultimate particles of matter, that the true theory of definite proportions must ultimately rest.'

It is difficult to imagine a greater contrast than existed between Davy and Dalton. Davy, of almost as humble origins as Dalton, was young, handsome and eloquent—a dazzling figure. He was the idol of London Society, and his fair admirers wrote him sonnets. Coleridge used to attend his lectures at the Royal Institution and said of him, 'If Davy had not been the first chemist, he would have been the first poet of his age'. He married a rich widow. He became an F.R.S. at 25, a knight at 34, a baronet at 40 and was President of the Royal Society at 42. Dalton was 56 when he became an F.R.S. His comment on Davy was as follows: 'the principal failing in his character as a philosopher is that he does not smoke'.

Scepticism about the existence of atoms continued as an undercurrent throughout the nineteenth century. As a framework, as an expository scheme, the atomic theory was indispensable. But not until the twentieth century were all doubts removed that atoms really exist.

*   *   *

The great weakness of the atomic theory, as it came from the hands of Dalton, was its inability to indicate how many atoms of each element combine together to form a molecule of a compound. Dalton gave the formula of water, for example, as HO instead of H_2O, and the formula of ammonia as NH instead of NH_3.

Several systems of chemistry arose in the second quarter of the nineteenth century, each with its own set of atomic weights and its own formulae for common chemical compounds. There was indeed great confusion; as Berzelius remarked, 'every few years the whole thing changes'. It is a strange fact that an hypothesis was put forward by Avogadro in 1811 which could have prevented all the turmoil and which, indeed, put an end to the confusion after 1860.

In 1808 Gay-Lussac published his discovery of the law that gases combine in simple proportions by volume. For example, one volume of nitrogen and one volume of oxygen produce two volumes of nitric oxide; two volumes of hydrogen and one volume of oxygen produce two volumes of steam. Dalton became aware of this law almost as soon as it was published because Berthollet sent him a copy of the *Mémoires de la Société d'Arcueil* in which Gay-Lussac's paper appeared. But Dalton refused to accept its accuracy; being a somewhat crude experimenter himself, he failed to appreciate the accuracy of Gay-Lussac.

In vain did Berzelius urge that it was necessary only to substitute the word atom for volume, when Gay-Lussac's law became Dalton's atomic theory. Dalton could see no reason for Gay-Lussac's law and he could not account for it on his theory; hence he denied it. Dalton's theory was concerned exclusively with combination by weight; it required adaptation to account for combination by volume.

Three years later Avogadro put forward the hypothesis that equal volumes of all gases (at the same temperature and pressure) contain equal numbers of *molecules*. The special significance of the word molecule will be apparent shortly. But it is well to make clear the distinction between the *law* of Gay-Lussac, which was simply an induction from the experimental facts, and the *hypothesis* of Avogadro, which was an interpretation placed upon Gay-Lussac's law.

An hypothesis similar to that of Avogadro, that equal volumes of all gases contain equal numbers of *atoms*, had already occurred to Dalton and he had rejected it. He argued in this manner. When a compound atom of nitric oxide is formed from one atom of nitrogen and one atom of oxygen, the number of atoms is halved. 'Thou knowst', said Dalton, 'no man can split an atom.' Hence the two volumes of nitric oxide contain only half as many atoms as the original two volumes of nitrogen and oxygen.

$$N + O = NO$$

| (1 atom) | (1 atom) | (1 compound atom) |
| 1 volume | 1 volume | 2 volumes |

But Avogadro had foreseen this difficulty. He drew a distinction between an atom and a molecule which was, in effect, the one accepted today. He suggested that the particles of which nitrogen gas is composed, now known as molecules, consist of two atoms bound together; thus the molecule of nitrogen is N_2. Similarly the molecule of oxygen is O_2. Thus

$$N_2 + O_2 = 2NO$$

(1 molecule)	(1 molecule)	(2 molecules)
1 volume	1 volume	2 volumes

Similarly

$$2H_2 + O_2 = 2H_2O$$

(2 molecules)	(1 molecule)	(2 molecules)
2 volumes	1 volume	2 volumes of steam

Avogadro obscured his luminous idea by a cloud of irrelevancies and by clumsy expression. Moreover, the idea of a molecule consisting of two atoms bound together was not credible to the chemists of that time because chemical affinity was considered to be electrical in nature and it was thought that two similarly electrified atoms should repel each other. Avogadro's hypothesis was ignored for half a century.

In 1860 a Congress of Chemists was held in Carlsruhe and, at the close, Cannizzaro distributed a pamphlet explaining how his fellow-countryman's hypothesis could clear up all the confusion about atomic weights and chemical formulae. Lothar Meyer relates that when he read this pamphlet, 'It was as though scales fell from my eyes, doubt vanished, and was replaced by a feeling of peaceful certainty'.

Cannizzaro's method was to use Avogadro's hypothesis to find the molecular weight of as many compounds of an element as were gaseous or volatile by weighing a volume of each gas and comparing it with the weight of an equal volume of hydrogen. The fraction by weight of the element contained in each gaseous compound can be determined by experimental analysis and these fractions of the molecular weights must represent the atomic weight of the element or a multiple of it. As a case in point, consider three compounds of nitrogen:

Compound	Molecular weight	Fraction of nitrogen in the compound
Nitrous oxide	44	28/44
Nitric oxide	30	14/30
Ammonia	17	14/17

The atomic weight of nitrogen is likely to be 14, three times Dalton's value. The molecule of the first compound, nitrous oxide, must contain two atoms of nitrogen.

Avogadro died, comparatively unknown and obscure, a year or two before his hypothesis was rescued from oblivion, but his posthumous fame has been enormous. Every schoolboy knows— or should know—Avogadro's hypothesis; it is the cornerstone of chemistry.

PLATE XI. (a) Newton's rings with yellow light. (b) Young's interference fringes with red light. (c) Thomas Young. (d) Augustin Fresnel.

(*a*)

(*b*)

PLATE XII. (*a*) Diffraction fringes due to a wire. (*b*) Diffraction fringes due to a circular obstacle. Note the bright spot at the centre of the shadow.

CHAPTER 12

THE WAVE THEORY OF LIGHT

THE WAVE THEORY of light is one of the major historic theories of physics. Its early history is linked with the controversy over Newton's investigations into the colours produced by prisms and lenses.

In 1672 Newton sent a paper to the Royal Society describing his experiments on the splitting up of a narrow beam of white light into a rainbow-coloured spectrum on passing through a glass prism. In one paragraph of the paper he mentioned that, to account for the breadth of the spectrum, he had considered whether the rays of light, on traversing the prism, might acquire a curved instead of a straight path, by analogy with the curved path of a tennis ball which has been given spin by a racket. One of the members of the committee appointed to assess the paper, Robert Hooke, seized on this and criticised Newton for suggesting that light consists of tiny corpuscles, like balls, with the necessary implication that there must be many different kinds of corpuscles in white light, responsible for the different colours into which it can be split. He himself believed that light consisted of vibrations or waves in the æther and he hinted that Newton should confine himself to a description of his experiments and leave discussion of the nature of light to those who, like himself, had worked out a satisfactory theory.

Newton, nettled and astonished, pointed out, quite rightly, that the conclusions he drew from his experiments with the prism were completely independent of any theory of the nature of light. He was fully aware that there were three possible mechanical hypotheses: that light consists of tiny corpuscles; or that light consists of waves in the æther, set up by the vibrations of the atoms of luminous bodies; or that light consists of a combination of corpuscles and waves.

In December 1675 he sent a memoir to the Royal Society in which he discussed these hypotheses, and showed that Hooke's wave theory of light could easily be adapted to explain the composite character of white light, revealed by the experiments with a prism, by assuming white light to consist of waves of many different wavelengths:

The most free and natural application of this hypothesis I take to be this—that the agitated part of bodies, according to their several figures, sizes and motions, do excite vibrations in the æther of various depths or sizes, which being promiscuously propagated through that medium to our eyes, effect in us a sensation of light of a white colour; but if by any means those of unequal sizes be separated from one another, the largest beget a sensation of a red colour, the least or shortest of a deep violet, and the intermediate ones of intermediate colours.

This was an incisive advance on Hooke's rather nebulous views. But Hooke's opinion of the memoir was 'that the main of it was contained in his *Micrographia*, which Mr Newton had only carried farther in some particulars'. The mischief-making secretary of the Royal Society, Oldenburg, duly reported this to Newton in Cambridge. Newton was deeply offended and retorted that Hooke's wave theory of light was merely a modification of the views of Descartes, with the result that Hooke, in his turn, was equally annoyed.

Newton favoured a combination of a corpuscular and wave theory, curiously prophetic of the modern view. He described it more fully in his *Opticks*, which was not published until a quarter of a century later, after the death of Hooke, and we shall therefore postpone its consideration.

* * *

The chief founder of the wave theory of light was the Dutchman, Christiaan Huygens (1629–95). Like Newton, Huygens was fond of making models when a boy and all his life he was engaged, with great success, in improving the telescope, which had been invented in Holland. As a young man he went to Paris, the centre of the cultivated world, where he remained for much of his life, and he became a close friend of Colbert, the minister of Louis XIV. He was made a pensioner of the *Académie Royale des Sciences*, and he retained this position while Louis XIV waged war on the Netherlands. Wars in those days were limited to professional soldiers, and were regarded by men like Huygens as minor inconveniencies, of less importance than the international activity of science.

In his book, *Traité de la Lumière*, giving the first detailed account of the wave theory of light, Huygens said nothing about colour because he found the subject difficult and perplexing. Strictly speaking, his theory was a pulse theory, rather than a wave theory, because he had no conception, as Newton had, of various wavelengths which determined the colours of the light.

He conceived of the æther as made up of tiny elastic spheres and imagined light to consist of pulses transmitted by the pressing of these spheres against their neighbours. The picture he gave is shown in Fig. 26. If a row of steel balls is given a smart blow

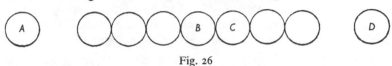

Fig. 26

at one end, the blow is passed on from each ball to the next, and the farthest ball moves away from the row. The æther particles must be very much harder than steel, to account for the enormous speed at which light is transmitted.

Since æther particles are not arranged in a row, but fill the whole of space, impulses must be transmitted in all directions. Figure 27 is another illustration from Huygens's book showing the waves or pulses radiating from three points, *A*, *B*, *C*, in a candle flame. Waves must radiate from every point on the candle flame and Huygens remarked on 'this prodigious quantity of waves which traverse one another without confusion and without effacing one another'. He pointed out that the property of the

Fig. 27

waves of passing through each other enables a man to look another man in the eye. If light consisted of particles, the particles from one eye would collide with those from the other. Huygens also commented on the great distances the waves of light can travel. 'But what may at first appear full strange and even incredible is that the undulations produced by such small movements and corpuscles, should spread to such immense distances; as for example from the sun or from the stars to us.'

He discussed the analogy between light and sound. Sound was known to consist of waves, or pulses of compression and rarefaction, in the air, set up by the vibrations of sounding bodies. Sound cannot pass through a vacuum, but light can; hence the need to postulate a medium for the transmission of light, the æther, which fills a vacuum and permeates the pores of transparent bodies. Sound is generated by the vibration of entire bodies, but light must originate from every part of a body; otherwise we could not see all parts of a body.

Huygens's most original contribution to the wave theory was a geometrical method of determining the positions of successive waves, known as Huygens's principle. He wrote of it: 'This is a matter which has been quite unknown to those who hitherto have begun to consider the waves of light, amongst whom are Mr Hooke in his *Micrographia*; and Father Pardies.'

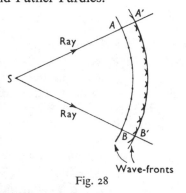

The principle is illustrated in Fig. 28. *S* represents a point source of light and *AB* the position at some instant of a wave, known as a wave-front. Huygens's principle enables us to determine the position of the new wave-front *A'B'*, at a short interval of time later. Huygens imagined that every point on the wave-front *AB* acts as a new centre of disturbance, from which a secondary wavelet spreads. The wave-front *A'B'* is the envelope of these secondary wavelets.

Fig. 28

The principle was applied in a satisfying manner to account for the laws of reflection and refraction. The theory required the velocity of light to be smaller in glass or in water than in air but unfortunately the experimental techniques of the time were quite inadequate to measure, or even to detect, such a difference in velocities.

The weakness of the principle was its failure to account for the fact that light travels in straight lines. In Fig. 28 *SA* is the direction of travel of the wave-front and represents a ray of light. Why should *SA* continue in a straight line to *A'*? According to Huygens's principle the light should spread in all directions from *A* and there is no obvious reason why it should be effective only at *A'*. If a small obstacle were placed in the path of the wave-front, the secondary wavelets should curve round the edges of the obstacle and obliterate the shadow. Thus light, like sound, should spread round corners. This was the decisive objection which led Newton to reject the wave theory. He wrote: 'To me, the fundamental supposition itself seems impossible, namely, that the waves or vibrations of any fluid can, like rays of light, be propagated in straight lines without a continual and very extravagant spreading and bending every way into the quiescent medium where they are terminated by it. I mistake if there be not both experiment and demonstration to the contrary.'

* * *

In 1704 was published Newton's *Opticks*, a work of originality second only to his *Principia*. It was written in English and not, like the *Principia*, in Latin. It had a remarkable fascination for the English poets of the eighteenth century, who made many references to the phenomena described in it; for example, the following couplet, from Pope's *Essay on Criticism*, shows the poet's familiarity with the composite nature of white light:

> When the ripe colours soften and unite,
> And sweetly melt into just shade and light.

The first part of *Opticks* deals only with experimental facts and laws. Newton realised, however, that the varied phenomena of optics cannot be co-ordinated without a theory of the nature of light and in the last part of the book he put forward the fruits of his imagination in the form of Queries. The Queries were framed as questions and, while they did not commit Newton to supporting any particular views, it is obvious which hypotheses he preferred. On the other hand, his attitude was that hypotheses are a matter of taste and, unlike his followers in the eighteenth century, he would have refused to be dogmatic about them.

Newton regarded light as consisting of tiny corpuscles emitted at great speed from shining bodies. Corpuscles travel naturally in straight lines, unless acted upon by a force, and hence the recti-linear propagation of light is accounted for. Reflection consists of the bouncing of these particles from a reflecting surface; refraction is explained by their attraction by the surface of a medium, such as glass or water, when they approach very closely to it. Newton's theory requires the velocity of light to be greater in glass or water than in air, in contrast to the theory of Huygens.

Experiment showed, in the middle of the nineteenth century, that the prediction of Huygens was correct.

The main reasons why Newton preferred a corpuscular theory are probably twofold; first, he could not see how the rectilinear propagation of light could be explained by a wave theory; secondly, it fitted well with the gravitational theory of particles, developed so successfully in the *Principia*. There were certain phenomena, however, which obliged him to supplement his corpuscular theory with a wave theory and these we must now consider.

Hooke in his *Micrographia*, published by the Royal Society in 1665, described what he calls the 'fantastical colours' of thin films, observed in soap bubbles, in very thin sheets of mica, in films on tempered steel, and in the film of air between two glass plates pressed together. He noted that the colours depended on the thickness of the film. He ascribed the formation of the colours to the combination of the beams of light reflected from the surfaces of the film. Here he was correct and he came near to anticipating, by a century and a half, the discovery of the principle of interference. But in this, as in all other matters, his interest was fugitive; his keen, darting mind did not dwell long enough upon it to develop it very far.

Hooke did not devise a method of measuring the thickness of his films and hence was unable to make any numerical observations. Newton, however, discovered a phenomenon, known as Newton's rings, which enabled such observations to be made. If a lens is placed on a flat glass plate (Fig. 29), tiny rainbow-

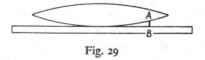

Fig. 29

coloured rings (Plate XI(*a*)) can be observed round the point of contact of the lens with the plate. Each ring corresponds to a particular thickness of the air film between the lens and the plate (such as *AB*), and the thickness increases by the same amount from ring to ring.

By using a lens having a surface of very slight curvature, so that the thickness of the air film changes slowly with horizontal distance, Newton was able to produce rings large enough to be measured with dividers. He observed the rings formed by red light and found that they were alternately red and black. Rings formed by blue light were alternately blue and black, and their radii were about half those of the corresponding rings formed by red light. The rainbow-coloured rings formed by white light were obviously caused by the superposition of red rings, of blue rings, and of rings formed by all the other colours of the spectrum.

Newton asked himself why, when using monochromatic light, alternate rings are black. Obviously, in these places, no light is reflected to the eye. 'And from thence the origin of these Rings is manifest', he wrote, 'namely that the Air between the Glasses, according to its various thickness, is disposed in some places to reflect, and in others to transmit, the Light of any one Colour.' But why should the thickness of the air film sometimes cause reflection and sometimes transmission? It was at this point that Newton invoked a wave theory. He imagined that the corpuscles of light on striking the surface of glass, gave rise to tremors or waves in the æther, rather as stones striking the surface of a pond give rise to ripples. These æther waves travel faster than the light corpuscles, overtake them, and at the boundaries of the air film put them into a 'fit' either of easy reflection or of easy transmission according to the state or phase of the wave. In the seventeenth Query Newton wrote: 'When a Ray of Light falls upon the surface of a pellucid Body and is there refracted or reflected, may not Waves of Vibrations, or Tremors, be thereby excited in the refracting or reflecting Medium at the point of Incidence...? And do they not overtake the Rays of Light, and by overtaking them successively, do they not put them into Fits of easy Reflection and easy Transmission described above?'

Thus, to explain the periodicity of the rings, Newton was obliged to postulate a periodicity in light, and this periodicity was introduced by the waves in the æther. The æther waves were not light; they merely served to insert a periodicity into the flow of corpuscles which formed the light.

Another optical phenomenon which required explanation was what is now called diffraction, and what Newton called inflexion. If sunlight is admitted through a small hole into a darkened room, the shadows of obstacles cast on a screen are found to have tiny rainbow-coloured bands round their edges. The shadow of a hair, or of a thin wire, has coloured bands inside the shadow as well as outside it (Plate XII (*a*)). Diffraction bands were first observed and carefully described by the Italian Grimaldi. They were subsequently studied by Hooke but neither he nor Newton seems to have observed the bands inside the shadow of a narrow obstacle.

Hooke explained the bands as due to the straying of light round the edges of the obstacle, rather as sound can pass round corners. Newton, on the other hand, explained it as a type of refraction, caused by a variation in the density of the æther near to the edge of a material body.

There was yet another optical phenomenon to be accounted for, what is now called polarisation. This can be most simply demonstrated by passing a beam of light through two crystals of tourmaline in turn (Fig. 30). If the second crystal is rotated, the

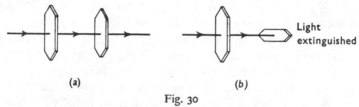

(*a*)   (*b*)

Fig. 30

intensity of the light emerging from it varies; when the axes of the two crystals are perpendicular to each other, no light emerges from the second crystal; when the axes of the crystals are parallel, the emergent light has its maximum intensity. Newton saw in this phenomenon an indication that light has different properties in the two perpendicular directions, represented by the axes of the tourmaline crystals in Fig. 30 (*b*). He could find no explanation for this on a wave theory, the waves being regarded as pulses or compressions travelling through the æther. He wrote: 'For Pressions or Motions, propagated from a shining Body, through an uniform Medium, must be on all sides alike; whereas by those Experiments it appears, that the Rays of Light have different Properties in their different Sides.... To me, at least, this seems inexplicable, if Light be nothing else than Pression or Motion propagated through Æther.' He was able to account for the phenomenon on his own theory by assuming that the corpuscles of light had a kind of asymmetry, with different properties in two perpendicular directions.

*   *   *

During the whole of the eighteenth century the corpuscular (or emission) theory of light held sway, backed by Newton's immense authority. The greatest achievement of physics at that time was the gravitational theory of particles, developed from Newton's work mainly by French mathematicians, and the climate of physical thought was much more favourable to particles than to waves. There was one notable critic of the corpuscular theory, the Swiss mathematician Euler (1717–83), but he produced no new experimental facts and his criticisms were disregarded.

The wave theory was revived in the opening years of the nineteenth century, by Thomas Young (1773–1829). Young was a universal genius, perhaps the last man who could take all knowledge for his province. He was a child prodigy. He could read at the age of 2 and had read the Bible through twice by the time he was 4. At the age of 14 he was asked to show how nicely he could write and, by way of a rebuke, he wrote out a sentence in fourteen different languages. Among the languages in which he was proficient were Latin, Greek, French, German, Spanish, Italian, Chaldee, Arabic, Syrian and Persian. One of his chief claims to fame is as an Egyptologist. In 1799 a tablet was discovered at Rosetta, near the mouth of the Nile, which bore an inscription in hieroglyphics, hitherto indecipherable, and also in Greek characters. The Greek provided the clue to the deciphering of the hieroglyphics, which was achieved by Young and independently by Champollion.

Young was educated at St Bartholomew's Hospital, at Edinburgh University, at Göttingen and at Emmanuel College, Cambridge, where he was known as 'Phenomenon Young'. In 1799, after leaving Cambridge, he set up in practice as a physician in London but his practice did not flourish. Two years later he became Professor of Natural Philosophy at the newly-formed Royal Institution in London.

Young's lectures at the Royal Institution covered the whole of the physical science known in his day, and were remarkable for their scope and originality; Sir Joseph Larmor has called them 'the greatest and most original of all lecture courses'. But Young was not a successful lecturer. His style was too condensed and laconic, more suited to research students than to a popular audience. His audience diminished daily while that of his dazzling colleague, Davy, who was Professor of Chemistry, increased to the utmost capacity of the premises in Albemarle Street.

Young's views on the wave theory of light were included in his lectures and also in four papers submitted to the Royal Society, two of which were read as Bakerian lectures, between 1800 and 1803. He criticised the corpuscular theory on the grounds that it did not explain satisfactorily why the velocity of light is the same, whatever the nature and temperature of the source. The velocity of waves would be expected to be constant, but why should the velocities of corpuscles be always the same? Again, reflection and refraction occur simultaneously. This is a natural occurrence with waves but why should some corpuscles be reflected while others are refracted?

Young examined Newton's contention that, if light consisted of waves, it would not travel in straight lines, and sharp shadows would be impossible. He maintained that if the wavelength of light were sufficiently short, the spreading of light round corners would be extremely small and shadows would appear sharp. He pointed out that high-pitched sounds spread round corners to a less extent than low-pitched sounds and that when a brass band goes round a corner, the low notes of the drum are heard after the high notes of the piccolos have died away.

Young had a wonderful flair—the hallmark of genius—for divining the essential clue to a problem. He discovered a principle, which had eluded both Newton and Huygens, and which is the keystone of the modern structure of wave theory. He called it the principle of interference; briefly, it is that two waves of light, which are half a wavelength out of step, so that the crests of the one coincide with the troughs of the other (Fig. 31 (*a*)) annul each other and produce darkness. Similarly, two waves which are exactly in step (Fig. 31 (*b*)) reinforce each other.

Young applied the principle to account for Newton's rings. In Fig. 29, light reflected at *A* interferes with light reflected at *B* and a dark ring results where these two sets of waves are half a wavelength out of step; a bright ring results where the waves are in step. The thickness of the air film, *AB*, increases by half a wave-length of light from each bright ring to the next bright ring (since the path difference of the two beams of light is 2*AB*). Thus Young was able to calculate the wavelength of light from a measurement of the rings; he found it to be about a hundred thousandth of an inch.

He took a further step. Newton had never linked diffraction fringes with his rings. Young explained diffraction fringes, as he explained Newton's rings, by the interference of two beams of light. He performed a simple but decisive experiment in support of this. He allowed a beam of light to enter a darkened room through a pinhole and to cast a shadow of a narrow obstacle,

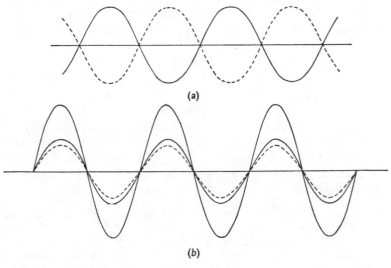

(a)

(b)

Fig. 31

consisting of a slip of card one thirtieth of an inch wide, on a screen. He obtained the diffraction fringes shown in Plate XII(a). He cut off the light from one side of the narrow obstacle and the fringes inside the shadow disappeared. These fringes must therefore be formed by the interference of the two beams of light passing on either side of the obstacle.

In the year 1802 the *Edinburgh Review* was founded and its liveliness gave it immense popularity. In this new journal, largely devoted to reviews of literature and science, the work of Young was assailed by an anonymous writer with repeated outbursts of malice and invective. The hostility of the reviewer was made clear in his opening sentence: 'As this paper contains nothing which deserves the name either of experiment or discovery, and as it is in fact destitute of every species of merit, we should have allowed it to pass among the multitudes of articles which must always find admittance into the collections of a Society which is pledged to publish two or three volumes every year.' But, said the reviewer, it was his duty to bring to the successors of Newton some sense of their responsibilities. Newton, 'great even in his most playful relaxations' (a reference to the fact that Newton gave his speculations on the nature of light in the form of Queries), 'amuses himself by conjecturing how the rays of light would act upon, and be affected by, an ætherial, subtile medium, were the existence of such a fluid ascertained'. 'But the clumsy hypothesis of Euler and Dr Young is that the æther itself constitutes light; and their object is to twist the facts into some sort of an agreement with what they conceive might be the laws of this fluid. From such a dull invention nothing can be expected....It has not even the pitiful merit of affording an agreeable play to the fancy.'

The lack of impartial interest in the merits of the theory by the reviewer, and his sole intention of destroying Young's reputation, is evident from his reply to Young's challenge concerning the experiment on the diffraction bands of a narrow obstacle. He refused to repeat the experiment himself and wrote:

The fact is, we believe the experiment was inaccurately made; and we have not the least doubt, that if carefully repeated, it will be found either that the rays, when inflected, cross each other and thus form fringes, each portion on the side opposite to the point of its flection; or that in stopping one portion, Dr Young in fact stopped both portions; a thing extremely likely, where the hand had only one-thirtieth of an inch to move in, and quite sufficient to account for all the fringes disappearing at once from the shadow.

The anonymous reviewer was the gifted and unstable Henry Brougham, later Lord Chancellor of England, and there is reason to believe that there was an old score to be paid off. Young had criticised, in just but tactless terms, a mathematical paper by Brougham, and the latter had nursed his resentment and awaited his opportunity.

Young's printed reply lacked the persuasive glitter of his adversary. Only one copy of it was sold, although he circulated it among his friends. The result of the affair was a damaging blow to Young's reputation and a set-back to the advance of the theory in Great Britain by twenty years.

Young's best-known experimental demonstration of interference was published shortly after the controversy. He passed light through a pinhole S (Fig. 32) and then through two pinholes S_1 and S_2 very close together. The intermingled light, falling upon a screen, gave rise to bright and dark bands (Plate XI(b)).

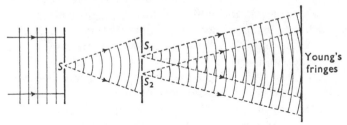

Fig. 32. Young's interference experiment.

* * *

About fifteen years later the concept of interference was put forward independently by Augustin Fresnel (1788–1827) in France. Unlike Young, Fresnel was backward as a child and at the age of 8 could scarcely read. Nevertheless he showed early signs of his future scientific powers; at the age of 9 he investigated the relative length and bore of popguns for maximum efficiency, with such effect that the toy became a dangerous weapon and had to be proscribed. He came of a family with intellectual interests, his cousin being Prosper Mérimée, the writer. He became a civil engineer and was sent to make roads in the Vendée, an occupation for which he was utterly unfitted by reason of his shyness, his delicate health, and his inability to handle men.

Passing through Paris in 1815, Fresnel had an interview with François Arago, a distinguished physicist and an influential figure in the scientific world of Paris, who later became his staunch friend and supporter. Fresnel told Arago that he believed in the truth of the wave theory, and shortly afterwards Arago wrote to him recommending the papers of Young. Fresnel could not read English and he replied anxiously: 'Please tell me if anybody has already determined this law of diffraction by exact measurement.' Arago urged him to prepare a paper on diffraction and to send it to the *Académie des Sciences*.

Fresnel's paper gave a critical account of Newton's theory of fits (p. 151) and showed that Newton's rings and other fringes could be accounted for by the principle of interference. He explained the diffraction fringes caused by a narrow obstacle, as Young had done, by the interference of the two beams of light passing on either side of the obstacle.

Arago was anxious that Fresnel, then road-making in Brittany, should obtain leave and come to Paris to continue his scientific investigations. By pulling strings in the French Civil Service he was able to arrange that Fresnel arrived in Paris in March 1816, where he met many of the celebrated French physicists of the day.

Fresnel found that all the members of the *Académie*, with the exception of Arago, were supporters of the corpuscular theory and opposed to his views. It was a magnificent period of French mathematical physics and the *Académie* included perhaps a dozen men with an international reputation. Arago has left an account of his admission to the *Académie* in 1809, and of the presentation of the members of the *Institut*, which included literary men and artists as well as scientists, to the Emperor. Napoleon passed down the rows of members, all clad in green uniform, all eager to catch his eye and to obtain the favour of a word with him. He spoke to Arago and commented upon the latter's youth. The old man Lamarck offered him a book on Natural History. Napoleon thought the subject was Meteorology; 'Do something in Natural History, and I should receive your productions with pleasure. As to this volume, I only take it in consideration of your white hairs. Here!' and he passed the book to an aide-de-camp. Lamarck, flustered and trembling, tried to explain, but his explanations were waved brusquely aside. It was to this body of men that Fresnel was yearning to be elected.

Fresnel's papers had been strongly opposed by Laplace, Biot and Poisson and it was decided, in 1817, to offer a prize for a study of diffraction, in the hope that the corpuscular theory would find a champion. Fresnel entered a *mémoire* and it won the prize. It was a masterpiece of mathematical physics, completing and correcting his earlier work. In it he explained diffraction by a combination of the principle of interference and of Huygens's principle.

His mathematical theory of diffraction was a complete justification of rectilinear propagation because its essential point was to show how far light actually departs from propagation in straight lines. Light does spread slightly round corners and in so doing gives rise to diffraction fringes.

Poisson pointed out that Fresnel's theory required a bright spot at the centre of the shadow of a small circular obstacle. When an experiment was performed and the bright spot was found (Plate XII(*b*)) a profound impression in favour of the wave theory was made upon the members of the *Académie*.

Fresnel now turned his attention to polarisation. An important discovery had been made by Étienne Louis Malus. One day in 1808 Malus was looking out from his house in the rue d'Enfer in Paris through a crystal with properties similar to tourmaline (p. 152), at the windows of the Luxembourg palace, which were reflecting the rays of the setting sun. He found, to his surprise, as he rotated the crystal, that the light was extinguished in certain positions. This suggested that light could be polarised by reflection, as well as by passing it through a crystal, a conjecture which he confirmed the same evening by reflecting candle-light from the surface of water.

The discovery of Malus made it quite clear that polarisation was caused by some inherent property of light as Newton had surmised. Like Newton, Young and Fresnel were at a loss to explain the phenomenon in terms of waves. They thought of the waves of light as similar to those of sound or to those passing through a row of steel balls as in Fig. 26. Such waves are called longitudinal waves and the vibrations of the medium occur along the direction of travel of the waves.

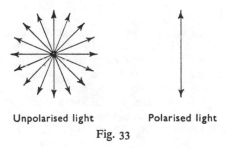

Unpolarised light   Polarised light

Fig. 33

Then Young hit upon the idea of transverse waves, in which the vibrations of the medium occur perpendicularly to the direction of travel, like the waves which can be set up in a long rope by moving one end periodically up and down with the hand. Suppose that the æther vibrations occur in all possible directions perpendicular to the direction of travel (Fig. 33), and that, after the light has become polarised by passage through a tourmaline crystal, they occur in one direction only. A second tourmaline crystal will allow the vibrations to pass through if its axis is parallel to the first, but not if it is perpendicular.

Fresnel too conceived the idea of transverse waves and decided to include it as an hypothesis in a paper which was being published in the joint names of himself and Arago. But Arago had not the courage to subscribe to so startling a conjecture and withdrew his name. Hitherto the æther had been regarded as a fluid, rather like a very tenuous gas. A fluid cannot resist distortion and hence cannot support transverse waves. The new æther must have the properties of an elastic solid. What about the resistance of such a medium to the motion of the planets? And, if transverse waves, why not longitudinal waves simultaneously?

Many years after Fresnel's early death, Biot, who had devoted a great deal of effort in trying to explain the phenomena of optics on the corpuscular theory, remarked to his brother, 'What a genius he had, to see as soon as he did the transverse nature of light vibrations!'

Fresnel was indeed one of the most brilliant scientific men that France has produced, a rare ornament in the intellectual life of his country, in whom all Frenchmen may take legitimate pride. Monsieur Boutry has recorded that, during the Second World War, he was living in the little town near Paris in which Fresnel died:

During the last war, I was living at Ville d'Avray and had one day to see the Mayor on business. Before leaving this worthy physician, it was only fitting to congratulate him on the many celebrities which the small town had some time known. I found this was his pet subject and listened with pleasure to anecdotes of Balzac writing endlessly in his small house, of Corot wandering along the lake, of Gambetta roaring in Les Jardies, of Mérimée strolling with great ladies in the Saint-Cloud woods. At last he paused for breath; I chanced it: 'And what of Fresnel?'—'Who was he?' said the Mayor.[1]

* * *

The theory that light consists of transverse waves in the æther provided a satisfying explanation of all known optical phenomena.

[1] G. A. Boutry, 'Augustin Fresnel: His Life, Time and Work, 1788–1827'. *Science Progress*, 1948, vol. XXXVI.

Newton and the supporters of the corpuscular theory had been obliged to invent arbitrary, unconnected hypotheses to account for several phenomena: fits of easy transmission and reflection to explain the colours of thin films, attraction and repulsion of the corpuscles by a variation in the density of the æther at the edge of a body to explain diffraction, and a two-sided property of the corpuscles to explain polarisation. The wave theory provided a single common principle by which all these diverse phenomena became intelligible and interconnected. As Sir John Herschel wrote: it is 'a theory which, if not founded in Nature, is certainly one of the happiest fictions that the genius of man ever invented to grasp together natural phenomena...a succession of felicities...'.

On the other hand, Sir David Brewster never found the wave theory congenial; 'his chief objection to the undulatory theory of light was that he could not think the Creator guilty of so clumsy a contrivance as the filling of space with æther in order to produce light!' But Brewster was wrong. It was not the Creator who had filled space with æther, but men of science.

An enormous amount of effort, which we now see to have been misguided, was expended in the nineteenth century in trying to explain the mechanical properties of the æther. Young had suggested that the æther could pass through bodies 'like the wind through a grove of trees'. Sir George Stokes pointed out that pitch and shoemaker's wax are rigid enough to transmit transverse vibrations, but plastic enough to allow bodies to pass slowly through them. He suggested that the æther might possess these two properties in an extreme degree. It was then understandable why the æther could transmit light and yet not impede the motion of the planets.

A new complication was added by the publication in 1865 of Maxwell's electromagnetic theory of light, according to which light consists, not of mechanical vibrations, but of waves of fluctuating electric and magnetic fields. It was then necessary to invent an æther which could transmit electric and magnetic energy.

We shall see, in chapter 21, that Einstein's fundamental postulate in the special theory of relativity is that it is impossible to detect absolute motion and hence that an æther, which would make this possible, cannot exist. We shall see also, in chapter 22, that certain phenomena have been discovered which make it necessary to supplement the wave theory with a modern type of corpuscular theory, known as the quantum theory. But although its mechanical basis has been removed, and although it must be combined with a new corpuscular theory, the wave theory still expresses, in some fundamental manner, the way in which light behaves.

CHAPTER 13

THE CONSERVATION AND DISSIPATION
OF ENERGY

THE PRINCIPLE of the conservation of energy is the most all-embracing generalisation in physical science. In the limited form of the impossibility of perpetual motion, it had been used in the development of mechanics by Galileo and his successors; in the middle of the nineteenth century it was extended beyond the realm of mechanics to include all the processes of nature.

The search for a conservation principle, which should hold in the transformations between mechanical, thermal, optical, electrical and chemical phenomena, was at first confused by the ambiguous use of the word 'force'. Force has a precise meaning in mechanics as the product of mass and acceleration but the term was also given by some writers to the cause of all physical effects. Mayer wrote: 'a force once in existence cannot be annihilated, it can only change its form'; and Helmholtz spoke of 'the persistence of force'. This was perplexing and illustrates the importance of a rigorous definition of scientific terms.

It was felt intuitively that something must be conserved or, as Joule put it, 'the grand agents of nature are, by the Creator's fiat, indestructible'. This something eventually was called energy, and it was defined in terms of its simplest manifestation as capacity to do mechanical work, measured by the product of force and distance, for example in foot-pounds.

The principle of the conservation of energy is that energy cannot be created or destroyed. Energy is merely transformed from one kind to another. The principle provides a potent method of attacking diverse problems. It is supplemented by another principle, equally fundamental and fertile, that energy, although never destroyed during a transformation, becomes less available for performing useful work, and this is known as the principle of the dissipation or degradation of energy.

Energy may be compared with wealth, and its transformation

with the conversion of wealth into different currencies. When we change our money into a different currency we normally lose on the process although the quantity of real wealth in the universe remains unaltered by our transaction. The energy locked up in coal or petrol represents a desirable, hard currency because it can be readily converted into some other useful form. The softest currency is heat at the temperature of its surroundings; indeed this currency is inconvertible.

* * *

The principle of the conservation of energy, although approached by Joule through his investigations of electrical phenomena, and by Mayer from a biological observation, is perceived most directly through the equivalence of mechanical energy and heat. Its recognition emerged simultaneously with the establishment of the kinetic theory of heat.

In the seventeenth century the kinetic theory of heat was widely accepted. Heat was regarded as the energy of motion of the tiny particles or molecules of which a body is composed. In the eighteenth century, however, the kinetic theory of heat lost favour and was replaced by the conception of heat as an imponderable, self-repellent, indestructible fluid, which was given the name caloric by Lavoisier.

Black mixed hot and cold bodies, such as water and mercury, and assumed that a quantity of caloric was transferred, unchanged in amount, from the hot to the cold body. Unit quantity of heat was taken as that quantity which would raise the temperature of unit mass of water through one degree. The caloric theory was useful at the early stage of the study of heat, but it became later a barrier to further progress.

The first serious attack on the caloric theory came in 1798 from the experiments of Benjamin Thompson, Count Rumford (1753–1814), one of the more colourful figures in the history of science. Rumford served for a time in the American army in the War of Independence, founded the Royal Institution in London, was Minister of War to the Elector of Bavaria, became a count of the Holy Roman Empire and married the widow of Lavoisier (followed quickly by a divorce). He had a practical turn of mind; he made an inquiry into the best way of making coffee and invented the Rumford stove. He wrote of what, to many Englishmen, is still familiar: 'the folly of sitting in a room which has a large blazing fire roasting one side of the body, while blasts of cold air are coursing the apartment.' He borrowed the house of Lord Palmerston in Hanover Square for his experiments on fireplaces.

During the boring of a cannon at the military arsenal at Munich, he was much struck by the large amount of heat produced. According to the caloric theory, the borer pressed out caloric from the barrel through the pores of the metal.

He placed a brass gun barrel in a wooden box containing about 19 lb. of cold water, used a blunt steel borer, and continued turning the barrel by means of two horses for $2\frac{1}{2}$ hr., at the end of which time the water boiled. 'It would be difficult to describe the surprise and astonishment expressed in the countenances of the bystanders', he wrote, 'on seeing so large a quantity of cold water heated, and actually made to boil, without any fire.' The supply of heat was apparently inexhaustible and he drew the following conclusion: 'anything which any *insulated* body, or system of bodies, can continue to furnish *without limitation*, cannot possibly be a *material substance*; and it appears to me to be extremely difficult, if not quite impossible, to form any distinct idea of anything capable of being excited and communicated in the manner the Heat was excited and communicated in these experiments, except it be MOTION.'

*　　*　　*

The experiments of Rumford did not make an immediate impression among physicists, but forty years later the idea of the conservation of energy was put forward independently by several men, foremost among whom was James Prescott Joule[1] (1818–89).

Joule began his first research, at the age of 19, with no thought in his mind of a new theory of heat nor of any comprehensive generalisation, but with the object of improving the electric motor, invented two years earlier by Sturgeon from the fundamental discoveries of Faraday. Joule constructed an electric motor

---

[1] The pronunciation is Joole, not Jowle. Professor H. S. Allen writes: 'I venture to suggest that the pronunciation "Jowle" may have originated through the sardonic humour of local workpeople who, having in mind such an expression as "cheek by jowl", spoke of "Jowle's brewery".' (*Nature, Lond.* 152, p. 354.)

weighing $7\frac{1}{2}$ lb. which he drove with a battery; he found that it could 'raise 15 lb. a foot high in a minute'.

He realised that here was a new prime mover which might, in due time, replace the steam-engine. He therefore compared the work done by his motor with the weight of zinc consumed in the battery. He was disappointed to find that, comparing the prices of zinc and coal, his motor was much more expensive to run than a steam-engine.

In 1840 he presented a paper to the Royal Society on the production of heat by an electric current. After a series of fundamental researches he took the crucial step of relating the heat produced in a circuit to the amount of chemical action in the battery. He gradually came to realise that the chemical energy disappearing in the battery is equivalent to the heat energy generated in the circuit.

He discovered that his electric motor when running had a greater resistance to the current than when it was at rest, and that the chemical energy released by the battery to overcome this increased resistance was proportional to the mechanical work done by the motor. He had now reached the stage of realising that chemical energy is equivalent to mechanical energy as well as to heat energy.

The relation between heat and work did not at first excite his curiosity; it was the phenomena of electricity that interested him. But he now perceived that the fundamental determination which required his attention was the quantity of mechanical work equivalent to unit quantity of heat, known as the mechanical equivalent of heat.

He compared the mechanical work done in driving a dynamo with the heat produced by the current generated; he measured the work done and the heat produced by the friction of water in its passage through fine holes in a reciprocating piston; by the compression of air with a hand pump; and by the churning of water with paddle wheels. The fact that he obtained approximately the same value for the mechanical equivalent of heat by these diverse methods was a strong indication that, whenever work is done, it results in the appearance of an equivalent amount of heat.

In some of his experiments, the rise of temperature by means of which he measured the heat generated was of the order of only $\frac{1}{2}°$ F. or $1°$ F. He obtained his accuracy by the use of an extremely sensitive thermometer. Ten divisions of his thermometer, of length about half an inch, represented about $1°$ F., so that each division represented about $\frac{1}{10}°$ F. He was able, by long practice, to estimate by eye to $\frac{1}{20}$th of a division; hence his readings were to $\frac{1}{200}°$ F. He showed confidence and boldness in making large conclusions from small observational effects. His critics said that he had nothing but hundredths of a degree by which to prove his case. We can now appreciate that he was an experimenter of the highest calibre.

The record of Joule's work is contained in his scientific papers, which are concerned with his experiments and their immediate implications for theory. Only once did he give a wide survey of his views on the conservation of energy, and this was in a popular lecture at St Ann's Church Reading Room, Manchester, on 28 April 1847. This lecture is one of the landmarks of science; it clearly states the equivalence of mechanical energy, heat, electrical energy and chemical energy.

During the years 1843–6 Joule presented to various meetings of the British Association an account of his determinations of the mechanical equivalent of heat. His papers were received with chilling apathy and silence. At the Oxford meeting in 1847, when Joule came forward with his customary contribution, the Chairman suggested that, as they had many important things to discuss, perhaps he would give a brief verbal description of his experiments instead of reading his paper. He did so and at the close no discussion was invited. But a young man rose and made some remarks which created a lively interest in Joule's ideas. The young man was William Thomson, who had taken his degree at Cambridge two years earlier. Thomson had been to Paris, where he had worked in the laboratory of Regnault and become acquainted with the work of Carnot. The theory of Carnot, which we shall discuss shortly, seemed to conflict with that of Joule. Thomson, although he was a convinced supporter of Carnot, found Joule's views of great interest and he secured for them a hearing.

A fortnight later Thomson met Joule at Chamonix. Joule was on his honeymoon and was carrying a long stick which looked in the distance like a walking stick. This was a thermometer with which he was measuring the temperature of the Alpine waterfalls.

Since the mechanical equivalent of heat is about 800 ft.-lb. per B.Th.U., the temperature of the water at the foot of a waterfall 800 ft. high should be 1° F. higher than at the top.

*   *   *

Sadi Carnot (1796–1832), a French military engineer trained at the École Polytechnique, wrote at the age of 23 or 24 *Réflexions sur la Puissance Motrice du Feu*, which revealed a scientific imagination of extraordinary prescience and founded the science of thermodynamics.

Carnot, like Joule and Mayer, was much interested in the steam-engines then spreading over Europe. Although England was the home of the steam-engine, her engineers were practical men, having skill and enterprise but also the limitations of their training and environment. It was left to a Frenchman, with a university education, to put forward a theory of what they were practising. Carnot's object was to provide a firm basis for the future improvement of the steam-engine. He asked, and answered, such questions as: On what does the motive power of a steam-engine really depend? Is there any limit to the efficiency of a steam-engine? Would another working substance be better than steam? Would a different type of engine be more efficient?

By the term motive power ('puissance motrice') Carnot meant work or energy—measured, for example, in ft.-lb. He suggested that the motive power of a steam-engine is derived from the fall of caloric (or heat) from a higher to a lower temperature, just as the motive power of a watermill is derived from the fall of water from a higher to a lower level. Wherever there is a difference of temperature a motive power can be produced.

Carnot imagined an ideal engine, consisting of a gas in a cylinder fitted with a frictionless piston. He conceived the idea of a 'cycle', in which the gas absorbs heat from a furnace, expands and does work by pushing the piston outwards, gives up heat to a condenser and contracts, coming back to exactly the same state as at the beginning. Since the gas comes back to the same state, any problems of heat required for internal work in the gas are avoided.

Carnot derived a simple mathematical expression for the efficiency of an ideal engine, in terms of the temperatures between which it works, and showed that, for a high efficiency, there must be a large difference of temperature. Substituting some other working substance for steam, or changing the type of engine, say from a reciprocating to a rotary type, would give rise to improved efficiency only if these changes resulted in the engine working between a bigger range of temperature.

* * *

After the first meeting of Thomson and Joule in Oxford in 1847, Thomson's thoughts were dominated by the apparent conflict between the theories of Carnot and Joule. Carnot held that work is done when heat falls from a higher temperature to a lower, without any conversion or annihilation of the heat. Joule, on the other hand, had shown by numerous experiments that work can be converted into heat, and he had determined the rate of conversion; but he had not demonstrated convincingly the inverse process, the conversion of heat into work.

One of the problems presented by Carnot's theory, which perplexed Thomson, was what happens to the work which should be done when heat passes from hot to cold water or when it is conducted along a metal bar. The heat has fallen through a difference of temperature but apparently it has done no work.

In 1849, James Thomson (William's brother) used Carnot's theory to make a quantitative prediction of the lowering of the melting-point of ice by pressure—imagining a small engine with ice as the working substance. William confirmed the accuracy of the theory by experiments in the laboratory and this afforded a tremendous reinforcement of his belief in the essential truth of Carnot's theory. He was beginning to perceive how Carnot's theory could be modified when, in February 1850, Clausius submitted a paper to the Berlin Academy of Sciences, showing that the most important conclusions in Carnot's theory remain unaffected by Joule's idea of the annihilation of heat to form work. Clausius developed thermodynamics from two laws: the first law of thermodynamics, which is the principle of the conservation of energy, that energy cannot be created or destroyed; and the second law of thermodynamics, which is essentially the principle of the dissipation of energy, that heat never passes of itself from a colder to a warmer body.

In March 1851 Thomson read to the Royal Society of Edinburgh a paper *On the Dynamical Theory of Heat*, in which he developed the theory of thermodynamics in his own manner, making the physical principles clearer than Clausius had done. He pointed out that, if an engine works between a small difference of temperature, only a small amount of heat can be converted into work, 'the remainder being irrevocably lost to man, and therefore "wasted", although not annihilated'. In a later paper on the dissipation of energy he developed the idea that, as heat tends to pass from a higher to a lower temperature, the energy in the world is falling to a uniform level, and hence is becoming unavailable for performing work. He forecast that the earth would cease to be habitable within a finite period of time, if still subject to the same laws of nature as at present.

Thomson lived to a ripe old age, was created a peer with the title of Lord Kelvin, and became the Grand Old Man of Victorian science. For fifty-three years he was professor at Glasgow where he created the first teaching laboratory of physical science in Great Britain. At the time of his appointment as professor, most physical constants were not accurately defined, and their values were unknown; scarcely any accurate measuring instruments existed. Thomson made a large contribution to the systematising of physics and invented several important instruments. It has been said that Victorian physics, without Thomson, would have been like Hamlet without the Prince of Denmark.

His weakness lay in his inability to read the work of others. He did not read right through a book for thirty years. His lectures, which he always opened with prayer, were not at all easy for the majority of his students. He was probably the only man who has ever made discoveries while lecturing. A new point would occur to him, he would digress and forget all about the topic of his lecture. During his lectures he would call continually upon his assistants for data of all kinds including, it is said, data from the ordinary multiplication tables.

His desultory methods of work, his want of sensitivity to the ideas of others, his lack of that intuitive gift to see beyond the immediate facts, prevented him from attaining to the very highest rank of scientific investigators, despite his strong clear mind, his outstanding mathematical technique, and his powerful personality.

*   *   *

Joule was anticipated in the statement of the principle of the conservation of energy and in the determination of the mechanical equivalent of heat by Julius Robert Mayer (1814–78). Mayer's work, however, attracted little notice; his ideas made no converts and were unrecognised until Joule's experimental researches had convinced the scientific world of their truth.

Mayer was a doctor in the little town of Heilbronn on the Neckar. His first appointment was as a ship's doctor on a Dutch East Indiaman and, during a voyage between February 1840 and February 1841, he had almost nothing to do since the health of the crew was good. He had with him plenty of scientific books and these he read and pondered.

In Java, owing to slight illness among the crew, Mayer was obliged to let blood, and he was surprised to find that the venous blood was bright red, instead of a bluish red as it is in colder climes. He was aware of Lavoisier's experiments on nutrition and he attributed the brightness of the blood to an excess of oxygen present because less combustion of food was required to provide body heat.

The voyage was as scientifically fruitful as that of Darwin in the *Beagle*, for, in the autumn of 1840 at Sourabaya, the germ of the idea of the conservation of energy came into his mind. During the journey home he made the first draft of his paper of 1842.

He submitted his paper for publication in *Annalen*, the foremost scientific journal in Germany, but it was rejected by Poggendorf, the editor, on the ground that it contained no experimental work. It was, however, accepted by Liebig for publication in his *Annalen*.

The paper received little attention and, later, other men began to put forward similar ideas. Mayer attempted to assert his claims to priority but his proffered contributions to German scientific journals were not published. He therefore wrote an article in a daily newspaper and this was answered in the same paper in a manner which made him appear ridiculous and unscientific. This so preyed upon his mind that it was thought advisable to put him in a mental home. The doctors found that his chief symptom was an obsession that he had discovered a great scientific principle and that no one else would recognise this. He was treated like a lunatic and put in a strait-jacket.

Some years later John Tyndall was preparing his lectures on heat and sought for information about Mayer. He wrote to Clausius, who sent him Mayer's complete works. He saw at once Mayer's greatness and warmly championed a man so harshly treated, claiming that 'the writings of Mayer form an epoch in the history of the subject'.

This was taken by British physicists as an attack on Joule and they rose to his defence. Joule himself wrote that it was wrong to apportion the credit for the theory to any one man, but added: 'I therefore fearlessly assert my right to the position which has been generously accorded me by my fellow physicists as having been the first to give a decisive proof of the correctness of the theory.' Tyndall replied: 'I do not think the public estimate of your labours can be in the least affected by any recognition which may be accorded to Mayer. There is room for both of you on this grand platform.'

It was undoubtedly the experimental work of Joule which established the theory. The difference between his work and that of Mayer is well summed up in Tyndall's words: 'True to the speculative instinct of his country, Mayer drew large and weighty conclusions from slender premises, while the Englishman aimed, above all things, at the firm establishment of facts.'

*　　*　　*

Thermodynamics, as developed by Carnot, Clausius and Kelvin, requires no assumptions about the nature of the constitution of matter and there is no need to postulate the existence of atoms or molecules or even the kinetic theory of heat. Side by side with it there grew up the kinetic theory of gases and the marriage of these two theories threw a flood of light on the nature of the second law of thermodynamics.

The kinetic theory was based upon the conception of a perfect gas, composed of a swarm of minute, hard, perfectly elastic, spherical molecules, like tiny billiard balls, of negligible volume compared with the total volume of the gas, and exerting no forces on each other except when they collided. The so-called perfect gases, such as hydrogen and helium, obey laws approximating closely to those for the theoretical perfect gas.

A gas exerts a pressure by the bombardment of its molecules. The velocity of the molecules can be calculated quite simply from a knowledge of the pressure and density of a gas at a particular temperature. The molecules of air at ordinary temperatures travel at about 1000 m.p.h., while those of the lighter gas hydrogen travel at about 4000 m.p.h. Each molecule makes about 10^{10} collisions every second.

In the earliest mathematical theory all the molecules were assumed to have the same velocity. But two perfectly elastic spheres, of equal mass and travelling with the same velocity, do not rebound with equal velocities unless the impact is direct or symmetrical. Clerk Maxwell, from the theory of probability, deduced a distribution of velocities similar to the distribution of shots round the bull's eye of a target. A few molecules travel very quickly and a few very slowly, but the majority travel at an intermediate speed.

Maxwell used his result to explain how it would be possible to violate the second law of thermodynamics—that heat cannot pass of itself from a lower to a higher temperature—but he required for the purpose a sorting demon. Imagine two volumes of gas, one at a higher temperature than the other, separated by a partition in which there is a frictionless trap-door, controlled by the demon. The average velocity of the molecules of the cooler gas will be less than that of the warmer gas, but there will be some molecules in the cooler gas which are travelling faster than the slowest molecules in the warmer gas. By suitably manipulating his trap-door, as the molecules of the gas bombard it, the demon can collect the fastest molecules from the cooler side on the hotter side, and the slowest molecules from the hotter side on the cooler side. In this way heat can be transferred from the cold to the hot gas, without the performance of work. Even without the assistance of the demon there is the remote possibility that such a sorting of the molecules could occur by chance. But, in view of the billions of molecules involved (27×10^{18} per cubic centimetre) the chance is about the same, in Eddington's metaphor, as that an army of monkeys, strumming upon typewriters, should write all the books in the British Museum.

Thus the second law of thermodynamics is a statistical law. When heat flows from a higher to a lower temperature, the molecules are merely changing from a less probable to a more probable state. By their collisions, the molecules tend, on an average, to equalise their velocities and hence to equalise the temperatures. The process is irreversible in the sense that the mixing of two powders by shaking is irreversible. Prolonged shaking could conceivably separate the powders completely again but the chance of this happening is infinitesimal.

The fact that the laws of gases and of thermodynamics are statistical laws, deducible from the laws of chance, and stating merely probabilities, prompted the question whether all physical laws are not also of this kind. A typical law of chance is that the probability of throwing a particular number with a six-sided dice is $\frac{1}{6}$. In the case of an assemblage of molecules the number of possible distributions of velocities and also the probability of one particular distribution can be calculated, using the same principles, but necessarily more difficult mathematics, as for the throwing of dice.

Two views are tenable as to the nature of the laws of statistical mechanics. It may be argued, on the one hand, that if we could follow the behaviour of a single molecule we should find that its behaviour is determined by rigorous laws, so that in principle we could predict its motion with certainty; we are obliged to have recourse to a statistical treatment giving conclusions of probabilities only, because of the vast numbers of molecules involved. On the other hand, individual molecules may not obey rigorous laws. A large number of molecules, through the operation of the laws of chance, must show a statistical uniformity, and chance may lie at the root of the laws of nature. This would have been regarded as a revolutionary point of view in the closing decades of the nineteenth century, but not so today (see p. 343).

*   *   *

At the beginning of the twentieth century Einstein took another step towards one of the ultimate goals of science, which is to reduce the number of its fundamental generalisations to a minimum; he combined the principles of the conservation of energy and of the conservation of mass. He deduced from his postulates of relativity that a body of mass m is equivalent to an energy E, where $E = mc^2$, c being the velocity of light. In some physical processes mass may disappear and energy appear, or *vice versa*, but the total of mass and energy is unchanged.

Einstein's mathematical deduction has been demonstrated by experiments on the largest scale—in the explosions of atomic bombs and in far greater cataclysms in the stars. The annihilation of a comparatively small mass releases a vast quantity of energy. The disappearance of 1 lb. of matter gives rise to the same energy as burning two million tons of coal.

The source of the energy of the sun was the subject of scientific speculation for over a century. In 1837 Sir John Herschel calculated, from his measurements of star radiation, that the heat reaching the earth from the sun per annum is sufficient to melt a crust of ice round the earth 100 ft. thick. Mayer estimated, in 1848, that if the sun were made of coal it would burn out in 5000 years. He suggested that the source of its heat might be the energy of meteors and asteroids striking its surface. But it was shown that the same proportion of meteors striking the earth would keep it permanently red hot. Helmholtz, in 1854, suggested that the sun's heat came from the release of gravitational energy during condensation and that a shrinkage of a few hundred feet a year would be sufficient. The theory indicated that the sun is 20–30 million years old and that its prospect of life is another 10 million years.

The present theory is that the sun's energy comes from an annihilation of mass and that it is losing mass at 4 million tons per second. The mass is lost during the transmutation of hydrogen into other elements; it is estimated that a transmutation of 10 per cent of its mass would give the sun a further lifetime of 10,000 million years.

CHAPTER 14

FIELD PHYSICS

THE MOST FUNDAMENTAL change in the outlook and basis of physics, after the time of Newton, originated from the work of Michael Faraday and James Clerk Maxwell. Newton's picture of physical reality was in terms of the movements of particles by the action of forces across empty spaces. Faraday conceived of the space surrounding a particle as an extension of the particle, developing his ideas from his researches in electricity and magnetism. Such a space is known as a field. A space in which electric, magnetic or gravitational forces can be detected is called an electric, magnetic or gravitational field respectively. Maxwell put Faraday's ideas into mathematical form and represented combined electric and magnetic fields, called the electromagnetic field, by a set of equations, which describe the changes in the field in space and time.

The most important developments of field physics after the work of Faraday and Maxwell were the replacement by Einstein of Newton's gravitational theory of particles and forces by the field equations of general relativity, and the development of quantum field theory, according to which every fundamental particle, such as the electron, has its own associated quantum field.

*   *   *

Michael Faraday (1791–1867), the third child of a London blacksmith, was brought up in poverty, with no formal education beyond the three R's, and became one of the greatest of experimental investigators. He was, however, extremely fortunate in his opportunities for further education. He began his working life as errand boy to a bookseller, and then was taken as an apprentice without premium. Inside the bookshop he came upon Mrs Marcet's *Conversations on Chemistry* and an article on electricity in the *Encyclopædia Britannica*; these gave him his first interest in science. He was impressed also with Watt's

PLATE XIII. Michael Faraday lecturing at the Royal Institution.

PLATE XIV. James Clerk Maxwell.

Improvement of the Mind, and he made serious and sustained efforts to teach himself to speak and to write correctly. He took lessons in drawing from a French *emigré*.

One of the customers of the bookshop took Faraday to hear Davy lecture at the Royal Institution. Faraday sat in the gallery and was entranced. He made copious notes, and wrote up a full account of the lectures, with diagrams, which he sent to Davy, begging for an opportunity of working at the Royal Institution. He said that trade was 'vicious and selfish' and later confided to Davy, who must have listened with an indulgent smile, that he thought that men of science were animated by higher moral feelings than other men.

Davy was kind and sympathetic, but told him that science was a harsh mistress and that he would do far better in bookselling. It so happened, however, that shortly afterwards the laboratory assistant at the Royal Institution was dismissed for misconduct, and Davy offered Faraday the post, which was accepted with alacrity.

Seven months after his appointment at the Royal Institution, in October 1813, Faraday set out with Davy, as secretary and valet, on a tour of France, Italy, Switzerland, Germany, Holland and Belgium, lasting one and a half years, during which he met most of the leading men of science in Europe. He was at that impressionable and receptive age at which some young men are university students. He had never before travelled more than twelve miles from London. He was excited by the journey to Plymouth; in Paris he had a glimpse of Napoleon in his carriage; he took part in experiments on electric fish at Genoa; he visited the laboratory of the *Accademia del Cimento* at Florence, and saw Galileo's first telescope; in Milan he met Volta, then an old man; he climbed Vesuvius; he accompanied Davy on fishing expeditions in the Alpine streams and acted as loader when Davy went shooting with de la Rive near Geneva; on his way home he slept in Brussels, two months before Waterloo. The travellers wandered through France and Italy without molestation during the Napoleonic war.

The only jarring note in the tour was that Lady Davy felt it her duty to keep Faraday in his place. He was, after all, only a laboratory assistant and valet. He was often not allowed to dine at the same table as Davy and his friends.

Back in London Faraday spent seven years as Davy's assistant. In Copenhagen in 1820 Oersted discovered that an electric current flowing in a wire caused a nearby magnetic needle to be deflected. This stimulated intense interest in electricity and magnetism, and, in 1821, Faraday wrote *Historical Sketch of Electromagnetism* which gave him a mastery of the work of other men in this field. In the same year he made his own first important contribution. He caused a magnet to rotate round a wire carrying a current, and also a current-carrying wire to rotate round a magnet—the first primitive electric motors.

In 1823 Faraday was proposed for election as a Fellow of the Royal Society; Davy, who was President of the Society, tried to have his candidature withdrawn. Davy was a sick man at the time, but he was also jealous. Never again were the relations between the two men the same. Faraday was elected a Fellow and two years later became Director of the Royal Institution.

Faraday did not forget the debt he owed to Davy. Twenty years after Davy's death the French chemist Dumas spoke slightingly of him to Faraday. After lunch Faraday took Dumas down to the library of the Royal Institution, paused and said: 'He was a great man. It was at this spot that he first spoke to me.'

Faraday married in 1819 and took his wife to live in a flat over the Royal Institution, where they remained for over forty years. During this time he performed thousands of experiments which are recorded in his diary, now published in seven large volumes. He was always full of ideas and used to say that he was content if one in a thousand of his experiments led to a really important discovery. He worked wearing an apron full of holes and assisted by an old soldier, Anderson, whose chief virtue was his blind obedience. One evening Faraday forgot to tell Anderson to go home, and found him still at work next morning.

Faraday's pleasures were simple. Occasionally in the evening he would take his wife to see Keen act or to hear Jenny Lind sing. He loved to watch the antics of the monkeys at the Zoo, and laughed until the tears ran down his cheeks. He sometimes rode his velocipede to Hampstead in the morning, or round the lecture theatre of the Royal Institution after the day's work was done.

He took the greatest pains to make himself a good lecturer. He wrote out rules for himself: for example, 'If at a loss for a word, not to ch-ch-ch or eh-eh-eh, but to stop and wait for it'.

He would lecture with a card before him bearing the word SLOW and, towards the end of the hour, Anderson would bring him a card with the word TIME. He had a gift for simple and striking experimental demonstrations. One of his audience related: 'When he threw a coal-scuttle full of coals, a poker and a pair of tongs at the great magnet, and they stuck there, the theatre echoed with shouts of laughter.'

The theatre held 700 persons and among its audience were sometimes to be found the Prince Consort and the Prince of Wales. Charles Dickens, then editor of *Household Words*, asked Faraday if he could report the lectures in his magazine. Faraday initiated the Christmas lectures for children, and the Friday evening discourses.

Faraday's salary, when he was married, was £100 per year with house, coals, and candles. By 1831 he was making £1000 a year as a consultant scientist, and Tyndall reckons that he could have raised this to £5000 a year. He found, however, that too much of his time was taken by these activities, and he deliberately sacrificed wealth to scientific research. In 1832 his income fell to £150 a year.

Faraday found it necessary deliberately to restrict his interests, as well as his finances, in order to pursue his all-absorbing investigations. At the age of 49 he had a mental breakdown and suffered from giddiness and loss of memory, which necessitated a four years' rest.

One outside activity Faraday permitted himself. He became scientific adviser to Trinity House, which is responsible for the lighthouses of the country. His visits of inspection to the lighthouses all round the coast provided him with a much needed change and relaxation. Once, at a Trinity House dinner, the Duke of Wellington advised him to give his speculations a more practical turn. He was often asked what was the use of his researches, for in those days electricity seemed only a scientific toy.

In 1858 Queen Victoria gave him a grace and favour house at Hampton Court. Here he spent his last few years sadly affected by loss of memory.

*   *   *

Faraday's most famous discovery, and the one which we will consider in a little detail in order to illustrate his views on the magnetic field, was the phenomenon known as electromagnetic induction. Oersted discovered in 1820 that an electric current creates a magnetic effect. Faraday looked for the inverse pheno- menon, the creation by magnetism of an electric current. He nearly found it in 1824 and 1825; he realised later how close he had been. If any date is to be ascribed to the discovery, it is 29 August 1831, although a full understanding was obtained only by numerous experiments performed during the next six months, and then by years of reflection.

Faraday wrapped two coils of copper wire round opposite sides of an iron ring (Fig. 34). The wire was insulated from the

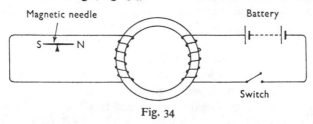

Fig. 34

iron by calico, and from neighbouring turns by string, since only bare copper wire was then obtainable. He connected the ends of one coil to a voltaic battery and a switch, and the ends of the other coil to a wire, lying north and south, under which lay a pivoted magnetic compass needle, as in Oersted's arrangement. On switching on the current in the first coil a deflection of the magnetic needle showed that a momentary current, now called an induced current, flowed in the second coil. While the current flowed steadily in the first coil, no induced current flowed in the second coil. On switching off the current in the first coil a deflec- tion of the magnetic needle occurred in the opposite direction, indicating a momentary, reversed, induced current in the second coil.

When the electric current from the battery flows through the first coil it causes the iron ring to be magnetised, and it is through this magnetism that it exerts its influence on the second coil.

Faraday now understood the reason for his earlier lack of success. He had not realised that an induced current would be set up only by a change of magnetism, and not by steady magnetism. From his diary he seems to have been slightly disappointed at the momentary character of the induced current. But he now had the essential clue.

By 17 October, seven weeks later, he had made considerable progress. On this day he obtained an induced current merely by plunging a bar magnet into a cylindrical coil (Fig. 35). The coil consisted of 220 ft. of copper wire wrapped round a hollow cylinder of paper. The induced current was detected by a galvanometer, invented by Schweigger in 1820, which consisted of a wire wound many times over and under a magnetic needle, thereby multiplying the Oersted effect. The fact that Faraday took some time to realise the value of a galvanometer illustrates the novelty and confusing nature of these early, electrical experiments. He had even a froggery in the cellars of the Royal Institution, where he kept frogs for detecting electric currents as Galvani had done.

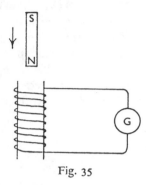

Fig. 35

On 28 October Faraday performed another experiment of major significance. His apparatus was the prototype of the dynamo, by which electric power is now generated for industrial and domestic use; the diagram which he drew of it in his notebook is shown in Fig. 36. He rotated a copper disc between the poles

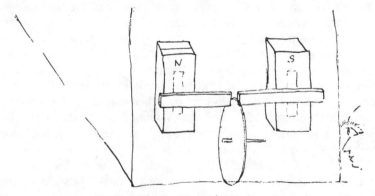

Fig. 36. Faraday's sketch of his 'dynamo'. (Courtesy: The Royal Institution.)

of the large, compound, horseshoe, permanent magnet belonging to the Royal Society. The induced current was led away to a galvanometer by means of grooved, slipping copper contacts at the rim and axle of the disc, and it was found to be continuous.

We have selected only three of Faraday's numerous experiments on electromagnetic induction and will now consider the theory he developed to explain them. It was not a mathematical theory—Faraday had learnt no mathematics beyond arithmetic—and yet it was quantitative and exact. He subjected it to all kinds of experimental tests and it never failed.

Long before his time it was known that iron filings, sprinkled over a card placed on a magnet, set themselves in lines as in Fig. 37. He called these lines *lines of magnetic force* and came to

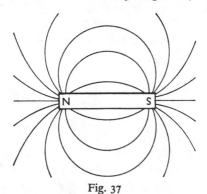

Fig. 37

believe that they existed in space independently of the iron filings. They represented, or indeed constituted for him, the magnetic field surrounding the magnet. Their direction represented everywhere the direction of the magnetic field; where they were crowded together the magnetic field was strong and, as they diverged, the magnetic field became weaker. By regarding them as being in tension, like threads of stretched elastic, and as tending to push each other sideways, Faraday accounted for the attraction of north and south magnetic poles and for the repulsion of two north or of two south magnetic poles.

The essential difference between Faraday's theory and those of his contemporaries, such as Ampère, was that it postulated the occurrence of magnetic action from point to point, between contiguous particles of the æther, and not at a distance. The lines represented a state of strain, or the transmission of some kind of stress, through the æther.

Faraday suggested that an induced current occurs whenever lines of magnetic force are cut by a circuit. In the experiments

represented in Figs. 34–6 magnetic lines of force are moving relative to the circuits while the induced current flows. The simplest
way in which we can obtain an induced current is to move a single
wire (connected to a sensitive galvanometer) near to a magnet.
Faraday argued that the mere motion of the wire could not be
responsible for the induced current. The space surrounding the
magnet must be affected by the presence of the magnet, and this
implied, he thought, the physical existence of the lines of
magnetic force.

Faraday performed a number
of experiments with the apparatus
indicated in Fig. 38. A wire
passes from *a* to *b* along the axis
of a magnet to its middle, and

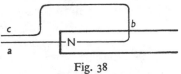

Fig. 38

then in a loop back to *c*. If the loop of wire is turned through one
complete revolution or, alternatively, if the wire is fixed and the
magnet turns through one revolution, the whole system of lines
possessed by the magnet is cut once by the wire and an induced
current flows. It does not matter how big the loop is, or what is
its shape; the same number of lines of force are cut, and the same
quantity of electricity is induced. Assuming the number of lines
of force to be proportional to the strength of the magnet, Faraday
stated the following law: 'The amount of current induced is precisely proportionate to the amount of lines of force intersected by
the moving wire.'

Faraday's theory led him to speculate whether the rotation of
the earth caused induced currents, by the cutting of the lines of
the earth's magnetic field. He tried, without success, to find
whether any induced currents were flowing in the Round Pond
in Kensington Gardens. He dipped two metal plates into the
pond, 480 ft. apart, and connected them by wires to a galvanometer. He performed a similar, fruitless experiment on Waterloo
Bridge, using the moving waters of the Thames. The induced
currents for which he was looking have since been detected by
their influence on submarine cables.

Faraday believed that lines of *electric* force with properties
similar to lines of magnetic force exist in the space surrounding
an electrically charged body. He regarded a line of electric force
as running from a positive charge to an equal negative charge,
and he often insisted, supporting his views as usual by experiments,

that it is impossible to produce one kind of electricity without producing at the same time an equal quantity of the other. For example, dry glass which has been rubbed with silk has a positive charge and the lines of electric force run from the glass to its surroundings—to the person holding it or to the walls of the room—where an equal negative charge will be found. This negative charge is said to be induced by the positive charge.

Faraday found that the force between two electric charges is changed by immersing them in a different insulating medium, for example in sulphur instead of in air. This reinforced his view of the part played by the intervening medium in all electrical effects. Indeed he tended to believe that electricity existed in the insulating medium surrounding a charged conductor, rather than in the conductor itself; the conductor merely served to tether one end of each line of electric force.

He extended his views to gravitation. He pointed out that matter in the sun exerts a gravitational effect on matter in the earth, at a distance of over 90,000,000 miles. Matter should not be regarded, therefore, as atoms of limited form and size. 'Matter is everywhere present', he wrote, 'and there is no intervening space unoccupied by it.' And again: 'This view of the constitution of matter would seem to involve necessarily the conclusion that matter fills all space, or at least all space to which gravitation extends.'

He was fond of quoting Newton on this subject. 'That gravity should be innate, inherent and essential to matter so that one body may act upon another at a distance through a vacuum and without the mediation of anything else...is to me so great an absurdity that I believe no man who has in philosophical matters a competent faculty of thinking can ever fall into it.'

One of Faraday's fundamental convictions was that all the manifestations of physical nature, such as gravitation, electricity, magnetism and light, have a common origin and are mutually related and dependent. He speculated that light might consist of vibrations transmitted along lines of gravitational force: 'The view which I am so bold as to put forth considers, therefore, radiation as a high species of vibration in the lines of force which are known to connect particles and also masses of matter together.' He pointed out that the æther was considered to be a medium of almost infinitely high elasticity and almost infinitely small density. He was therefore inclined to dismiss the æther as materially non-existent and to substitute for it his lines of force.

*   *   *

Faraday's conception of the physical existence of lines of force did not find immediate acceptance. Sir George Airy, the Astronomer Royal, wrote: 'I can hardly imagine anyone who knows the agreement between observation and calculation, based on action at a distance, to hesitate an instant between this simple and precise action on the one hand and anything so vague and varying as lines of force on the other.' It was not until Maxwell had put Faraday's ideas into mathematical form that the new concepts of field physics began to prevail. Einstein has pointed out that the relation between Faraday and Maxwell was not unlike that between Galileo and Newton. In each case the earlier investigator provided many of the intuitive ideas, and his successor gave them mathematical expression.

James Clerk Maxwell (1831–79) was born in the year that Faraday discovered electromagnetic induction. He was the son of the laird of Glenlair, Kirkcudbright, and was brought up to enjoy a country life. As a child he was interested in everything mechanical and he was always asking 'What is the go of that?'

Maxwell studied for three years at Edinburgh University, and then, in 1850, entered Peterhouse, Cambridge. After one term at Peterhouse he moved to Trinity because there were many able mathematicians at Peterhouse, and there seemed to be a better chance of a Fellowship at Trinity. Like Kelvin and J. J. Thomson, he graduated as Second Wrangler, Routh being the First, and in 1855 he obtained his Fellowship at Trinity.

Maxwell had a keen sense of humour which was betrayed by a peculiar twinkle in his eyes. He took with him to Cambridge in 1857 one of his spinning tops—he was particularly interested in rotating bodies and the gyroscope. He showed the top to his friends in his room one night and it was spinning when they left. In the morning he saw one of these friends coming across the court to his room. He sprang out of bed, started the top spinning and returned to bed, pretending to be asleep when his friend arrived.

All his life Maxwell played with great skill the game of diabolo, or the devil on two sticks. The devil is a double cone which is spun by a string connected to two sticks, thrown into the air and caught again on the string. He also retained his interest in mechanical toys. He used to fly some mechanical butterflies, with revolving antennae, round his drawing room in Cambridge when he was a professor; he made a toy ophthalmoscope with which he would examine the eyes of his friends and of his dog; he had a zoetrope, a primitive machine for showing moving pictures.

In 1856 Maxwell became Professor of Natural Philosophy at Aberdeen. After two months he wrote to one of his friends: 'No jokes of any kind are understood here.' Between 1860 and 1865 he was professor at King's College, London, and these were the most fertile years of his life.

In Cambridge plans were afoot for the institution of a professorship and demonstratorship in Heat, Electricity and Magnetism, and for the building of a physics laboratory. Before this time no provision had been made by the University for experimental training or research; Newton did his experiments in his rooms at Trinity, and Stokes in his rooms at Pembroke. The cost of the laboratory was estimated at £6300 and the money was generously provided by the Chancellor of the University, the Duke of Devonshire. Maxwell was appointed as the new professor in 1871 and he supervised the building and equipping of the laboratory, which was given the name of the Cavendish Laboratory.

Cambridge was fortunate in having as its first Cavendish professor a man of the highest genius, although Maxwell did not live long enough to found a school of research and the magnitude of his greatness was not appreciated until some years after his death.

* * *

Maxwell was trained as a mathematician but he had great insight into physical phenomena. When he was an undergraduate his Cambridge coach said 'it appears impossible for Maxwell to think incorrectly on physical subjects'. Maxwell studied Faraday's *Experimental Researches* and absorbed Faraday's way of looking at electrical and magnetic phenomena. He set to work to see how Faraday's ideas could be represented mathematically and in 1855 published a paper *Faraday's Lines of Force*, in which he showed that Faraday's conceptions led to the same results as action at a distance.

The prevailing attitude to the physical world at that time was to regard it as a machine. Lord Kelvin said that he could not understand a phenomenon until he could make a mechanical model of it. Maxwell, as a young man, wrote many letters to Kelvin for his advice, and it is small wonder that he set himself the task of making a mechanical model of Faraday's lines of magnetic force. He wrote: 'I think we have good evidence for the opinion that some phenomenon of rotation is going on in the magnetic field, and that this rotation is performed by a great number of very small portions of matter.' He visualised a line of magnetic force as a string of beads, each bead being a tiny, spinning vortex of æther. The speed of rotation was proportional to the strength of the magnetic field. Centrifugal force caused the beads to contract longitudinally and to expand laterally, and this accounted for the property of the lines of force of being in tension and of repelling each other laterally.

The direction of rotation of the beads was determined by the direction of the magnetic field. Now two adjacent beads, if geared together, must revolve in opposite directions. Maxwell therefore assumed that between each adjacent pair of beads there was a sphere of æther, which acted like an idle wheel and geared the beads together, enabling them to rotate in the same direction (Fig. 39). He identified these intermediate spheres as particles of

Fig. 39. The shaded spheres (called beads in the text) are magnetic vortices; the unshaded spheres are electric vortices.

electricity and their linear motion as an electric current. Motion of the spheres through a short distance represents a very brief electric current and Maxwell termed this an electric displacement. If the speed of one of the beads is changed, representing a change in the strength of the magnetic field, the spheres next to it will move linearly, representing an electric displacement. This accounts for electromagnetic induction. Faraday's discovery was that a change in a magnetic field gives rise to an induced current in an electric circuit. Maxwell's model indicated that, if no circuit is present, the effect of a changing magnetic field is to cause a slight displacement of electricity in the æther.

Suppose that the beads are not spinning and that a force is applied to the spheres. The linear motion of the spheres will cause the beads to rotate. Hence the production of an electric displacement creates a magnetic field. This, fundamentally, is Oersted's discovery, that an electric current creates a magnetic field.

An electric displacement is an interpretation in terms of an elastic æther of an electric field. Hence two basic ideas emerged from the model: a changing magnetic field gives rise to an electric field: a changing electric field gives rise to a magnetic field.

Maxwell's superiority to his contemporaries like Kelvin was that when his model had served its purpose he was able to discard it. His final attitude was somewhat ambivalent but, in the construction of his mathematical theory, it was fundamentally modern. The modern view is that it is impossible to represent the electromagnetic field in mechanical terms; electric and magnetic fields must be accepted as irreducible facts of experience. A mechanical model may be suggestive, but it is ultimately stultifying because it cannot be verified by any evidence independent of the phenomena it represents. Moreover, a mechanical model, designed to cover a wide range of magnetic and electrical phenomena, becomes so complicated as to give little satisfaction.

On 8 December 1864, Maxwell read to the Royal Society his most important paper, *On a Dynamical Theory of the Electromagnetic Field*. The word 'dynamical' in the title is perhaps misleading and it may have been a relic of his mechanical model. He justified it by saying that electromagnetic phenomena are caused by matter in motion.

Maxwell opened the paper by remarking that the mathematical theories accounting for the phenomena of electricity and magnetism, then in existence, had been developed without consideration of the surrounding medium; they postulated action at a distance. The most complete theory, that of the German physicist Weber, required the force between two electric charges to depend on their relative velocity, as well as on their distance apart, and Maxwell said that this assumption was such as to prevent him from accepting the theory as a final one.

Maxwell's own theory was essentially a theory of the field, that is to say, of the condition of the space surrounding bodies in electric or magnetic conditions. All space must be filled with an ætherial medium to account for the propagation of waves of light. The passage of waves of light represent a transfer of energy through the medium. The energy is passed on from one part of the medium to the next, and alternates from kinetic energy, that is energy of motion, to potential energy, that is energy stored up by displacement of the medium. To illustrate this, consider ripples spreading over the surface of a pond and a cork on the surface bobbing up and down. At one moment the cork is moving up and has kinetic energy; shortly afterwards it is at rest on the crest of a ripple, when its energy is potential; the restoration of this energy causes it to move down and again possess kinetic energy.

A similar alternation of kinetic and potential energy occurs in the mutual creation of magnetic and electric fields, indicating the possibility of the propagation of electromagnetic waves. A change in a magnetic field gives rise to a displacement of electricity in the æther, like an elastic yielding, and this provides potential energy. The restoration of this displacement gives rise to a magnetic field and this provides kinetic energy. The creation of the magnetic field causes a further electric displacement and so on. In this way an electromagnetic wave may be propagated consisting of alternating magnetic and electric fields.

Maxwell then derived his famous equations which, in their modern form, are four in number. They express mathematically the fundamental laws of electricity and magnetism with beautiful conciseness. They predict the existence of electromagnetic waves and they show that the velocity of these waves must be equal to the velocity of light. They lead to the obvious deduction that light must be electromagnetic and hence to the unification of the sciences of optics and of electricity and magnetism. Maxwell was able to calculate the optical properties of a medium from its magnetic and electrical properties.

Interpreted in terms of Faraday's lines of magnetic and electric force, the equations show that these lines can move through space, at right angles to themselves, with the velocity of light.

Maxwell expanded his paper into his *Treatise on Electricity and Magnetism*, published in 1873. This was a difficult book to understand, mainly because Maxwell had no clear idea of the nature of electricity itself. One French physicist is reported to have said: 'I understand everything in the book except what is meant by a body charged with electricity.' The establishment of the modern view of the atomic nature of electricity was delayed for half a century by the influence of Faraday.

The first continental physicist to accept Maxwell's theory was Heinrich Hertz (1857–94). He took the view that the theory was embodied in the equations, which described the structure of the electromagnetic field in space and time, and that the ideas of lines of force and strains in the æther were best regarded as the scaffolding by means of which the equations were constructed. These ideas, he thought, were superfluous and confusing accretions. The real crux of the theory, by which it could be tested against its rivals, was whether or not electromagnetic waves existed. In 1888 he produced electromagnetic waves in the laboratory, thereby removing all doubt of the fundamental truth of the theory and providing the basis of radio broadcasting.

Students of electricity and magnetism are still taught to interpret phenomena as Faraday did because lines of force provide a simple model or mental picture. Physicists cannot do without their models. A model in science is like a metaphor in language. Physicists speak of the flow of heat, of waves of light, of bouncing atoms, and of ideal heat engines. A model interprets complex phenomena in terms of something simpler and acts as a link between theory and experiment. It is useful so long as it does not become too complicated.

Hertz, in rejecting a model as an ultimate explanation, anticipated the modern positivistic view of physical theories. A physical theory does not describe or explain the intrinsic nature of the world of sense observation. It merely depicts structure and represents, usually mathematically, functional relations between phenomena. Its validity rests upon its success in making predictions about future phenomena.

CHAPTER 15

THE RISE OF ORGANIC CHEMISTRY

THE DISCOVERY of molecular architecture, or the way in which the atoms are arranged in the molecules, provided the basic idea of organic chemistry and has led to one of the greatest achievements of science, the synthesis in the laboratory of some of the complex materials of living organisms and of hundreds of thousands of organic compounds not found in nature.

The distinction between organic and inorganic chemistry, the former dealing with substances found in living things and the latter with inanimate matter, was drawn towards the end of the eighteenth century. The analysis of organic compounds, such as sugars, fats, and vegetable oils, showed that they had a more complicated composition than inorganic compounds. It was thought that the production of organic compounds required the action of a 'vital force', to be found only in living things, and that the chemical laws of the organic world were different from those of the inorganic. This belief died slowly and became untenable when organic substances were synthesised in the laboratory from inorganic materials.

The immense complexity of the molecules of some organic substances, such as proteins, and the limitless number of possible organic compounds spring, as we shall see, from the remarkable and unique property of the carbon atom of being able to link up with other carbon atoms and to form long chains or rings of atoms. Organic chemistry is now defined as the chemistry of carbon.

One of the chief methods of investigation available to the early chemists was to find the composition of a substance by experimental analysis. Many organic compounds contain only the elements carbon, hydrogen and oxygen; Lavoisier determined their compositions by burning them in a known quantity of oxygen gas, when the carbon was converted to carbon dioxide (CO_2), and the hydrogen to water (H_2O). By measuring the quantities of carbon dioxide and water formed, and the quantity of oxygen used, Lavoisier was able to calculate the percentage of carbon, hydrogen and oxygen in the organic compound. The method was simplified and improved, notably by Liebig, who burnt the organic compound with a solid substance yielding oxygen instead of in oxygen gas. In ways similar to this the compositions by weight of many organic compounds were determined.

A remarkable fact of fundamental significance emerged. Two or more substances, with quite different chemical properties, were found to have identical compositions. Silver fulminate (CNOAg), analysed by Liebig in 1823, had the same composition as silver cyanate, analysed by Wöhler. At first it was thought that an error must have been made, but it was then found that the same applied to other substances. In 1828, for example, Wöhler found that he could prepare urea by evaporating ammonium cyanate, and that both substances had the same composition. Urea is found in urine, and its preparation from the inorganic substance, ammonium cyanate, was one of the first blows sustained by the theory of vitalism. 'I must tell you', wrote Wöhler to Berzelius, 'that I can prepare urea without requiring a kidney of an animal, either man or dog.'

Berzelius coined the word isomerism (Gk. *isos*, equal; *meros*, part) to describe the phenomenon of several substances of different chemical properties having the same composition, and it was realised that the arrangement of the atoms in a molecule must have an importance comparable with the number and nature of the atoms. It is now known that a fairly simple substance like decane, $C_{10}H_{22}$ has seventy-five isomers; the hydrocarbon $C_{16}H_{34}$ has more than ten thousand.

* * *

It is not possible, in a brief space, to recount the long and involved story of the development, during the first three-quarters of the nineteenth century, of the modern idea of molecular structure, but we shall outline some of its most significant features.

The daunting complexity of the composition of organic substances, which prompted Wöhler to compare organic chemistry to the impenetrable jungle of a primeval forest, was somewhat simplified when it was realised that groups of atoms in organic compounds could retain their identity during chemical changes and behave like a single atom. From this was developed the radical theory of Liebig, Wöhler and Dumas, a radical being a group of linked atoms.

Liebig and Wöhler published in 1832 an account of their joint research in a classic paper *Researches on the Radical of Benzoic Acid*. From oil of bitter almonds they obtained a series of compounds all containing the benzoyl radical C_7H_5O, for example, benzoyl chloride $C_7H_5O.Cl$, benzoyl bromide $C_7H_5O.Br$, benzoyl cyanide $C_7H_5O.CN$ and benzamide $C_7H_5O.NH_2$. Other radicals such as methyl CH_3, ethyl C_2H_5, acetyl C_2H_3O and cacodyl AsC_2H_6 were discovered and they seemed to justify the generalisation of Liebig and Dumas that the radicals 'are the real elements with which organic chemistry operates'. They provided a much needed method of classification in the bewildering welter of organic compounds.

Fig. 40. Liebig's laboratory at Giessen. Redrawn from a contemporary drawing by Trautschold. (From J. Read: *Humour and Humanism in Chemistry*, by courtesy of G. Bell and Sons Ltd.)

Justus von Liebig (1803–73) was one of the most influential chemists in the whole of the nineteenth century. The status of chemistry in his youth can be judged from the reaction of his classmates when the Rector of his gymnasium inquired about his future career; his reply that he wanted to be a chemist touched off immoderate and uncontrollable laughter. His apprenticeship to an apothecary came to an abrupt end when he blew out the attic windows with his experiments. Nowhere in Germany could he obtain an adequate training in chemistry, but he received permission to work in the private laboratory of Gay-Lussac in Paris. On his appointment at the small German university of Giessen he built up a school of chemistry to which students came from all over the world. His influence as a teacher became immense.

Liebig had a lifelong friendship with Friedrich Wöhler (1800–82), who was trained in the private laboratory of Berzelius at Stockholm and who also built up a famous school of chemistry—at the university of Göttingen. It was Wöhler who suggested to Liebig that they might undertake a joint research: 'If you are so minded we might, for the humour of it, undertake some chemical work together.' Wöhler was of a placid and equable temperament, and he had often to restrain his impetuous and irritable friend. The two men laid the foundations of experimental organic chemistry.

In 1852 Edward Frankland (1825–99), a former student of Liebig at Giessen, put forward the important idea that every atom has a limited, definite, although sometimes variable, capacity for combining with other atoms. He pointed out that phosphorus, nitrogen, antimony and arsenic always combine with three or five atoms of hydrogen or chlorine. Before this time chemists had no clear conviction of the limited combining capacities of the elements, and they made no attempt to investigate the effects of the constituent atoms on the combining capacities of the radicals.

Combining capacity is now called valency. A univalent atom, such as hydrogen or chlorine, combines with one univalent atom, as in hydrochloric acid, HCl; a divalent atom, such as oxygen, combines with two univalent atoms, as in water H_2O; the atom of phosphorus may be trivalent or pentavalent since it can give rise to such compounds as PCl_3 and PCl_5.

Frankland's views did not meet with immediate acceptance, partly because there was still uncertainty about the atomic weights of the elements and about the formulae of chemical compounds. Wöhler and Liebig, for example, gave the benzoyl radical the composition $C_{14}H_{10}O_2$, not C_7H_5O as we do today. The confusion was dispelled after 1860 at the instigation of Cannizzaro, as we have already explained on pp. 143–4.

The greatest name in the history of theoretical organic chemistry is that of August Kekulé (1829–96). It was he who launched organic chemistry into the highly deductive, architectural form that it holds today. It is no doubt significant that he began his university career at Giessen as a student of architecture; coming under the spell of Liebig, he transferred his interest to chemistry. In later middle age his health broke down through overwork. 'If you want to be a chemist', Liebig said to him when he was working in his laboratory, 'you will have to ruin your health; no one who does not ruin his health with study will ever do anything in chemistry nowadays.'

Kekulé complained of the 'total lack of exact scientific principles' in organic chemistry and set about their provision. He recognised the significance of valency and established two fundamental facts: that carbon is quadrivalent, that is it combines with four atoms of hydrogen, and that atoms of carbon can link together. The idea of the linking of carbon atoms into chains came to him when he was travelling on the open top of a London horse bus. It was the last bus on a fine summer evening, and, as he travelled from Islington to Clapham, the streets were deserted. He fell into a reverie; the atoms seemed to gambol before his eyes, whirling as in a dance, until they formed themselves into a chain.

Kekulé's graphical symbols for molecules were not very suitable and resembled strings of sausages. The following are the first three members, CH_4, C_2H_6, and C_3H_8, of the endless homologous series of hydrocarbons, given in modern symbols:

Of the same family as ethane is ethyl alcohol C_2H_5OH; it has an isomer of a different family, dimethyl ether $(CH_3)_2O$:

The strength of Kekulé's theory was that it explained isomerism and predicted the number of isomers.

The idea of chains of carbon atoms was wonderfully successful in the case of the aliphatic compounds, so called because they include the natural fats, but it required modification to explain the aromatic compounds, derived from essential oils. The simplest example of the aromatics is benzene, C_6H_6, discovered by Faraday in 1825. When the more complex aromatic compounds are broken down by reagents, there remains a residual product containing six carbon atoms.

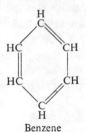

Benzene

Once more inspiration came to Kekulé in a waking dream. He was sitting in a chair by the fire when the atoms began their dance; they formed into chains like snakes, when one of the snakes seized hold of its own tail and formed a ring. Such was the genesis of that basic concept of organic chemistry, the benzene ring, in which the carbon atoms are linked in a closed chain by alternate

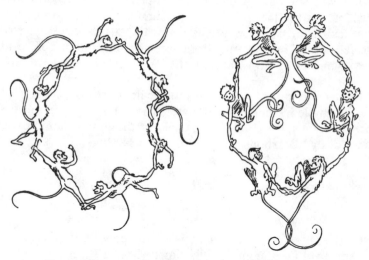

Fig. 41. Humorous drawings of the benzene ring, showing the alternate single and double bonds.

single and double bonds, thereby accounting for the four bonds of each carbon atom. Fig. 41 is a humorous drawing of the benzene ring which appeared in a journal prepared for a beer-drinking meeting of German chemists in Berlin in 1886.

Kekulé suggested that his ring concept could be tested by an investigation of the numerous substitution derivatives of benzene and an examination of their isomers. The field of aromatic chemistry, until then almost untouched, was invaded by numerous researchers and the new facts fell into place in Kekulé's theory.

One further step was needed to complete Kekulé's work. Methylene chloride CH_2Cl_2 should exist as the following two isomers, whereas only one is known:

$$
\begin{array}{ccc}
& Cl & \\
& | & \\
H-&C&-Cl \\
& | & \\
& H &
\end{array}
\qquad
\begin{array}{ccc}
& H & \\
& | & \\
Cl-&C&-Cl \\
& | & \\
& H &
\end{array}
$$

If one assumes, however, that the four bonds of the carbon atom are distributed symmetrically in space, instead of in a plane, there can be only one substance CH_2Cl_2.

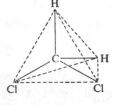

The origins of stereo-chemistry, dealing with molecular architecture in three dimensions, must be traced back to Louis Pasteur. Pasteur, in his first scientific investigation as a young man, discovered that crystals of sodium ammonium tartrate which, in solution, have the property of rotating through an angle the plane of vibration of polarised light, possess unpaired facets and are asymmetrical. He suspected that the asymmetry of the crystals might represent an asymmetry of molecular structure and hence be linked with the property of rotating polarised light. He therefore examined the crystals of sodium ammonium racemate, isomeric with sodium ammonium tartrate and optically inactive, expecting to find the crystals symmetrical. But in fact there were two kinds of crystals, both asymmetrical and each the mirror image of the other. He carefully separated the two kinds of crystal and found that when dissolved in water, one kind caused a right-handed rotation, and the other a left-handed rotation. He related how, when the experiment was demonstrated to Biot, 'the illustrious old man took me by the arm and said: "My dear child, I have loved science so much throughout my life that this makes my heart throb."'

Pasteur's suggestion of the asymmetric structure of optically active molecules was combined with the conception of a carbon atom situated at the centre of a tetrahedron with its bonds towards the four vertices, in the theory put forward independently in 1874 by Van't Hoff and Le Bel. They pointed out that there is an asymmetric structure only when a carbon atom is attached to four different atoms or groups. Such a case is the molecule of lactic acid, which exists in the following two stereo-isomeric forms:

$$CH_3 \quad\quad\quad CH_3$$
$$C\text{---}CO.OH \quad\quad CO.OH\text{---}C$$
$$H \quad OH \quad\quad OH \quad H$$

The theory that atoms were arranged in space came as a shock to many chemists because they were not really convinced that atoms existed. The theory was eminently satisfactory, however, as an explanation of the chemical facts.

* * *

The creation of theoretical organic chemistry led to the development, between 1856 and 1880, of the synthetic dye industry. It was in 1856 that Perkin discovered the first aniline dye, mauve, and this was followed by aniline red and aniline blue. It was realised that development would proceed far more quickly if the chemical constitutions of the dyes were understood and there arose a demand for organic chemists. A method was discovered of manufacturing from coal tar the dye alizarin, hitherto obtained from the root of the madder plant and employed from time immemorial for dyeing cotton a bright red. In consequence the cultivation of madder in the South of France, Alsace and Algiers came to an end. The German firm Badische Anilin und Soda Fabrik developed a method of synthesising and manufacturing from a coal product the dye indigo—a research project which cost them a million pounds. This ruined the Indian plantations where the plant, from which natural indigo had been extracted, was grown.

Synthetic drugs and perfumes were also manufactured, although not on so large a scale as dyes. In the twentieth century industries sprang up for the production of synthetic fibres, such as rayon and nylon, of plastics, such as bakelite, perspex and alkathene, of a substitute for rubber, of insecticides, weed killers, detergents, explosives, lubricants and fuels for internal combustion engines. These developments were based on chemical theory and research.

At the present time work is proceeding all over the world to elucidate the structure of enzymes, hormones, vitamins, alkaloids and genes. Many of these molecules contain scores, hundreds, or even thousands of atoms, and they exist in thousands of isomeric forms. The chemist must find the exact structure of the molecule—where each atom is fitted in the chains or rings of carbon atoms—and only when he has done this can he attempt the task of synthesis, which is the building up of the complex molecule from its simpler constituents.

It is possible that, just as in the nineteenth century the continuity between the inorganic and organic worlds was established, so, in the twentieth century, the gap between non-living and living matter will be bridged. It seems to be only a question of time before the organic chemist will be able to synthesise, in the laboratory, life.

CHAPTER 16

EVOLUTION

THE THEORY of evolution postulates that all living organisms have developed, over the course of aeons, from a common primitive germ of life. A century ago it required an effort really to believe that an oak, a haddock, a spider, a crocodile, a robin and a man had all descended from a common ancestor. The evidence for it was marshalled so convincingly by Charles Darwin, however, that its truth became generally accepted. No reputable biologist today would deny that the process of evolution is a fact, though some find unsatisfying the currently accepted explanation of its cause.

Until the nineteenth century it was believed that every different form of life had been created separately and simultaneously. In Milton's picture of the creation, the earth opened and all land animals emerged, the lion pawing his way out of the ground, shaking his brindled mane.

Attempts to classify living organisms showed that it was possible to arrange them into groups, the members of a group having marked similarities to each other. In the animal kingdom there are about a dozen of these major groups called phyla: for example, the Arthropods are animals possessing an external skeleton, and they include crabs, insects, spiders and centipedes; the Vertebrates are animals with a backbone and they include the fishes, amphibians, reptiles, birds and mammals. Each phylum can be divided into a system of subgroups, of which the ultimate unit is the species, representing a set of individuals which interbreed.

Why should such classification be possible and what lies behind it? The answer of the school of German *Naturphilosophie* in the eighteenth and early nineteenth centuries was that, although the Creator gave very free play to his fancy at the creation, He subjected Himself to the discipline of an archetypal plan for all organisms. Parallelisms of structure in widely different organisms were sought, but no satisfactory theory of form was achieved.

A different kind of explanation was suggested by the Frenchman, Buffon (1707–88). His immense encyclopaedic work, *Histoire Naturelle*, appeared at intervals between 1749 and the end of his life in the form of twelve volumes on quadrupeds and nine on birds. As he collected his information Buffon was struck by the similarities between quadrupeds, for example, between the dog, wolf, fox and jackal, and between the horse and zebra, suggesting that the different species were branches of one family tree. It appeared unlikely that each species had been created separately because of small but noticeable anatomical features which seemed to serve no useful purpose. What, asked Buffon, is the purpose of the toes of a pig?

Buffon had a superb style, a fine presence, and a majestic self-esteem. When asked how many really great men there had been he answered: 'Five—Newton, Bacon, Leibnitz, Montesquieu and myself.' The pressure of orthodox opinion forced him to throw out his hints on the evolution of species in a guarded and ironic manner. When he speculated that all quadrupeds might have been derived from a single ancestral species, he added, 'But no, it is certain, by revelation, that all animals have equally enjoyed the grace of creation'.

The theory of evolution was put forward in a more forthright fashion by Dr Erasmus Darwin (1731–1802), Charles Darwin's grandfather, in a book with the curious title, *Zoonomia*. Charles read *Zoonomia* and was not impressed; he felt that the mass of speculation was too great to be carried by the small number of facts. The conclusion of *Zoonomia* was that all forms of life had descended from a single 'filament' or spermatozoon. The arguments adduced were the unity of plan of organisms as revealed in systems of classification; the fact that animals can be changed by selective breeding, as with dogs, sheep and horses; that they can be changed by climate—for example the sheep of warm climates are covered with hair instead of with wool, and partridges in northern climes become white in winter; and finally the growth of a tadpole into a frog, which gives some idea of what nature can do in the way of metamorphosis.

A far more important precursor of Charles Darwin was the Frenchman, Lamarck (1744–1829). Lamarck's theory was rejected by his contemporaries and he had small influence in establishing the fact of evolution, although his ideas on its cause influenced Darwin, and survive today among the advocates of neo-Lamarckism. He is regarded by Frenchmen as the greatest of their biologists.

Lamarck arranged all animals in a linear series, beginning with simple, primitive organisms and continuing to the most complex. He found that the graduation from simplicity to complexity was sustained but irregular, and he held that the more complex animals had evolved from those that were simpler.

He accounted for the process of evolution by the striving of animals to adapt their bodies to new habits. This involved two principles which he laid down as laws: that use and disuse cause modifications of organs; that these modifications are transmitted to offspring.

In his *Philosophie Zoologique*, published in 1809, he gave numerous examples to illustrate his theory. The long legs of the heron, he maintained, were the result of its efforts to stretch and lengthen them, in order to adapt itself to a wading habit. Stretching, too, was responsible for the long neck of the giraffe and for the long tongue of the ant-eater. The habit of snakes of gliding along the ground and of crawling through narrow spaces caused their bodies to elongate and their legs to disappear. Plants also evolve through changes in their environment. A seed of a meadow plant, blown by the wind to dry and stony ground, gives rise, if it germinates, to an ill-nourished plant. Its descendants, said Lamarck, will be dwarfed and adapted to drought. All these examples are plausible but Lamarck's fellow biologists did not find them sufficiently convincing to support the tremendous postulate of evolution.

Lamarck was harried by poverty all his life. In his last years he became blind, and his remains, like those of Mozart, were thrown into a common grave. It fell to Cuvier, after his death, to read a eulogy before the *Académie des Sciences*. 'What a eulogy!' exclaimed one of poor Lamarck's friends. In sarcastic tones Cuvier poured ridicule upon Lamarck's theory, stigmatising it as worthless, unscientific speculation.

Lamarck's explanation of evolution, in contrast to the mechanistic view put forward fifty years later by Darwin, was teleological, postulating in organisms a striving or will to evolve. His law of the inheritance of characters, acquired by use and disuse, is now rejected by most biologists since, despite repeated attempts, it has never been satisfactorily confirmed by experiment.

* * *

Charles Darwin (1809–82), by reason of his immense influence on human thought, must be classed with the greatest men of science— with Aristotle, Galileo, Newton, Lavoisier and Einstein. He was the son of a doctor in Shrewsbury and his boyhood interests were country pursuits. From his early years he had a strong passion, not shared by his brother and sisters, for collecting: he was a born naturalist. He loved dogs and shooting. 'How well I remember', he wrote, 'killing my first snipe, and my excitement was so great that I had much difficulty in re-loading my gun from the trembling of my hands.' After leaving Shrewsbury school he went to the University of Edinburgh to train as a doctor, but found that he had no taste for the profession. His father then suggested that he should become a parson and he was sent to Cambridge. His academic career was undistinguished; he spent much of his time shooting, hunting and collecting beetles, so much so that his father feared he would become an idle sporting man.

He became friendly with Henslow, the professor of botany at Cambridge, and, at the latter's suggestion, attended Professor Sedgwick's lectures on geology. During one vacation he accompanied Sedgwick on a tour of North Wales, which taught him how to investigate the geology of a country. On revisiting the area later, when he was an accomplished geologist, he realised how much they had failed to observe. The evidence for glaciation, he said, was as obvious as that of a burnt house. This reinforced his conviction that observation is fruitful only if guided by speculation and theory.

The decisive event of his life came at the end of his Cambridge career; it was the invitation to join, as unpaid naturalist, the expedition of H.M.S. *Beagle*, commissioned to chart some of the world's unknown waters. This five-year voyage, which took him to many parts of the world and introduced him to the teeming life of the tropics, was his true education. How nearly he missed it he has himself recorded. His father was at first hostile to the project and was only induced to change his mind by Charles' uncle, Josiah Wedgwood. At the interview with the commander of the *Beagle*, Captain Fitzroy, a fine but exceedingly aristocratic naval officer, Darwin was nearly rejected because of the shape of his nose; Fitzroy, aware that they would be together in a small ship for years, thought that the nose indicated insufficient determination and character.

The *Beagle* set sail from Plymouth on 27 December 1831. At Henslow's suggestion Darwin took with him the newly published first volume of Lyell's *Principles of Geology*, which he rapidly assimilated. At the very first port of call, St Jago in the Cape Verde Islands, he saw unmistakable evidence of Lyell's theory of uniformitarianism, that the earth's surface has reached its present condition by gradual changes occurring over a vast period of time. Along the cliffs was a horizontal white band of limestone with numerous shells embedded in it lying between layers of volcanic rocks. The upper layer of volcanic rock must have poured into the sea while the limestone was lying at the bottom. Darwin's observations as a geologist during the voyage of the *Beagle* were perhaps more remarkable than his work as a naturalist. The theory of evolution was, in a sense, merely the transfer of Lyell's theory of uniformitarianism from rocks to living organisms.

The tangled, green luxuriance of the forests of Brazil, gleaming with orchids and butterflies, made a lasting impression on Darwin's mind. Everywhere he collected insects, flowers, rocks and fossils; he shot birds; and crates of specimens were shipped to England from different ports during the *Beagle's* voyage. Fitzroy, engaged in charting the seas, left him ashore for weeks or months at a time, and Darwin hired horses and guides to explore the hinterland.

The *Beagle* made her way slowly south. Darwin rode with the gauchos and watched them hurl their bolases. Across the Salado river he noted that the coarse high grass of the pampas changed to fine green pasture as a result, not of a change of soil, but of the introduction of cattle, which cropped and manured it. In Uruguay he found that a European plant, the prickly cardoon, had changed the balance of nature. Unhindered, apparently, by any check, it had spread over hundreds of square miles, making the country impassable to man or beast. From the mud deposits of the Parana, he dug up the fossil remains of a gigantic, extinct armadillo-like animal.

In the desolate Tierra del Fuego he looked with wonder on the savages, and saw the sleet melting on the naked bodies of a woman and her baby. Off the coast of Chile, he watched the volcano of Osorno in eruption, throwing up dark masses of lava in a red glare. While resting in the forest in Valdivia he experienced the tremors of a severe earthquake, and when, a fortnight later, the *Beagle* reached Concepcion, the town lay in ruins. The land in the Bay of Concepcion had risen two or three feet. Rocks, formerly under water, were now 10 ft. above it and were covered with putrefying mussels. In the Andes Darwin found petrified trees at 7000 ft., and sea-shells at 13,000 ft.

Large Ground Finch   Medium Ground Finch

Cocos Finch   Warbler Finch

Fig. 42. Galapagos finches. (From D. Lack: *Darwin's Finches*.)

The *Beagle* now approached the Galapagos Islands about six hundred miles west of the coast of Ecuador. It was here, more than anywhere on the long voyage, that the idea of evolution insinuated itself into Darwin's mind. The fauna of these volcanic islands consisted mainly of reptiles; there were no mammals, except mice, probably brought in a ship. The huge, aboriginal tortoises, munching cacti growing in the black lava, differed from island to island so much that they could be recognised at a glance. Why should they differ? But the most remarkable evidence of evolution was to be found in the Galapagos finches. These included over a dozen separate species, all with different kinds of beaks, from a large and parrot-like beak to one as small and fine as that of a chaffinch, adapted for different types of food—insects, seeds, leaves and buds. Darwin wrote, 'One might almost fancy that from an original paucity of birds in this archipelago, one species had been taken and modified for different ends'. The original species must have come from the South American mainland.

The *Beagle* crossed the world via Tahiti to New Zealand and Australia; then, calling at Keeling Island, Mauritius and Ascension Island she made her way towards England. Darwin's journals and specimens, sent home from South America, had already created a most favourable impression. He wrote in his autobiography: 'Towards the close of our voyage I received a letter, whilst at Ascension, in which my sisters told me that Sedgwick had called on my father, and said that I should take a place among leading scientific men....After reading this letter, I clambered over the mountains of Ascension with a bounding step, and made the volcanic rocks resound under my geological hammer.' He had set out on the voyage a rather indolent, intellectually un-disciplined, young man; he returned an acute observer, both as a geologist and naturalist, inured to habits of constant reflection on his observations.

Back home in England, Darwin turned his thoughts to marriage; he wrote down the pros and cons on paper. He proposed to his cousin, Emma Wedgwood, and she accepted him. They settled, eventually, at Down House in Kent, sixteen miles from London. Here he spent the rest of his life, absorbed in his biological studies, virtually in retirement from the world, a semi-invalid, subject to a chronic digestion complaint. Whether the cause of his illness was the continual seasickness he suffered on the *Beagle*, or an hereditary weakness, or a neurotic condition caused by his dominating father, has been disputed. He made frequent visits to hydropathic establishments to take the 'water-cure'—two shower-baths a day. His best medicine, however, was his work.

His long hours of sleeplessness in the middle of the night enabled him to ponder over his problems. Indeed, he found it difficult to stop his mind from working. He wrote to his friend Hooker, 'It is an accursed evil to a man to become so absorbed in a subject as I am in mine'.

To train himself as a serious biologist he devoted eight years to a study of barnacles. His interest had been aroused in Chile by a curious form he had observed burrowing in a shell. A dissecting-table was let into the window of his study and, as his children grew up, his work on barnacles seemed to them so much part of the established order of things that one of them asked of a neighbour, 'Where does he do his barnacles?' In 1852 Darwin wrote, 'I hate a Barnacle as no man ever did before'.

It was in 1837, the year following his return from his voyage in the *Beagle*, that he started his first notebook on the evolution of species. He confided the idea of the theory to his friends Lyell and Hooker as though he were confessing a crime. The idea of evolution may have been in the air, but he did not know a single biologist who believed in it.

In 1838 he came upon a book by Malthus, *An Essay on the Principle of Population*, which argued that population tends to increase in geometrical ratio and is checked only by the means of subsistence. Malthus maintained that the result, among mankind, was misery and vice, and his central theme was that young men should postpone marriage until they had the means of supporting a family. Darwin obtained from the book the idea of the mechanism of evolution: the intense competition among organisms to survive, he realised, must lead to a natural selection of beneficial, random variations.

In 1844, he wrote out, in 231 pages, a statement of his theory of evolution by natural selection. All the time he was collecting facts of every kind which had any bearing on the theory. Characteristically he took especial care to record unfavourable facts, because he found that he was apt to forget them much more readily than those that were favourable. He had an elaborate filing-system, consisting of thirty to forty large portfolios, in cabinets with labelled shelves, in which he entered references to all the information he had obtained by reading, by correspondence, by observation and by experiment. In addition to his own considerable library of books, he had a drawerful of abstracts taken from other books.

To obtain first-hand information of the variations among domestic animals he decided to breed pigeons. He built a pigeon-house and stocked it with every breed of pigeon he could obtain. He cultivated the acquaintance of pigeon fanciers. He also had conversations with horse-breeders and consulted stud-books.

The years passed by and it is possible that he might never have published his theory had he not received, in June 1858, a manuscript from a young naturalist, Alfred Russel Wallace, who was collecting specimens in the Moluccas, putting forward precisely the same theory as his own, that of evolution by natural selection. Deeply disturbed, he consulted his friends. He said that he would rather burn his own statement of the theory rather than that Wallace or anyone else should think he had behaved in a paltry spirit. On the advice of his friends he wrote a short paper which, with that of Wallace, was read before the Linnaean Society on 1 July 1858. There was no discussion. As Hooker said, opponents did not dare to 'enter the lists before armouring'.

Darwin then set about writing his epoch-making work, *The Origin of Species*. The delay of over twenty years, since he first conceived the theory, had been a great advantage. He had amassed an enormous amount of evidence from which to select and over which he had brooded continually. The book was published in 1859. The first edition of 1250 copies was sold out on the first day and in the next twenty-five years about 30,000 copies were printed.

Like all the Darwins, he could write; not that composition came easily to him, but his style was vivid, and gave the impression of an attractive candour. His punctilious sense of honesty of expression is illustrated by an incident which occurred one evening after he had been discussing with his friends the sense of the sublime. He retired to bed early, before they did, but reappeared in his dressing-gown. 'Since I went to bed', he said, 'I have been thinking over our conversation in the drawing-room, and it has just occurred to me that I was wrong in telling you that I felt most of the sublime when on the top of the Cordillera; I am quite sure that I felt it even more when in the forests of Brazil. I thought it best to come and tell you this at once in case I should be putting you wrong. I am now sure that I felt most sublime in the forest.'

Another of his endearing traits was his complete modesty. When he found one of his ideas to be wrong he wrote to a friend, 'I am the most miserable, bemuddled, stupid dog in all England, and I am ready to cry with vexation at my blindness and presumption'.

The greatest thing about him was his judgement—as sound in finance as it was in science. His private income of some £1200 per annum at his marriage had been increased, by the end of his life,

(a)

(b)

PLATE XV. (a) The *Beagle* in the Straits of Magellan. (b) Archaeopteryx lithographica.

PLATE XVI. Charles Darwin.

to £8000 per annum, his investments being mainly in railway stock. He kept a record of every penny he spent and received, just as he kept the aggregate scores, over the course of many years, of his nightly games of backgammon with his wife.

* * *

The *Origin of Species* is a sustained argument, supported by a wealth of evidence, all of it circumstantial. No other naturalist, either before or since Darwin's time, had his range of knowledge and world-wide experience.

The book begins with two chapters to establish what Darwin had to accept as an empirical fact, without any understanding of its cause, that marked variations among animals or plants of the same species are constantly being thrown up. The breeders of domestic animals have selected these variations with the object of accentuating and perpetuating them. The result is that the wild dog has given rise to such diverse varieties as the greyhound, the pug, the bloodhound and the bulldog; the wild rock-pigeon has been developed into the runt, with a long massive beak and large feet, the barb with a short broad beak, and the pouter with an elongated body and enormously developed, inflatable crop; from the wild strawberry have come the domestic varieties, with a much larger fruit. The common oak has so many varieties that a German author, said Darwin, classified them into over a dozen species.

The next step in the argument is to explain how these naturally occurring variations could have given rise to the evolution of species. Animals and plants produce far more offspring than can possibly survive. Many plants produce hundreds or even thousands of seeds per year, but suppose that they produced only two per year, and that each of these produced two more the next year, and so on. The result of this increase in geometrical ratio would be, in twenty years, over a million plants. This is no mere speculation, for Darwin had seen an increase of this order, where there was no check, in the prickly cardoon in Uruguay. He considered the case of the elephant, which is well known to be one of the slowest breeders of all animals. He calculated that, if all the offspring originating from a single pair survived, in about 750 years there would be nearly nineteen million elephants.

Under normal conditions, the number of individuals in a species does not increase in geometrical ratio, but remains approximately stationary, and hence there must be intense competition to survive. Any slight variation which is beneficial in the struggle for existence will enable its owner to remain alive and this variation will be perpetuated in the next generation; those not possessing the advantage will become extinct. This process Darwin called natural selection, in contrast with the artificial selection of breeders of domestic animals and plants. Whereas the effects of artificial selection have been produced in a few thousand years, natural selection could have been operating, so geologists like Lyell affirmed, for many millions of years.

To illustrate the frightful mortality in nature Darwin cleared a piece of ground 3 ft. by 2 ft., and counted the seedlings of wild plants that sprang up. Out of 357, no fewer than 295 were destroyed, mainly by slugs and insects. In the winter of 1854–5 he estimated that four-fifths of the birds in his grounds were destroyed by the severity of the weather.

Besides the struggle for life among individuals in the same species, there is a complex of relationships between all animals and plants living in the same region. Darwin gave several examples of this, including the effect of cats on the survival of flowers; cats destroy mice, mice destroy the combs of humble bees, and humble bees pollinate flowers.

Suppose that there were a change of climate—and it was known that there had been enormous changes of climate in the world's history during the appearance and disappearance of the Ice Ages—some species might become extinct, the complex of relationships would be disturbed, and the proportional numbers of the different species might change drastically. When the number of individuals in a particular species increases, the amount of variation for natural selection to work upon increases.

If a species becomes dominant it may tend, in course of time, to become adapted to different environments and to evolve into other species and genera of strikingly different habits and appearance. The process is now known as adaptive radiation. In this way a convincing explanation can be given of what Darwin's contemporary, Richard Owen, called homologous organs, which are organs constructed on the same pattern in different animals, but having every variety of form and function. Thus the hand of a man, the leg of a horse, the paddle of a porpoise and the wing of a bat are made up of similar bones in the same relative positions

(Fig. 43). Likewise the hind-feet of the kangaroo, of the leaf-eating koala bear, of the insect-eating bandicoot and of other Australian marsupials, although adapted for such different modes

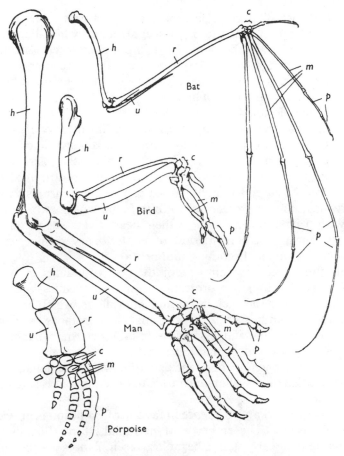

Fig. 43. A comparison of the homologous bones in the forelimb of a man, a porpoise, a bird and a bat; *h*, humerus; *r*, radius; *u*, ulna; *c*, carpals; *m*, meta-carpals; *p*, phalanges. (From *Proc. Soc. Psychical Res.* 50 (1953), 117, by courtesy of Sir Alister Hardy.)

of life, are of the same characteristic type. Crustaceans with elaborate mouthparts have fewer legs, and those with simpler mouths have more numerous legs, suggesting that the jaws are metamorphosed legs. The sepals, petals, stamens and pistils of flowers, although serving such different purposes, are all constructed on the same pattern. Owen regarded all this as evidence of a plan or plans in creation; Darwin saw in it powerful evidence for evolution.

On the view that creation took place on a preconceived plan one would expect that those parts of structure which determine habits of life would be of decisive importance in classification. But no systematist would class a mouse with a shrew, or a whale as a fish. The two individuals in each of these pairs, although adapted for a similar way of life, must be, said Darwin, of different genealogical descent.

Darwin pointed out the significance of rudimentary or atrophied organs, which are common in nature. Boa constrictors have rudimentary limbs; there are birds, such as the ostrich and penguin, and certain insects, whose wings are so atrophied that they cannot be used for flight. 'What can be more curious', wrote Darwin, 'than the presence of teeth in foetal whales, which when grown up have not a tooth in their heads?' He suggested that disuse, during the long period of evolutionary change, could account for organs becoming rudimentary.

He considered the mimicry of their surroundings in bodily structure and colour, as a protection from their enemies, by certain animals, such as the stick insect and caterpillars, which hang like dead twigs from bushes. This is a remarkable instance of adaptation to environment and it is understandable in terms of haphazard variations preserved by natural selection over a long period of time. It is a phenomenon, incidentally, whose explanation on a Lamarckian theory of an inner urge to development is singularly unconvincing.

Further remarkable evidence in favour of evolution is provided by embryology. The embryos of mammals, birds, lizards and snakes, in their early stages, bear close resemblances to each other. Darwin quoted a story told by von Baer, the founder of embryology, that he had preserved two embryos in spirit without labelling them, and then was unable to tell whether they were lizards, birds or mammalia.

In Fig. 44, representing the embryos of a man and of a dog, the gill clefts in the neck of both embryos, corresponding to the gill slits of a fish, can be seen, and also the incipient tail of the human embryo. Darwin's cautious comments were boldly developed by

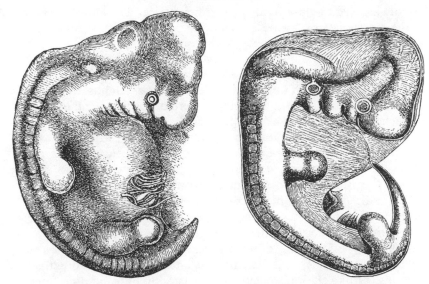

Fig. 44. Embryos of man and dog, taken from Darwin's *Descent of Man.*

his followers into the theory of recapitulation, that the embryo, during its development, passes through the adult stages of its evolutionary ancestors. This is true only in outline and not in detail.

The geographical distribution of animals was, for Darwin, one of the most convincing lines of evidence for evolution. He suggested that a new species probably, evolved at one place and then fanned out, but not equally in all directions, because of the barriers constituted by mountains, deserts and seas, and because of differences of climate. Sometimes barriers appeared after the spread of a species had occurred, and then evolution proceeded independently in the separate regions. This would account for the very different forms of life found today in various parts of the world where the climate and physical conditions are much the same. It would also account for the remarkable affinity of the different species inhabiting the same continent. The rhea or American ostrich is quite different from the African ostrich or the Australian emu; one species of rhea is found near the Straits of Magellan and another allied species in the plains of La Plata. The South American animals corresponding to European rabbits and hares have a typically American structure. The tiger is found only in Asia, and the jaguar in South America. The kangaroo and the marsupials of Australia belong to a type dominant long ago; they have been preserved by their isolation.

It is understandable why, on oceanic islands, whole groups of animals are missing, and why the animals which are found there are endemic or peculiar. In the Galapagos Islands there are virtually no mammals, and the numerous species of Galapagos finches exist nowhere else in the world.

The most direct evidence of evolution is provided by palaeontology, the study of fossils. In the lowest and oldest strata the fossils represent simple, primitive forms of life; passing consecutively to higher and more recently formed strata, we find fossils of more and more complex organisms. In the case of the vertebrates, fishes appear in the older rocks, then amphibia, then reptiles, then mammals and birds.

Many species and groups of species not alive today are found among the fossils, such as trilobites and ammonites and extinct monsters like the pterodactyl, the mastodon and the dinosaurs. 'No one can have marvelled more than I have done', wrote Darwin, 'at the extinction of species.' He rejected the theory of successive catastrophes, which postulated that species were wiped out by convulsions and floods. He pointed out that some species today are rare; their conditions of life must be unfavourable. They are like a sick man, whose death should cause no great surprise.

To the objection that, if evolution had really occurred, the rocks should contain many forms of life intermediate between existing species, he replied: 'I have found it difficult, when looking at any two species, to avoid picturing to myself forms *directly* intermediate between them. But this is a wholly false view; we should always look for forms intermediate between each species and a common but unknown progenitor.'

He emphasised the imperfection of our geological record: the remote chance of an animal's remains becoming embedded in the rocks, the small part of the earth's surface that had been explored, the submergence beneath the sea, and the elevations and denudations which had occurred during the course of evolution. The abrupt appearance of some species in certain rocks could be explained as the result of migrations of the species from other parts of the world.

With characteristic candour Darwin dealt faithfully with the difficulties of his theory. 'Some of them are so serious', he wrote, 'that to this day I can hardly reflect on them without being in some degree staggered.' How, for example, could the blind process of natural selection, operating on fortuituous variations, have resulted in so intricate and delicately adjusted an organ as the human eye? Darwin described how the simplest eye among living creatures consists merely of an optic nerve coated with pigment and has no lens. In the eyes of certain starfishes a gelatinous, primitive lens is found, which does not focus an image but merely concentrates the light. Since these gradations in eye-structure occur in nature it is possible to assume that the eye has evolved and that the human eye could have been formed, over a vast period of time, by gradual changes preserved by natural selection.

Another problem which Darwin considered was the way in which organs could be so modified that they were adapted for an entirely new mode of life. How, for example, could a fish have evolved into a land animal? He suggested that the swimbladder of a fish, originally developed for flotation, became modified into a primitive lung for respiration. The fins of fishes living in shallow water evolved gradually into legs.

Darwin did not discuss the evolution of man in *The Origin of Species*. But he realised that this would be regarded as the most significant aspect of his theory and in order that 'no honourable man' should accuse him of concealing his views, he included a sentence on the penultimate page: 'Much light will be thrown on the origin of man and his history.'

* * *

Samuel Butler remarked that Darwin was lucky, in that the theory of evolution was a ripe fruit which he had only to pluck. The immediate reception of *The Origin of Species* hardly bore this out however. Darwin's former teacher of geology, Sedgwick, was deeply offended by the theory, and even Lyell considered that it should have taken into account the guidance of a Higher Power. Owen, the foremost English authority on palaeontology and comparative anatomy, made a show of sympathetic deliberation but worked secretly and inveterately to oppose the theory. Jane Carlyle said that Owen's sweetness reminded her of sugar of lead. Darwin had one doughty champion, Thomas Huxley, who wrote a long and favourable review of *The Origin of Species* in *The Times*. The book created a sensation and it was talked about everywhere.

In June 1860 the British Association met in Oxford and there was lively anticipation of a discussion on the theory arranged to take place during the meeting. The debate was opened by the Bishop of Oxford, Samuel Wilberforce, who had acquired the nickname of Soapy Sam. Wilberforce, ignorant of biology, had been coached by Owen, and spoke for half an hour in a florid, scoffing manner. He made the mistake of turning to Huxley and asking, with jocular sarcasm, whether it was from his grandfather or his grandmother that he claimed descent from a monkey. Huxley whispered to his neighbour, 'The Lord hath delivered him into my hands'.

When Huxley rose, he said quietly that he was there in the interests of science. Darwin's theory was the best explanation of the origin of species which had so far been advanced. In reply to the bishop, he would not be ashamed to have a monkey as an ancestor, but he would be ashamed to be connected with a man who used great gifts to obscure the truth.

The atmosphere became electric. A lady fainted and had to be carried out. At the back of the hall Fitzroy, Darwin's old commander in the *Beagle*, now an admiral (retired), stood up shouting, and started waving a bible in the air. A story went round, after the meeting, that Huxley had declared that he would rather be a monkey than a bishop; this he disclaimed.

Huxley became Darwin's bulldog. He rejoiced, in his own phrase, in 'smiting the Amalekites'. One day he saw Thomas Carlyle, an old, forlorn figure, and crossed the road to greet him. 'You're Huxley, aren't you?' grunted Carlyle, 'the man who believes we are descended from monkeys', and walked on.

Von Baer rejected the theory on the grounds that evolution, without a purpose, is meaningless. The mechanical operation of natural selection, on variations arising apparently by chance, seems completely aimless. Darwin foresaw this criticism. What could be the aim, he asked, of the serrated sting of the bee which, when used, causes the bee's viscera to be torn out? Of the hundreds of drones, only one of whom can serve a purpose, when it mates with the queen? Of the blind alleys into which creatures have evolved and of the thousands of species which have become extinct? To his wife's distress, Darwin lost his religious faith.

Darwin's theory of a struggle for survival as the guiding force of evolution was in tune with the economic doctrine of unrestricted competition or *laissez-faire*, current in his time. Indeed, the idea came to him after reading Malthus. It has likewise been suggested that Lamarck's theory was inspired by the ideals of individual liberty and self-expression disseminated by the *philosophes* and that Cuvier's belief in geological catastrophes can be traced to the political convulsion of the French revolution. Such theories of the origin of ideas are plausible but not always convincing. An equally plausible case could be made that the atomic theory was developed in the nineteenth century because of Dalton's delight in bowls, and because of the growth in popularity of ball games.

* * *

The sheer weight of the evidence which Darwin had put forward, and the large number of facts which could be explained so much more economically and convincingly by evolution than by special creation, soon began to tell. The younger biologists eagerly embraced the Darwinian theory and it gave them a completely new outlook on the study of living organisms. Much work was done on the evolution of particular families of species and there sprang up what has been called a forest of family trees.

The search for fossils was enormously stimulated, and the finds were astonishing. As early as 1862, in the Jurassic slates of Bavaria, the remains were found of *Archaeopteryx*, a link between the reptiles and the birds (Plate XV (*b*) and Fig. 45). It had teeth, claws on its wings and a long vertebrated tail. In North America Professor O. C. Marsh, and his rival E. D. Cope, discovered amazingly rich fossil beds and they dug up hundreds of species and tens of thousands of specimens of extinct reptiles, including pterodactyls and huge dinosaurs.

Fossil links were found tracing the ancestry of the horse from *Hyracotherium*, a creature the size of a large dog, with four toes on its front limbs and three on its hind limbs; the horse today has only one toe on each limb, with relics of two more in the form of thin, functionless splint bones. The fossil pedigrees of tapirs, elephants, camels, sea urchins, starfish and ammonites, were also pieced together. There is now, for example, an almost complete set of fossils showing the development of the elephant's trunk. Indeed, the main evidence for evolution today is the fossil record.

Fig. 45. *Archaeopteryx*, a link between the reptiles and the birds.

The approximate dates at which many species evolved have been determined by estimating the age of the rocks in which their earliest fossils are found. These estimates have been made by calculating, from a reasonable average rate of deposit, the time required for the thickness of the sedimentary rocks to be formed, and also by measuring the amount of radioactive change which has occurred in the rocks since their deposition. The radioactive element uranium decays at a very slow, known rate into a series of other elements, of which the end product is a particular form of lead. The time during which the decay has been taking place can be calculated from the ratio of the uranium to the lead in the rocks.

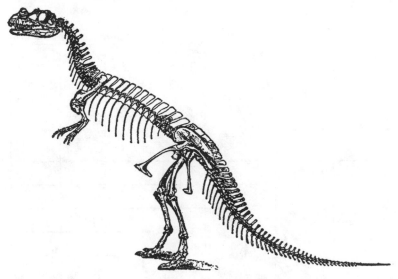

Fig. 46. Reconstruction by O. C. Marsh of the extinct dinosaur, *Brontosaurus excelsus*. (From C. Schuchert and C. M. Le Vene: *The Life of O. C. Marsh*, by courtesy of Yale University Press.)

Fig. 47 shows the geological time-scale and the duration of the existence of some important types of animal. The dates must be regarded as provisional and may well be modified when more accurate determinations have been made.

The earliest discoveries in the search for 'missing links' between the apes and man were primitive human remains at Abbeville in Northern France, and at Neanderthal in Rhenish Prussia. The Neanderthal skull was simian, with a receding forehead and protuberant eyebrows, and it is now thought to represent a distinct hominid species which has died out. In 1891 the remains of an early man, *Pithecanthropus erectus*, with simple tools and evidence of fires, were found in Java and, some years later, the bones of a number of similar primitive men were found near Peking. During the present century Broom and Dart have discovered in South Africa many fossils of a species which they have called *Australopithecus transvaalensis*, more primitive than Java or Peking man. No tools or fires were associated with the fossils and it is likely that the tongues of these ape-men were too big for articulate speech. The species appears to be a genuine link between the apes and man.

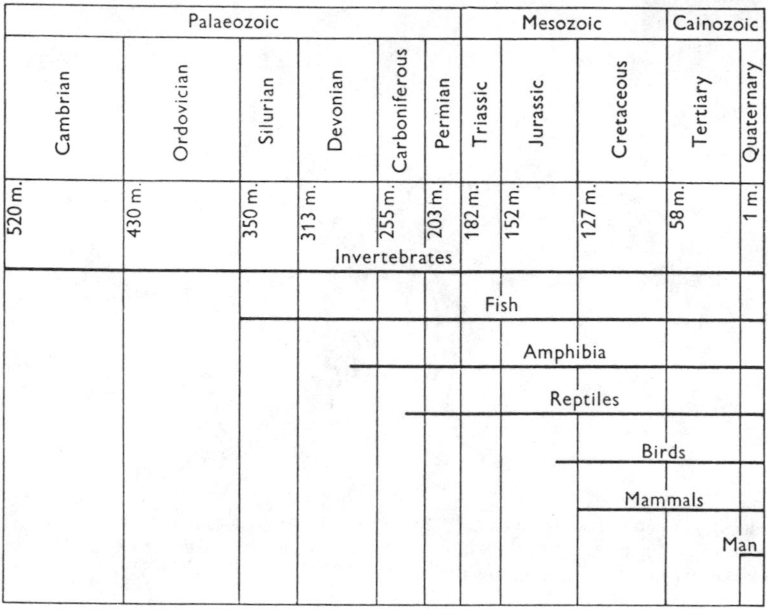

Palaeozoic						Mesozoic			Cainozoic	
Cambrian	Ordovician	Silurian	Devonian	Carboniferous	Permian	Triassic	Jurassic	Cretaceous	Tertiary	Quaternary
520 m.	430 m.	350 m.	313 m.	255 m.	203 m.	182 m.	152 m.	127 m.	58 m.	1 m.
				Invertebrates						
					Fish					
						Amphibia				
							Reptiles			
								Birds		
								Mammals		
										Man

Fig. 47. Time-scale of animal evolution from the fossil record (after Zeuner). The top line represents the era: the Palaeozoic or Age of Ancient Life lasted about 340 million years, the Mesozoic or Age of Intermediate Life about 125 million years, and the Cainozoic or Age of Modern Life about 58 million years. In the second line are geological periods or systems and below are ages in millions of years. The length of the line denoting the one million years of the existence of the genus Man is out of scale; it should be only 1/520 of the width of the table. The science of dating the rocks is still in its infancy and the above figures must be taken as provisional.

One of the rare instances of deliberate fraud in science is provided by the story of Piltdown man. In December 1912 Charles Dawson, an amateur geologist and archaeologist, announced the discovery of a skull in the ancient gravels of the

Sussex Ouse, together with some crude flint tools. Although this was received with great enthusiasm, it produced a certain amount of uneasiness among the experts. When, however, a missing canine tooth was discovered at the same spot in 1913 and the remains of another primitive man two miles away in 1915, all criticism was stilled.

The remains in Java and Peking did not tally with Piltdown man. The obvious deduction seemed to be two lines of evolution. In 1949 the age of the Piltdown skull was estimated by a new technique. A chemical determination of the fluorine content, which had seeped in from the soil water, showed it to be much more modern than had been supposed. It was then discovered that the bones had been stained and the teeth had been filed. The evidence pointed to deliberate forgery on the part of Dawson—who could not be cross-questioned because he had died in 1916.

Some series of fossils, showing the development of particular organisms, seemed to suggest that evolution occurred in a straight line, as it were, supporting Lamarck's idea of a predestined path controlled by the inward directive force of the organism, and the theory of *orthogenesis* (Gk. *orthos*—straight, *genesis*—a becoming) was put forward. At the end of the nineteenth century Darwinism seemed to be in decline for another reason also. Information, unknown to Darwin, on the nature of variations and the way they occurred, seemed to suggest that natural selection played a very much smaller part in evolution than he had believed. By the First World War, Darwin's theory of the mechanism of evolution was apparently dead; by the Second World War it was once more vigorously alive. We shall consider the rise of neo-Darwinism in chapter 19.

CHAPTER 17

THE GERM THEORY OF DISEASE

DISEASE HAS been ascribed, during its long affliction of mankind, to three main causes: demons, miasmas and germs.[1]

The demonic theory of disease prevailed in New Testament times, when healing was described as casting out devils and the expelled devils were thought to be sufficiently real to take possession of a herd of swine. Healing was effected by religious faith and it was also attempted by magic and charms. Marcellus the Empiric in the fifth century gave the following treatment for stomach and intestinal complaints: 'Press abdomen with left thumb and say "Adam, bedam, alam, betur, alem, botum". Repeat nine times, touch the earth with the same thumb and spit, say charm nine times more, again for a third series of nine, touching the ground and spitting nine times also.'[2]

In more recent times the royal touch was considered a cure for the 'King's Evil' (scrofula). Charles II is said to have touched about one hundred thousand people; Samuel Johnson was touched by Queen Anne. The witch-doctor and the medicine-man still employ charms, talismans, incantations and other forms of powerful psychological pressure.

The theory that disease is caused by miasmas can be traced back to the Greek Hippocratic school in the fifth century B.C. Hippocrates, the father of medicine, thought of disease, rather than of specific diseases, and considered one of its main causes to be impure air.

Leprosy, the Black Death or bubonic plague, and syphilis taught the men of the Middle Ages the facts of infection and contagion. Infection is the communication of disease, particularly by the atmosphere and water; contagion is its communication from body to body by direct contact.

Leprosy became widespread in Europe after the Crusades; it

---

[1] A germ is a unicellular micro-organism or microbe.
[2] C. E. A. Winslow, *The Conquest of Epidemic Disease.*

was realised that the disease was contagious, and it was kept in check by the isolation of lepers.

For this there was the authority of scripture: 'And the leper in whom the plague is,...shall cry Unclean, Unclean,...he shall dwell alone; without the camp shall his dwelling be.'

The Black Death was responsible for a European catastrophe in the fourteenth century. Introduced into Constantinople from Asia in 1347, it spread throughout Europe during the next three years. Estimates of the mortality range from one-quarter to three-quarters of the population. The land lay uncultivated and the cattle trod the wheat and the vines. The conception of the cause of the disease, as given in the many plague tractates, which instructed the people how to protect themselves, was a corruption of the air, arising from decaying organic matter in marshy and putrid waters, and from unburied bodies. The plague continued to rage intermittently in Europe for the next three centuries. The Great Plague of London of 1665 wiped out almost one-fifth of the population.

Syphilis was a main subject of *De Contagione* by the Italian, Fracastorius (1478–1553), a contemporary of Copernicus at Padua, and one of the outstanding figures in the history of the study of disease. Fracastorius explained contagion in terms of agents of disease which he called *seminaria*, translated as seeds or germs, although it is doubtful whether he considered them to be alive. He thought that they were transmitted through the atmosphere and he compared contagion to the exhalation of an onion.

The miasma theory reached its hey-day in the mid-nineteenth century when the slums of the industrial cities were choked with revolting filth and exuded a nauseating stench. In the absence of any system of sanitation the poor emptied their chamber pots into the street—there was nowhere else. The large dunghills which collected in courts and alleys swarmed with flies and were carted away periodically by farmers for use as manure. What London sewers there were discharged their contents untreated into the Thames and in 1858, the year of the Great Stink, Parliament carried on its work behind curtains soaked in chloride of lime, while travellers made long detours to avoid the Thames bridges. It was prophesied that the overpowering smell would give rise to pestilence on an immense scale. But the death-rate that year was below average.

The theory that disease was caused by the foul emanation or

gas from decomposing filth was a most fruitful error. It gave rise to what has been called the great sanitary awakening, by which London and other cities were provided with a reasonable system of sanitation and a pure water-supply. The death-rates from the prevailing epidemics of cholera, typhoid and typhus fever were strikingly reduced.

John Snow investigated the London cholera epidemics of 1854 and proved that their origin lay, not in the atmosphere, but in the water-supply. The epidemic in Broad Street, Golden Square was traced to the Broad Street pump. In neighbouring places, with independent water supplies, there were hardly any cases, while a lady in Hampstead, whose drinking water was specially transported from the Broad Street pump, died of the disease. William Budd made a similar investigation of typhoid fever and showed that the contagious element passed, like that of cholera, from the excreta of infected persons to the water-supply.

* * *

The man who was primarily responsible for establishing the germ theory of disease was Louis Pasteur (1822–95). He was born at Dôle in the Jura, the son of one of Napoleon's conscripts, who rose to the rank of sergeant-major and was decorated on the field of battle. The old soldier was a fine man; he had an 'admiration for great men and great things' and he inspired in his son a devotion to the greatness of France.

Pasteur was trained as a chemist and he had no medical qualification. He was led to the germ theory of disease by an astonishing variety of practical problems involving micro-organisms; some of these problems were solved in a blaze of controversy, in which he was a tiger. His work, despite its diversity, had an underlying unity; for most of his life he was occupied with the origin of disease.

The first of these problems was that of fermentation, employed from antiquity in the production of wine. If yeast is added to a solution of grape sugar the liquid begins to froth and this is caused by the release of carbon dioxide gas, as the sugar is broken down to alcohol.[1] Pasteur's attention was drawn to the pheno-

[1] $C_6H_{12}O_6 = 2C_2H_6O + 2CO_2$

Grape sugar	Ethyl alcohol	Carbon dioxide

menon in 1856, when he was professor at Lille, by one of his students whose father had been having difficulties in the manufacture of alcohol from beetroot. Pasteur examined the fermenting liquors under the microscope; he saw that the healthy ferments were round globules, and that the unhealthy ones were elongated.

The first man to observe yeast under the microscope was Leeuwenhoek and he saw it as tiny spherical granules. In 1836 Cagniard de la Tour in France and Schwann in Germany discovered that yeast was a living organism which reproduced itself by budding or division. Yeasts are, in fact, unicellular fungi and, since they lack the chlorophyll normally possessed by plants, they feed on organic matter.

Liebig and most other chemists maintained that fermentation was caused not by living agents but by chemical materials. They believed in explaining a chemical phenomenon by a chemical cause. Liebig pointed out that fermentation could be caused by adding almost any albuminous matter to grape sugar. He suggested that yeast, at its death, liberated an albuminous material which imparted a kind of molecular movement to the sugar, thereby breaking it up. This view Pasteur vigorously contested. He believed that the albuminous material, almost certain to be contaminated with wild yeasts, merely provided food for the yeasts. After many experiments he demonstrated that the albuminous material was not essential to fermentation; he prepared a solution containing only sugar and mineral salts in which the yeast multiplied and the sugar fermented. He showed that fermentation ceased if the living yeast was destroyed by boiling or removed by filtration.

There were other fermentations in which yeast played no part, such as the production of lactic acid in sour milk and of butyric acid in rancid butter. Pasteur discovered living ferments in both these cases and made the important observation that the ferments could live without oxygen. He found that yeast could live with or without oxygen and he concluded that, if no free oxygen were provided, the yeast took its oxygen from the sugar and caused fermentation.

Pasteur's theory that fermentation was effected by living microorganisms and not by their death or putrefaction was fruitful although it was essentially wrong and Liebig's theory was the correct one. In 1897 Buchner showed that alcoholic fermentation is caused by the chemical zymase which can be isolated from yeast, and he introduced the term enzyme for an organic catalyst of this kind. Enzymes play a vital part in all plant and animal metabolism.

Pasteur was urged by Napoleon III to investigate the diseases of wines. Each variety of wine had its own peculiar maladies; champagnes were apt to become ropy, burgundies to become bitter, and clarets turbid and flat. Pasteur turned his microscope on to the diseased wines. He saw that, in addition to the wild yeasts originally present on the skins of grapes and responsible for the healthy fermentation of grape sugar, there were other micro-organisms, and he was convinced that these gave rise to the products causing the bad flavour. He first tried to kill the contaminating micro-organisms by antiseptics, but eventually found that heating to 55° C. would do so without destroying the flavour and bouquet of the wine. This heating process, now called pasteurisation, was tested by a commission. Five hundred litres of wine, half of it heated and half unheated, were put aboard a ship at Brest and taken on a ten months' cruise. At the end of the cruise the heated wine had an excellent flavour, while the unheated wine was acid and sour. Pasteur had an experimental cellar of his own where heated and unheated wines were kept for many years.

He made similar investigations into the diseases of vinegar and beer. In 1871 he visited the Whitbread Brewery in London and examined the yeasts from the fermentation vats with his microscope. He was able to explain the spoiling of beer as the work of harmful micro-organisms and it was not long before microscopes were installed in breweries in London and in Burton-on-Trent.

One of the most famous controversies in which Pasteur was involved was that concerning spontaneous generation. From ancient times it was believed that life could arise spontaneously: maggots from putrefying meat, fleas, bugs and lice from moisture and filth, caterpillars from leaves, bees from the entrails of dead bulls, and even mice from dirty linen. By the time of Pasteur the real point at issue was whether micro-organisms could be spontaneously generated. At some point in the process of evolution, micro-organisms might well have originated from inanimate matter.

During the previous two centuries several investigators had

tackled the problem. Francesco Redi (1626–97) placed meat behind gauze and left other meat uncovered. He found that maggots appeared in the uncovered meat but not in the covered. He thought, correctly, that the maggots came from eggs deposited by flies. John Needham (1713–81) took some boiling mutton gravy and corked it in a flask. In a few days it was teeming with micro-organisms. He repeated the experiment with an almond infusion and various others; on each occasion he obtained a multitude of microscopic life. He believed that he had destroyed all micro-organisms by the preliminary boiling and hence claimed that spontaneous generation was a fact. His experiments were repeated by the abbé Spallanzani (1729–99). Spallanzani did not use corks like Needham; he drew out the air from his flasks with an air-pump and sealed them hermetically. No micro-organisms appeared. His superb technique was not equalled until the experiments of Pasteur.

Pasteur realised, like Spallanzani, that if an infusion were subjected to prolonged boiling, so that all the micro-organisms in it were killed, any subsequent appearance of life probably had its origin in germs carried by the atmosphere. He placed an infusion in a flask and drew out the neck into a sinuous, swan-neck

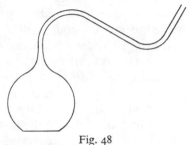

Fig. 48

tube (Fig. 48). During boiling the air was driven out of the flask by the steam; on cooling, the air re-entered but the germs were trapped on the wet walls and the liquid in the flask remained sterile. A drop of the infusion shaken into the neck soon became cloudy and, if this drop were then shaken into the flask, the whole infusion became alive with micro-organisms.

Pasteur demonstrated that the air high up in the mountains, far away from the smoke and dust of towns, is uncontaminated by germs. He transported sealed flasks, containing boiled infusions, to the Mer de Glace above Chamonix in the Alps. He broke the necks with pincers, which had been sterilised by passing through a flame, and the air hissed in. Most of the flasks remained sterile, showing the air to be almost free from germs.

Pouchet, the director of the Museum of Natural History at Rouen and an eminent biologist, was a convinced believer in spontaneous generation. He repeated Pasteur's experiments in the High Pyrenees, and found, when he opened his flasks, that life began to germinate in his infusions. This confirmed his belief that putrescible organic material, when activated by pure, sterile air, could spontaneously generate life.

Pasteur imperiously issued a challenge and suggested that he and Pouchet should together perform experiments before a commission to settle the issue. Pouchet delayed and dallied and, in the end, his nerve failed him. He never appeared before the commission. This was a pity because his experiments were carried out with scrupulous care. Pasteur used yeast infusions which can be readily sterilised by boiling; Pouchet, on the other hand, used hay infusions, which are very difficult to sterilise and his crop of micro-organisms originated from unkilled spores in the hay.

Incomplete sterilisation was also the explanation of a claim made ten years later by Bastian in England that urine, after being heated to 110° C., could still give rise to micro-organisms. At this point John Tyndall, the physicist, took a hand in the controversy. His interest in the dust in the atmosphere sprang from its effect in the scattering of light. He passed light through side windows in a closed chamber with a glass front, whose inside walls were coated with glycerine. After several days the floating matter in the air had all settled in the glycerine and the beam of light in the chamber was invisible. Tyndall found that, in this dust-free atmosphere, boiled organic liquids such as urine and infusions of meat or vegetables could be exposed and remain sterile indefinitely. He also discovered that spores, which are bacteria in a resting and highly resistant state, could be destroyed by repeated short periods of boiling. The modern practice for thorough sterilisation is to heat materials to 120° C.

The baffling problem of spontaneous generation, with its experimental difficulties and pitfalls, was thus finally settled, and it was established that all life comes from other life. Pasteur gained from his experiments the realisation that the germs carried by dust particles in the air might provide an explanation for some of the diseases of animals and men.

In 1865 Pasteur was begged to investigate the disease of silkworms, which had brought the French silkworm industry to the verge of ruin. He knew nothing about silkworms and even had to ask the naturalist Fabre what a chrysalis was. But he was recognised as the only man in France likely to be able to find a cure for the disease.

The silkworm is born from an egg laid by a moth. After eating a quantity of mulberry leaves it spins a cocoon, which is the source of the silk, and within this it changes into a chrysalis. Some of the chrysalides are allowed to develop into moths, each of which lays about 400 eggs, to provide a continuing supply of silkworms.

A symptom of the disease was the presence of black spots on the worms, resembling pepper grains, and hence the disease was given the name pébrine from the patois word for pepper, *pébré*. Pasteur travelled down to Alès in the south of France to collect information about the disease and to view the diseased worms under the microscope. He came to the conclusion that it was necessary to breed from a pair of moths, male and female, both of which were free from the spot-like microbes. In 1866 healthy eggs were obtained and in the spring of 1867 the new broods of worms were examined. Many of them were diseased. Pasteur became more and more dismayed and at length he said to his assistants: 'Nothing has been accomplished. There are two diseases.' He had discovered numerous bacteria in the alimentary canals of the worms responsible for a second disease, flacherie. It was necessary to select moths for breeding that were free from both harmful microbes. There were again the long periods of waiting between the breeding seasons but by 1869 the problem was completely solved. When asked for eggs by the Lyons Silks Commission he sent, with characteristic confidence, four packets: the brood from the first packet, he predicted, would be healthy; that from the second would die of pébrine; that from the third would die from flacherie; and that from the fourth would contract both diseases.

While pursuing his silkworm investigations he had a stroke which partially paralysed his left side for the remainder of his life.

He began to preach the doctrine that many human diseases had their origin in microbes, with a vehemence that was exacerbated by the resistance of the doctors, who regarded him as a quack. At a discussion in March 1879 at the Paris Academy of Medicine on the cause of puerperal fever, a disease responsible for the deaths of many women in childbirth, the opening speaker kept talking of the 'puerperal miasma'. Pasteur stood up and interrupted: 'The cause of the epidemic is nothing of the kind. It is the doctor and his staff who carry the microbe from a sick woman to a healthy woman.' He went up to the blackboard and drew a picture of a microbe. 'There! That is what it is like', he said.

A great discovery resulted from his investigation of chicken cholera (a disease having no relation to human cholera), which decimated many poultry farms. The microbe responsible for the disease was found, and it was cultivated in a suitable broth. A drop of this culture, inserted under the skin of a chicken, proved quickly fatal. It was Pasteur's practice to sow new cultures, perhaps every day, by inserting a drop of the old culture into fresh broth, when a rapid multiplication of the microbes took place. It so happened that some chickens were inoculated with a culture several weeks old. They contracted the disease in a mild form and then recovered. When subsequently inoculated with a fresh and virulent culture they did not die although all other chickens, inoculated with the same culture, did so.

Pasteur at once saw the similarity with vaccination against smallpox, a practice of some antiquity, which was introduced into England by Edward Jenner (1749–1823). As a student, Jenner heard that milkmaids who had caught cowpox, a mild disease manifested by pustules or sores on the hands, from pustules on the udders of cows, never contracted smallpox. To test this he inoculated a small boy in the arm with matter from a pustule on the hand of a milkmaid. The boy contracted cowpox and he was then inoculated—a horrifying experiment—with matter from a person suffering from virulent smallpox. The smallpox did not take. Jenner's method of vaccination against smallpox soon spread throughout the world.

As a tribute to Jenner, Pasteur used the term vaccine (Latin *vacca*, a cow) for his attenuated culture of chicken cholera microbes. He pondered the problem of the immunity induced by the attenuated culture. He found that the microbe grew well in fresh chicken broth but did not multiply in a medium in which a culture had already been grown. He came to the conclusion that the microbe produced a chemical substance which inhibited its growth, and he speculated that it might be possible to produce an effective vaccine containing this chemical instead of living microbes. In the modern anti-toxin treatment, particularly successful against diphtheria, Pasteur's speculation has become a fact.

Dead or living microbes are injected into an animal, such as a horse, and they cause the production of protective antibodies in the animal's blood. The animal's blood serum is then used for human vaccination.

Pasteur applied his discovery of vaccination with attenuated cultures of microbes to anthrax, a deadly disease of sheep and cattle. There had been several investigators of this disease before Pasteur and it was known that the blood of diseased animals contained little rods, the anthrax bacilli. Pasteur sowed them in various culture media and saw the rods multiply into tangled filaments, visible to the naked eye.

Many attempts were made by Pasteur and his assistants to produce an attenuated anthrax culture. The difficulty was that the rod-like bacilli formed highly resistant spherical spores whose virulence could not be diminished. In the infected pastures of Beauce and of Auvergne anthrax spores lay dormant for years, ready to spring to life when swallowed by some unfortunate sheep. Pasteur found that the anthrax bacilli could be cultivated without the formation of spores by keeping the culture medium at a temperature just below that at which reproduction stopped altogether. The virulence of these bacilli could then be attenuated by keeping them for a week or a fortnight exposed to the air.

Pasteur's vaccine was tried on a dozen or so laboratory sheep and it was successful. Its efficacy was then tested in a famous experiment, organised by a Melun veterinary surgeon. This same surgeon had been one of Pasteur's many critics. 'Micro-biolatry is the fashion', he wrote, 'it reigns undisputed; it is a doctrine which must not even be discussed, especially when its Pontiff, the learned M. Pasteur, has pronounced the sacramental words, "I have spoken".'[1]

At the farm of Pouilly le Fort, near Melun, on 5 May 1882, twenty-four sheep, six cows and a goat were inoculated with a living attenuated culture of anthrax bacillus, and on 17 May, with a less attenuated culture. On 31 May they were inoculated with a highly virulent culture and at the same time a control group of animals which had not previously been inoculated were also given the virulent culture. As always, Pasteur was supremely confident. He predicted that the control group would die, and that the others would survive. There was excitement in France during the days

[1] R. Vallery-Radot, *The Life of Pasteur*, translated by Mrs Devonshire.

of waiting and the fulfilment of his prediction brought him immense public fame and prestige.

As a result of his work on rabies, Pasteur's fame became worldwide and reached a height attained by no other scientist before Einstein. The devotion of so much of his time to rabies may have been the result of vivid boyhood memories of a mad wolf in the Jura, which had bitten men and beasts; he had seen the wounds cauterised by a blacksmith with red-hot irons.

The virus of rabies is invisible under the microscope. Pasteur was dealing with an unseen, elusive agent that could not be cultivated in his usual broths. He tried injecting the blood of a rabid dog into a healthy animal, but the disease was not transmitted. The disease attacks the nervous system, causing mania, convulsions and paralysis. He therefore opened the skull of a live healthy dog and placed some virulent saliva in its brain. The dog contracted rabies. Pasteur found that he could cultivate the virus in the brains of rabbits and monkeys. By extracting the spinal cord of a rabbit which had died of the disease and drying it, he was able to obtain an attenuated virus. When animals were inoculated with the attenuated virus they were found to be immune from the disease.

Pasteur was reluctant to inject his vaccine into a human being, but in July 1885 his hand was forced. A mother brought to him her little boy of nine, Joseph Meister, who had been bitten by a mad dog. The period of incubation of the disease is between two and twelve weeks, depending on which part of the body has been bitten; the virus takes time to work its way to the brain. Pasteur began to inject vaccines of increasing strength into the boy. Madame Pasteur wrote to one of her children: 'Your father has had another bad night; he is dreading the last inoculation on the child; and yet there is no drawing back now.' Joseph Meister survived to become the gatekeeper of the Pasteur Institute in Paris.

Another boy, J. B. Jupille, who had courageously driven off a mad dog with a whip to save his companions but had been bitten in the act, was also successfully treated. Children were sent to Pasteur from as far afield as Russia and the U.S.A., and in 1886 over two thousand persons were vaccinated. The Pasteur Institute in Paris, completed in 1888, was built by subscriptions from all over the world.

Pasteur's amazing record of scientific sucess was the result of intense powers of concentration and tenacity. He would spend hours absorbed in meditation, poring through his microscope or observing infected animals in the basement of his laboratory, quite unaware of the presence of other people. All his results, and those of his assistants, were recorded in his notebook. During discussion, if he said 'It is in the notebook' there was no question of further argument. His thoroughness and meticulous attention to detail gave him absolute confidence in his results.

He had certain idiosyncrasies. He avoided shaking hands for fear of infection. At table he carefully cleaned his plate, knife and fork and spoon with his napkin. 'He minutely inspected the bread that was served to him and placed on the tablecloth everything he found in it: small fragments of wool, of roaches, of flour worms', said his assistant, Loir. 'Often I tried to find in my own piece of bread from the same loaf the objects found by Pasteur, but could not discover anything. All the others ate the same bread without finding anything in it. This search took place at almost every meal and is perhaps the most extraordinary memory that I have kept of Pasteur.'[1]

Towards the end of his life, looking back on one of his earlier publications, Pasteur said to Loir: 'How beautiful, how beautiful! And to think that I did it all. I had forgotten it.'

* * *

With Pasteur, as co-founder of the germ theory of disease and co-initiator of the golden age of bacteriology, must be associated Robert Koch (1843–1910). Koch's range was more limited than Pasteur's; he was, however, the first to isolate pure strains of microbes by means of a new bacteriological technique and, while director of the new Institute of Infectious Diseases in Berlin in the eighteen-nineties, helped to make Germany supreme in medicine until the First World War.

Koch began his career as a young district doctor in the province of Posen, Silesia, where he created a laboratory by curtaining off part of his surgery, equipping it with a new microscope—a birthday present from his wife. In his district anthrax was rife, and his investigations of this disease forestalled those of Pasteur. He examined under his microscope the blood of infected animals; he

[1] R. J. Dubos, *Louis Pasteur*.

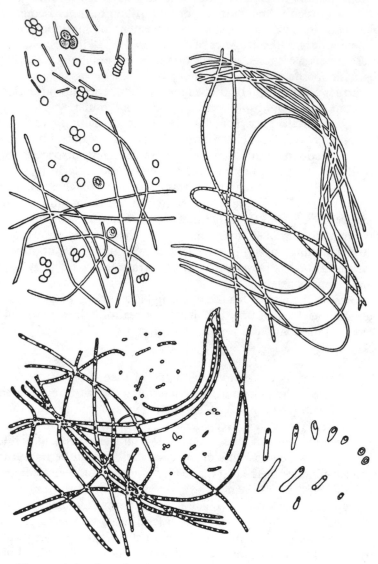

Fig. 49. A drawing, after Koch, of the development of anthrax spores.

saw the rods of the anthrax bacillus, some of them connected together in long threads (Fig. 49), and he observed also the formation of round spores from the rods.

He used mice for his experiments, inoculating them under the skin, by means of a sliver of wood, with the blood of an infected sheep. He took a tiny fragment of the spleen of a mouse which had died from anthrax, placed it in a drop of watery fluid from the eye of an ox, and suspended it from a cover slip in a concave hollow of a glass block (Fig. 50). In this way he watched the anthrax bacilli grow. He transferred the disease from mouse to mouse, demonstrating the enormous multiplication of the bacilli in each animal. He learnt to stain the bacilli with coloured dyes, fitted a camera to his microscope, and took photographs of them.

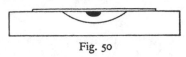

Fig. 50

His method of preparing pure cultures of microbes was to use a solid medium of gelatine mixed with sterile meat infusion. He is said to have conceived the idea from a cut potato, left by accident on his laboratory table overnight; in the morning it was covered with different coloured specks, each of which proved to be a pure culture of a specific microbe.

There was no love lost between Koch and Pasteur. Pasteur, with his intense devotion to his country, could not forget the humiliation of the Franco-Prussian war. Koch claimed that Pasteur's cultures were badly contaminated. He bought on the open market some of Pasteur's vaccine for anthrax and said that it was dangerous to use.

When he went to Berlin, to explain his preparation of a pure anthrax culture, Koch was received coolly. He was always considered rather an outsider because he did not rise through normal academic channels. A post was found for him, however, in the Imperial Health Office. There he discovered, in 1882, the tubercle bacillus. It was the first time that a specific microbe was proved to produce a specific human disease. Koch laid down three rules for identifying a microbe as the cause of a disease: (1) the microbe must invariably be present in the diseased animal; (2) it must be capable of cultivation outside the body; (3) it must, if injected into a healthy animal, produce the disease. He made the one great error of his career when he announced his discovery of a cure for tuberculosis, namely tuberculin, the product of dead tubercle bacilli. Tuberculin did not cure tuberculosis; patients and doctors, in their natural disappointment, heaped abuse upon him.

Koch was a dogged, reserved and crusty character. He created a scandal in Germany when, as an old man, he divorced his second wife and married a young actress. People felt that they could understand Koch's point of view, but were quite unable to divine what the young woman saw in him.

* * *

The application of the germ theory of disease to surgery was the work of Joseph Lister (1827–1912). The discovery of anaesthetics, made about twenty years before Lister's investigations, had done away with the agonised screams of patients and the necessity to hold them down forcibly on the operating table. Even so, before undergoing an operation, a prudent man made his will, since the chances of survival were smaller than on the field of battle. At least one-third of amputation cases died.

The main danger lay in various forms of blood poisoning, in particular hospital gangrene, which caused the tissues round the wound to become grey; these tissues died and the process spread slowly throughout the body. When a wound putrefies there is suppuration, or the formation of a yellow substance called pus, consisting, as we now know, of dead white corpuscles, which the blood manufactures in quantity to engulf and destroy invading bacteria. It was thought, in accordance with Galen's theory of 'laudable pus', that suppuration was desirable and assisted healing.

Lister perceived the falsity of this doctrine when, in 1856, he came upon Pasteur's paper, *Recherches sur la Putréfaction*. He realised that suppuration was caused by germs from the air and from the surgeon's hands and instruments. He looked for some means of destroying the germs. After trying various disinfectants he selected carbolic acid and dressed the wounds with lint soaked in the crude liquid. He found that the carbolic acid was an irritant of the skin; he began to use weaker solutions and to place a pad between the wound and the carbolic.

In the old operating theatres sawdust was scattered on the floor to absorb the blood, and surgeons operated in frock coats, often stiff with dried blood and encrusted with pus. Lister insisted on cleanliness and on the sterilisation of instruments, and used a carbolic spray for sterilising the air. He developed his antiseptic methods while he was professor of clinical surgery at Glasgow University.

Doctors were slow to recognise the value of his work. Sir James Simpson, the discoverer of the use of chloroform as an anaesthetic, spoke of Lister's 'mythical germs'. When Lister returned as professor to London, where he had been educated, he found a wall of polite obstruction set up by the nurses at King's College Hospital and he was starved of patients. His striking successes soon changed all this. He removed a tumour from a young woman on whom Sir Stephen Paget refused to operate because of the danger of blood poisoning. When he opened a fractured knee-cap and wired the pieces of bone together, one of his colleagues, convinced that the patient would die, exclaimed that he ought to be prosecuted. Operations became possible that had hitherto been unthinkable.

Lister was the first medical man to be made a peer. He represented the Royal Society at the celebrations in December 1892 of Pasteur's seventieth birthday and, in the large crowded lecture theatre of the Sorbonne, he spoke of the debt surgery owed to Pasteur. Pasteur, filled with emotion, embraced him.

Lister's antiseptic methods have now been replaced by aseptic surgery. Germs are excluded rather than destroyed. A modern operating theatre is completely sterilised; the surgeons and nurses wear sterilised clothes and rubber gloves, and breathe through masks.

*   *   *

The great triumvirate, Pasteur, Koch and Lister, laid the foundations for a transformation of medicine and surgery. Doctors no longer waited in helpless passivity for a disease to take its course; they mounted an offensive against the germs which were causing it, and they safely devised new and more daring operations.

In the wake of Pasteur and Koch there followed a whole army of workers. Some of them investigated the life-history of germs and the way they invaded the human body. It was found, for example, that the bacillus of the plague bred in the stomach of a flea, which lived on the body of the black rat; that the protozoa of malaria and the virus of yellow fever were transmitted to man through the bite of an infected mosquito; that the rickettsia of typhus was carried by lice; that the amoeba of dysentery was conveyed on the bodies of flies from the faeces of diseased persons. Knowledge of this kind enabled preventive medicine to be developed.

Other workers sought for chemical substances which would destroy the germs of disease without producing harmful effects on the body and they built up the science of chemotherapy. Paul Ehrlich discovered a cure for syphilis, called salvarsan or 606, being the six hundred and sixth substance he had tried. More recent well-known drugs are the sulphonamides and penicillin.

The expectation of life for a man in Great Britain has risen during the present century by nearly twenty years, from 48 to 67.

CHAPTER 18

THE NINETEENTH CENTURY

THREE MAIN scientific generalisations emerged in the nineteenth century: the conservation of matter, the conservation of energy, and evolution. The first two pointed unmistakably to materialism and the third, by displaying man and his moral sense as part of the evolving cosmos, produced a revolution in outlook comparable with that caused by the physical synthesis of the seventeenth century.

The increase in knowledge was such that science began to split up into the sub-branches that we know today (pp. 360–1). Specialisation in a comparatively narrow field became more and more necessary.

A notable feature of the intellectual life of the nineteenth century was the growing prestige of science. The French philosopher Comte (1798–1857) propounded what was called positivism, asserting that every branch of knowledge passed through three stages; a theological or imaginative stage, a metaphysical or abstract stage, and a scientific or positive stage. The third stage, in which science replaced theology and metaphysics, was the most advanced, and was that which civilisation had reached in the nineteenth century. Comte had a wide influence, despite a certain unpopularity among theologians and metaphysicians; one of his more famous adherents was George Eliot.

Attempts were made to apply the scientific method to almost all branches of thought; to treat, for example, history as a science rather than as an art, and to reduce religion to a system of spiritual laws. Walter Bagehot wrote *Physics and Politics, or Thoughts on the Application of the Principles of Natural Selection and Inheritance to Political Society* (1872), and Henry Drummond, *Natural Law in the Spiritual World*. Drummond added to the plant and animal kingdoms a third—the kingdom of God. Just as the biologists found that life is not generated spontaneously but arises from other life, so Drummond found, or thought he had, that there is no spontaneous generation of spiritual life and that the sole source of it is Christ.

Deeper thinkers, like Coleridge, contrasted the method of science, which splits experience into parts for the purpose of classification and control, with 'that intuition of things which arises when we possess ourselves as one with the Whole', the method appropriate to the world of aesthetic and spiritual realities. Indeed, in the nineteenth century the gulf between science and the humanities widened and became apparently unbridgeable.

* * *

The deterministic materialism of the eighteenth-century *philosophes* reappeared, strengthened by new scientific evidence, in the setting of staid and devout Victorian England. Science, in its conflict with theology, went over to the attack and, after the great controversies of the 1870's and 1880's, found itself permanently established in some of the territory hitherto held by its rival. 'Extinguished theologians lie about the cradle of every science', said Huxley, 'as the strangled snakes about that of Hercules.'

For the first time men of science became openly contemptuous of the main stream of philosophy. In contrast to Descartes and Locke, who had been scientists as well as philosophers, and drew no sharp dividing line between the two activities, the leading philosopher at the begining of the nineteenth century, Hegel (1770–1831) was ignorant of science. He held that everything in the universe, including matter, was essentially mental or spiritual, and part of the unfolding Absolute Idea. His philosophy could be made plausible in history and morals but, when applied to the natural sciences, it appeared to men of science, in the words of Helmholtz, as 'absolutely crazy'.

The great scientific principles of the indestructibility of matter and of the conservation of energy led inevitably to the view that the abiding reality in the universe was matter, and that the behaviour of this matter was rigorously determined by scientific law. Organic matter, as shown by the chemists, was just as much under the reign of law as inorganic matter, while life itself, according to evolution by natural selection, had developed in a mechanistic manner.

John Tyndall, the physicist, was one of the foremost exponents of the new materialism. He maintained that, just as atoms and

PLATE XVII. Louis Pasteur in his laboratory, from the painting by Edelfeldt. The glass at which he is looking contains the spinal cord of a rabbit which has died from rabies.

PLATE XVIII. Faraday, Huxley, Wheatstone, Brewster and Tyndall.

molecules can arrange themselves in the beautiful and often complicated forms of crystals, so, in a further stage of elaboration, they give rise to living matter, to plants and to animals. But how did consciousness arise? Every fact of consciousness appears as the result of chemical activity in the brain. When the brain is damaged or diseased the mind becomes deranged and deluded. The mind is thus a by-product of matter.

Tyndall's most famous exposition of his views, infuriating to many churchmen, was his address as President of the British Association for the Advancement of Science at Belfast in 1874. He asked what had prevented the rapid advance of science until comparatively recent times. He answered that it was the obscurantism of the Church. He went on to make an arrogant claim: 'The impregnable position of science may be described in a few words. We claim, and we shall wrest from theology, the entire domain of cosmological theory. All schemes and systems which thus infringe upon the domain of science must, in so far as they do this, submit to its control and relinquish all thought of controlling it. Acting otherwise proved always disastrous in the past, and it is simply fatuous today.' In his presentation of the argument that life and mind depend upon matter he admitted that we are ignorant how consciousness can arise out of atoms but he stated his belief in the primacy of matter.

Many of the greatest men of science, for example Maxwell, Kelvin, Faraday and Joule, did not allow scientific theories to affect their religious faith. Maxwell made playful fun of Tyndall's address in a humorous poem. He commented on Tyndall's view that the mind, like the body, has evolved and that it inherits the experiences of its pre-human ancestors:

We honour our fathers and mothers; grandfathers and grandmothers too;
But how shall we honour the vista of ancestors now in our view?

The issues involved in the conflict between science and religion were debated during the eleven years 1869–80, by the Metaphysical Society, which included many of the most eminent thinkers of the Victorian age. The society owed its origin to a young architect, James Knowles, the founder of the influential magazine, *The Nineteenth Century*. Debating societies provided one of the chief intellectual recreations of the Victorians, and almost every town had its Philosophical Society. Knowles was a friend of Tennyson who, throughout his life, made an effort to read and understand the latest scientific books.

The society included the Archbishop of York, the Bishops of Peterborough and of Gloucester and Bristol, the Roman Catholic Archbishop Manning, W. G. Ward—the friend and disciple of Newman and one of the earliest converts to Roman Catholicism from the Oxford Movement, the Duke of Argyll, the historians Froude and Bagehot, the scientists Huxley and Tyndall, Tennyson and John Ruskin. Notable thinkers who declined the invitation to join were John Stuart Mill, Spencer, Newman and Carlyle.

The society met once a month for nine months in the year and, at each meeting, one of its members read a paper, which was followed by a discussion. The two most brilliant debaters in the society were Ward and Huxley. At the first meeting it was suggested that moral disapprobation should be avoided in debate. There was a pause and Ward said: 'While acquiescing in this condition as a general rule, I think it cannot be expected that Christian thinkers shall give no sign of the horror with which they would view the spread of such extreme opinions as those advocated by Mr Huxley.' After another slight pause, Huxley replied: 'As Dr Ward has spoken, I must in fairness say that it will be very difficult for me to conceal my feeling as to the intellectual degradation which would come of the general acceptance of such views as Dr Ward holds.'

The two men represented the main opposing viewpoints among the members of the society: Ward held that knowledge is gained by intuition and revelation, and believed in the supernatural; Huxley held that knowledge can be gained only by experience and the scientific method, and disbelieved in the supernatural. Huxley coined the word 'agnostic' to describe his own religious position. Dr Wace said that 'infidel' would be a more suitable term, and Huxley agreed that 'infidel' had the advantage of being somewhat offensive.

Huxley's first paper had the title, 'Has the Frog a Soul?' In it he attacked one of the major assumptions of the supernaturalists, that man has an immortal soul. Assuming that the mind or soul has evolved, it is difficult to deny immortality to a frog if one grants it to a man. Huxley described experiments on a frog deprived of the front part of its brain, or with its spinal cord severed, which still reacted under certain stimuli. He asked whether the frog was an automaton, as Descartes maintained. If it had a soul, was this distributed along its spinal marrow? He could give no answer, but suggested that the discussion should throw light on the supposed relations between the human soul and body.

His next paper was the most notorious of all those read to the society and was the only one granted the honour of a second meeting to continue the discussion. Because of his rejection of the miraculous, his opponents invited him to write a paper on a definite miracle and to explain why he disbelieved it. He selected one of the greatest of all miracles and entitled his paper 'The Evidence of the Miracle of the Resurrection'. When Newman heard of this discussion he was filled with horror and thanked God that he had refused the invitation to join the society.

Huxley's opponents were by no means always on the defensive. Ward read a paper on 'Memory as an Intuitive Faculty' in which he attacked the major assumption of science that experience, largely based on memory, is trustworthy. Those who believed that experience is the sole source of knowledge resisted this attack, while the intuitionists pressed their view that man has intuitive faculties which provide him with deeply important knowledge.

Ward read another paper, 'Can experience prove the Uniformity of Nature?' The Principle of the Uniformity of Nature was stated by John Stuart Mill in the form, 'What happens once will, under a sufficient degree of similarity of circumstances, happen again, and not only again, but as often as the same circumstances recur'. The principle is the justification for predicting the future from scientific generalisations, which are necessarily based on present or past events. Ward asked whether scientists ever showed anxiety to examine asserted exceptions to the Uniformity of Nature. A miracle is merely an exception and does not invalidate the assumption of general uniformity. Huxley admitted that the Principle of the Uniformity of Nature cannot be proved and must be taken merely as a working hypothesis. In answer to the criticism that scientists showed no anxiety to examine miracles he replied that, if they did so, and denied them, they would wound many amiable people; if they accepted them they would lose their scientific reputations. Archbishop Manning maintained that the Uniformity of Nature rested on the wisdom of God, while Ward, in conclusion, said how much better our civilisation would be if we could prove that the Divine Will was an efficient cause in nature.

The society came to an end when it was found that old arguments were being repeated and that the great debate was a stalemate.

But controversies still flared up in print. In the issue of *The Nineteenth Century* of November 1885 there appeared an article by Gladstone on 'The Dawn of Creation and Worship'. The article was hardly worth a reply, but Huxley was goaded by the thought that 'all orthodoxy was gloating over the slap in the face which the G.O.M. had administered to science'. Gladstone maintained that the order of creation, given in Genesis, agreed with that discovered by science, whereas reptiles, as shown by the geological record, appeared earlier than stated in Genesis. Huxley's reply was scathing: 'Still, the wretched creatures stand there, importunately demanding notice; and, however different may be the practice in that contentious atmosphere with which Mr Gladstone expresses and laments his familiarity, in the atmosphere of science it really is of no avail whatever to shut one's eyes to facts, or to try to bury them out of sight under a tumulus of rhetoric.'

Symptomatic of the time were the new religions, or substitutes for religions, which sprang up. In 1875 the Theosophical Society was founded by Madame Blavatsky in New York, in 1879 the first church of Christ Scientist by Mary Baker Eddy in Boston, U.S.A., while in England, in 1882, the Society for Psychical Research was formed, supported by many distinguished members, to attempt to apply the scientific method to psychical phenomena.

There was a renewal of controversy at the end of the century with the publication in 1899 of Ernst Haeckel's *The Riddle of the Universe* (*Die Welträthsel*), the sale of which was enormous, hundreds of thousands of copies being sold in Germany, England, France and Italy. Haeckel was a philosophical monist, believing that only matter is real. He held that the mind is completely dependent on a material substratum, and therefore does not survive the body. He was a distinguished biologist and claimed that any animal with a centralised nervous system possessed consciousness. He professed himself a pantheist, believing that God is not distinct from the world, and quoted Schopenhauer with approval: 'Pantheism is a polite form of atheism.' The controversy aroused was full of asperity, and theologians were driven to quoting Huxley in their own defence: 'the materialistic position that there is nothing in the world but matter, force and necessity, is as utterly devoid of justification as the most baseless of theological dogmas.'

Huxley, normally aggressive and optimistic, was subject to moods of depression which resulted, so his opponents said, from the unresolved paradoxes in his intellectual life. He and Tyndall loved to sing hymns together on a Sunday evening.

The years of controversy between science and religion left the Roman Catholic position unmoved. As a result of the dogma of papal infallibility, announced in 1870, all utterances by the Pope *ex cathedra* were beyond argument: Catholics guilty of modernism were excommunicated. Among Protestants there was a minority, the fundamentalists, that clung to a belief in the literal accuracy of the Bible and rejected evolution. Others became modernists, stressing the beauty and ethical teaching of Christianity, but rejecting some of the miraculous elements that seemed to conflict with science.

The difficulty with modernism is to know where the pruning process should stop. Moreover, the cardinal tenet of Christianity, that there was a unique intervention by God in the world nearly 2000 years ago, is so tremendous a miracle that, once accepted, all other miracles present little difficulty. Many of those who no longer found Christianity credible became scientific humanists, believing in science and in man, and rejected the supernatural. We shall touch upon this further in our discussion of the influence of the theory of evolution.

* * *

Newton and his contemporaries believed that God created the universe in substantially the same form as it is today. The nineteenth-century view was very different. Laplace postulated that the solar system began as a glowing, rotating gas from which the planets condensed and cooled; Hutton and Lyell described the gradual transformation of the surface of the earth during millions of years, how mountains had been raised and wasted away, how continents had sunk beneath the sea and risen again; and Darwin drew a picture of the emergence of primitive forms of life which developed, over the vast periods of geological time, into a profuse variety of complex plants and animals, culminating in man.

Its success in astronomy, geology and biology suggested that evolution might be a fundamental principle applying to all branches of knowledge. Studies were made of the evolution of society, of religion, of morals, of marriage, of language and of war. Herbert Spencer (1820–1903), an influential if not a great philosopher, made evolution the guiding theme of his system. As a young man of twenty he read Lyell's *Principles of Geology* and was so little impressed by its arguments against Lamarckism that he became a convinced Lamarckian. He was the originator of the word 'evolution' and he was already beginning to sense its all-embracing significance seventeen years before the publication of Darwin's *Origin of Species*.

Spencer sought a formulation of the principle of evolution which would apply universally. He held that evolution necessarily follows from universal causation. The fact that every event is the cause of succeeding events must result in a continuous process of change. He obtained the clue to his universal formula, in a flash of illumination, from the embryology of von Baer, which revealed how the complex embryo develops from the simple fertilised egg cell. In the embryo different types of tissue and different organs appear, co-ordinated in a single individual. This led Spencer to the formula that all evolution is a process of integrated differentiation.

He applied his formula to sociology, psychology and ethics. He traced the organisation of society from the small homogeneous family and tribe, through the early civilisations, differentiated into rigid classes and integrated by a monarch, to the nineteenth-century industrial States with their increasing differentiation of labour, successfully integrated, as he believed, quite independently of centralised government controls. He did not regard socialism as the next stage in evolution, believing that socialism would result in a bureaucratic class more powerful and pernicious than the aristocracies of old, a conclusion which found a ready response from the *laissez-faire* liberals of his time.

The general application of an evolutionary formula is subject to several fallacies. One is the implication that the more developed can be explained and interpreted in terms of the less developed—what has been called the 'nothing-but' fallacy. Because man has descended from an ape-like stock it does not follow that he is nothing but a sophisticated monkey; and because religion has evolved from primitive animism we cannot assume that theology is merely intellectualised superstition. Evolution is a creative process in which novel characters and entities emerge. When hydrogen and oxygen combine, the water which results has properties which are new, and this applies to all chemical compounds. At a certain stage of chemical elaboration life emerges and, at a further stage, mind. But with any foreseeable increase in scientific knowledge it would be impossible to predict the properties of life and mind from chemical principles.

Spencer's attempt to apply an evolutionary formula, derived from biology, to psychology and sociology, was also vitiated by the fact, not appreciated until the twentieth century, that human evolution proceeds in a fundamentally different manner from that of other organisms. The physical evolution of man is now subordinate to cultural evolution, operating through the transmission of knowledge and tradition. Cultures reproduce themselves and it is they which compete for survival rather than individual men. All societies have some apparatus for mutual aid or welfare, organised to arrest the process of natural selection among its individuals.

The theory of natural selection was put forward to account for biological evolution and can have no relevance to astronomical or geological evolution. Hence the word evolution, when used to cover other fields outside biology, can mean little more than change, since there is no unifying principle to account for its cause.

Spencer's approach to ethics led him to define good and bad conduct in terms of evolution. In *The Principles of Ethics* he wrote: 'the conduct to which we apply the name good is the relatively more evolved conduct, and bad is the name we apply to conduct which is relatively less evolved'. But we can conceive of evolution as proceeding in an undesirable direction, from good to bad, and hence our judgement of what ought to happen is based on criteria other than what does happen. Indeed, outraged by the picture of nature red in tooth and claw, Huxley declared that ethically good conduct consists in opposing the predatory operation of biological evolution.

Evolutionary ethics have been revived in the twentieth century, notably by Huxley's grandson Julian. Julian Huxley expounded his ethical views in the Romanes lecture of 1943, delivered exactly fifty years after a similar lecture by his grandfather.[1] He maintained that our ethical ideas have evolved and are still evolving; that they should be based as far as possible on our knowledge of evolution; and that they can be used to guide the process of evolution itself in the future.

Julian Huxley rejected traditional theories of ethics, based largely on the compulsive sense of obligation imposed by our consciences, because modern psychology has shown that our consciences are not an absolute or reliable authority. They depend on heredity and on environment in infancy. He rejected religious revelation because he believed it to be relative to the age in which its founder lived; he rejected the absolute values of philosophy because he found them intellectually unconvincing. He fell back on the undoubted fact of cosmic evolution: 'We thus arrive at the apparent paradox that a process of change provides the only certainty to which a man can hope to attain.'

When, however, we look at Huxley's ethical principles, we find that they are based far more on intuitions of what is right and good than on the facts of evolution. 'In the broadest possible terms evolutionary ethics must be based on a combination of a few main principles: that it is right to realize ever new possibilities in evolution, notably those which are valued for their own sake; that it is right both to respect human individuality and to encourage its fullest development; that it is right to construct a mechanism for further social evolution which shall satisfy these prior conditions as fully, efficiently, and as rapidly as possible.'

We may ask why is it right, on evolutionary grounds, to respect human individuality?

The weakness of this approach to ethics is that the process of cosmic evolution is ultimately aimless and meaningless except by reference to factors outside itself. The evolution of the universe can have no goal or purpose unless there is some non-physical, eternal reality. To the religious mind the advocates of evolutionary ethics are like travellers in a train who endeavour to plot their journey by observations confined to the train itself, refusing to look out of the window because they regard what they see there as illusion.

* * *

[1] T. H. Huxley and Julian Huxley, *Evolution and Ethics 1893–1943* (London, 1947).

The chief significance of the nineteenth century is to be found not so much in its science as in its technology. The immense development of industry in Europe and in the U.S.A. produced a rise in the average standard of living greater than that achieved in the previous two thousand years. By the end of the century Europe had reached its apogee of world power. Its population, almost static until about 1700, rose from 187 millions in 1800 to 400 millions in 1900.

The industrial revolution, begun in Britain in the eighteenth century, proceeded at an ever-increasing tempo and spread to the other European countries. Methods for manufacturing cheap steel were invented by Sir Henry Bessemer in 1856 and by Gilchrist and Thomas in 1879. British coal production rose from 10 million tons in 1800 to 60 million tons in 1850 and 210 million tons in 1900.

The prime mover of the age was the steam-engine which, like the water-wheel, was invented and developed by men who owed little to theoretical science. Although science had but a small influence on industrial progress in the first half of the century, it was responsible in the second half for the creation of entirely new industries.

Transport was revolutionised by railways and steamships. The building of railways was stimulated by the needs of industry for the transport of raw materials and of manufactured goods; industry, in turn, was stimulated by the provision of rapid and cheap transport. Railways made possible the expansion over a continent of the inhabitants of the eastern seaboard of the United States of America, and of the immigrants that poured in from Europe.

Two industries which owed their origin entirely to science were the electrical and fine chemical industries. The first London telephone exchange was set up in 1879 and the first generating station for supplying electricity to private consumers began operation in New York in 1882. From 1870 there was considerable scientific research, mainly in Germany, on synthetic dyes, and many new ones were discovered. The Germans, because they led the world in theoretical chemistry, acquired a complete domination in the fine chemical industry, producing dyes, perfumes, drugs and explosives. The First World War forced other countries to set up chemical industries of their own, with the result that the output of Britain and the U.S.A. overtook and passed that of Germany.

The Germans became pre-eminent too in the optical industry, again because of their superior science and because of the research laboratories which they set up in connection with industry. They produced the finest cameras, microscopes, spectrometers and other scientific instruments.

Towards the end of the century industrialised Western Europe could no longer produce enough food to feed its greatly increased population. Food was imported from North America, where the methods of agriculture were transformed by such machines as the combine harvester. Refrigeration and canning enabled meat, fish and fruit to be transported to Europe over the oceans of the world. The first cargo of frozen mutton reached Britain in 1880.

Industry tended to congregate in the coalfields. Places which, in the eighteenth century, had been little more than villages developed with mushroom growth into large cities—for example, Liverpool, Manchester, Leeds, Sheffield and Birmingham. A squalid region of grime and ugliness spread over the Midlands and the north of England.

* * *

The nineteenth century was a period of public optimism and, among some intellectuals who had lost their religious faith, of private despair. The public optimism was based on the immense scientific and technological advances and the rising standard of living. It expressed itself in a belief in automatic and inevitable progress.

The origin of the idea of progress can be traced back about three hundred years, to the time of Francis Bacon and the beginnings of modern science. It was completely foreign to the thought of classical antiquity. In the Middle Ages men looked back to a Golden Age rather than forward to the millennium. In the eighteenth century, however, the idea of the progress of knowledge and of human welfare, resulting from the reign of reason and science, became widespread. Joseph Priestley wrote: 'Whatever was the beginning of this world, the end will be glorious and paradisiacal beyond what our imagination can conceive.'

Evolution seemed to provide a scientific basis for the idea of progress. Everything is evolving, said Spencer, and the direction of evolution is, by definition, good. Spencer was convinced that mankind was becoming better and better and would ultimately reach perfection, as opposed to the Christian doctrine that man is inherently sinful and that, although he can apprehend perfection, he can never attain it.

The two world wars of the twentieth century, the horrors of the Nazi concentration camps and events such as the extermination of six million Jews have dealt a severe blow at the idea of progress. Modern men behave just as wickedly as at any time in recorded history.

In the public mind, however, the idea of progress survives; not progress as understood by the biologist, with his vista of millions of years, but material progress provided by the technologist, in periods of five year plans. Science, as a body of cumulative knowledge, continually progresses. Art, on the other hand, does not; a work of art is changeless and ultimate.

* * *

Most scientific advances of the nineteenth century were made in France, Britain and Germany, and the sciences of the three countries had, to some extent, distinctive characteristics, French science tending to be perfect in form and logic, British science individualistic and often highly original, German science thorough and superbly organised.

For the first quarter of the century Paris was the scientific centre of the world. In 1794 the *École Polytechnique* and *École Normale Supérieure* were founded and these trained many of the finest French scientists throughout the century. In 1795 the *Académie des Sciences*, suspended during the Revolution, was reopened as part of the *Institut*. Although later in the century France produced an outstanding genius in Pasteur, and many other distinguished men of science, she never regained the preeminence which she had held earlier.

From about 1830 until 1865 the major scientific developments came from Britain, in the work of Lyell, Faraday, Joule, Darwin and Maxwell. Of these five men only Maxwell, and Lyell for a period, were professors at a university. Joule, Darwin and Lyell were amateurs with private means and Faraday worked at the Royal Institution. It is indeed surprising how small a part the universities played in the progress of British science in the nineteenth century. At Oxford and Cambridge some geology and botany was taught, but no experimental physics, chemistry and zoology until after the middle of the century. The amount of science teaching in the public schools and grammar schools, even at the end of the century, was negligible.

By contrast, Germany had a well-organised State system of scientific education and this largely accounts for the dominance of German science in the second half of the century. In 1824 the first technical school was established in Berlin, an example soon followed by other German towns, and by 1870, the year in which Germany became a unified State, the technical colleges were training a host of engineers and chemists. There were over twenty German universities all subsidised by the State and together turning out more than five times the number of graduates as Britain. At most of the German universities there were flourishing schools of scientific research: for example, the schools of chemistry of Liebig at Giessen, of Bunsen at Heidelberg and of Ostwald at Leipzig; the schools of physics of Helmholtz at Berlin and of Weber at Göttingen; and the school of cellular pathology of Virchow at Würzburg. These attracted foreign students, particularly from America, many of whom returned to their own countries to become professors, imbued with Germanic ideals and methods. One American biologist, Dr H. P. Osborn, regretted this, pointing out that although Germany produced most of the generals and the rank and file among the army of biologists, England had produced the commanders-in-chief. The same could be said of physics and geology in the nineteenth century, but not of chemistry nor medicine in which Germany was supreme.

By the end of the century the intellectual output of Germany was almost equal, at least in quantity, to that of the rest of the world, and the conviction of many Germans of the cultural superiority of the Teutonic race was understandable, if ironic.

Some landmarks of science in the nineteenth century

1800–3 Thomas Young, Wave theory of Light.
1808 John Dalton, *A New System of Chemical Philosophy*.
1809 J. B. Lamarck, *Philosophie Zoologique*.
1811 Amedeo Avogadro's Hypothesis.
1817 Georges Cuvier, *Le Règne Animal*.

1817   William Smith, *Stratigraphical System of Organised Fossils.*

1818   Augustin Fresnel, *Mémoire sur la Diffraction de la Lumière.*

1824   Sadi Carnot, *Réflexions sur la Puissance Motrice du Feu.*

1830–3   Charles Lyell, *Principles of Geology.*

1831   Michael Faraday's discovery of Electromagnetic Induction.

1832   Justus von Liebig and Friedrich Wöhler, *Researches on the Radical of Benzoic Acid.*

1842   Julius Mayer's paper on the Conservation of Energy.

1847   James Joule's lecture on the Conservation of Energy.

1850   Rudolf Clausius, Second Law of Thermodynamics.

1852   Edward Frankland's Theory of Valency.

1854   Lord Kelvin's Absolute Scale of Temperature.

1856   Louis Pasteur, *Recherches sur la Putréfaction.*

1858   Stanislas Cannizzaro reintroduces Avogadro's Hypothesis.

1858   August Kekulé's Theory of Molecular Structure.

1859   Charles Darwin, *Origin of Species.*

1864   James Clerk Maxwell, *A Dynamical Theory of the Electromagnetic Field.*

1865   Joseph Lister, Antiseptic system first used for compound fracture of a leg.

1874   Van't Hoff and Le Bel, Theory of Stereo-chemistry.

1881   Robert Koch, discovery of solid-culture media for microbes.

1888   Heinrich Hertz, discovery of electromagnetic waves.

CHAPTER 19

GENETICS AND NEO-DARWINISM

ONE OF THE MOST serious early criticisms of Darwin's theory of evolution was that the variations preserved by natural selection would rapidly become diluted and swamped by 'blending inheritance'. Thus a creature which is unusually fleet of foot, or has some other variation which favours its survival, on mating with a normal individual, will produce offspring which have only about half the advantageous variation, while the next generation will have about one quarter, and so on until the variation has virtually disappeared.

Darwin investigated every available field of information in his study of variation and he took to breeding pigeons himself. He knew that curious things happen in heredity. For example, there are two phenomena known as 'reversion' and 'prepotency': an individual appears sometimes who is a reversion to an earlier ancestral type; and sometimes the offspring resemble one parent much more than the other, showing this parent's 'prepotency'. Nevertheless Darwin did not seriously challenge the general truth of the theory of blending inheritance, in the sense that a particular variation is normally diluted in the offspring. He therefore tended, towards the end of his life, to give more weight than formerly to the effects of the environment and of the use and disuse of organs in producing adaptive variations, and rather less weight to natural selection of random variations.

He put forward a theory of the mechanism of heredity, designed mainly to account for the inheritance of characters acquired by use and disuse. This theory, known as pangenesis, postulated that minute, invisible particles were continually produced in all parts of the body, that some of these reached, and were absorbed by, the germ cells, and hence affected heredity. It was not one of Darwin's happier speculations, and it proved to be valueless.

*   *   *

The study of variations led, at the close of the nineteenth century, to two schools of thought: those who believed that evolution proceeded by the natural selection of small, continuous variations, in accordance with Darwin's views; and those who stressed large, discontinuous variations, regarding natural selection as of minor importance.

One of the chief representatives of the former school was Francis Galton (1822–1911), a cousin of Charles Darwin. He applied statistical, mathematical methods to the continuous variation of characters such as stature in man or size of seed in plants. He showed that the variations follow the laws of chance and can be represented by the normal curve of error, shaped like a cocked hat. In a crowd of spectators at a football match there will be a few men who are very tall, a few who are very short, and a large number of intermediate heights. Similarly in a large enough sample of bean seeds, a few seeds will be unusually large, a few unusually small, and the majority of average size. If evolution proceeds by the natural selection of these variations, we should expect that the offspring of an unusually large seed would be represented by a curve of error having a higher mean value than normal. Galton postulated that a permanent shift of the mean value could be achieved by progressive selection.

This postulate was shown by experiment to be untrue, notably as a result of the work of the Danish botanist, Wilhelm Johannsen (1857–1927). Johannsen's experiments were conducted with a race of beans which, by self-fertilisation, bred true; such a race is known as a 'pure line'. He caused some bean plants to produce small seeds, by giving them inadequate nutrition, and other bean plants, by liberal manuring, to produce large seeds. The next generation, grown under uniform conditions and with uniform nutrition, showed no signs of the treatment accorded to their parents. The plants from the small seeds produced seeds whose average size was the same as those grown from the large seeds. Johannsen was led to distinguish sharply between the effects of heredity and of environment; between hereditary constitution, handed down from generation to generation, and the visible characters of an individual, the result of both heredity and of environment. His experiments showed that variations caused by nutrition and environment are not transmitted to offspring. In Galton's work no attempt had been made to distinguish between variations due to heredity and those due to environment.

The theory that the variations which are transmitted to offspring occur in comparatively large, abrupt changes was put forward in 1900 by the Dutch botanist, Hugo de Vries (1848–1935). He gave the name *mutations* to these discontinuous variations.

There was a field near Hilversum, bordered by canals and inaccessible by road, which no one could cultivate. It became a jungle of plants, the height of a man, consisting mainly of the evening primrose, *Oenothera lamarckiana*, originating from seeds blown from a cultivated bed in a nearby park. De Vries studied this field for three years in the 1880's and found that the rapid multiplication of the plants had given rise to a high degree of variability. He cultivated the new varieties in his own garden and, in eight generations, raised 53,000 plants, among them eight new types, differing from the original type in size and shape of petals, in smoothness of leaves, and in the height of the plant. He found that each new type appeared in an abrupt change, without intermediate steps. He came to the conclusion that species do not originate by the natural selection of small continuous variations, as Darwin maintained, but by comparatively large, discontinuous mutations. A new species appears suddenly, he maintained, without transitional steps.

Six years before de Vries put forward his mutation theory, there appeared in England a book by William Bateson (1861–1926) entitled *Materials for the Study of Variation*, containing a wealth of examples of discontinuous variations. The book made little impression on English biologists and for some years Bateson was a lone voice, although an ardent and outspoken one, in favour of discontinuity in evolution. He believed, like de Vries, that evolution proceeded in sudden jumps, determined by the inner nature of the organism, rather than by a process of gradual adaptation to the environment.

* * *

In the spring of 1900 de Vries in Holland, Correns in Germany and Tschermak in Austria, rediscovered independently the papers of the Austrian monk Mendel on the breeding of peas, published thirty-four years earlier. Bateson read about Mendel's work in the train between Cambridge and London on 8 April 1900 and recognised at once its fundamental significance.

Mendel was, in fact, one of the outstanding scientific investigators of the nineteenth century, and yet his work was neglected and forgotten. The papers were read before an obscure body, the Brünn Society for the Study of Natural Science, but copies of the Proceedings of the Society were sent to Vienna, Berlin, Rome, St Petersburg, Uppsala and London. Mendel was in correspondence with Nägeli, one of the leading botanists of the day, about his experiments and Nägeli completely failed to realise their importance. Those responsible for making digests of scientific papers either did not trouble to understand Mendel's results or were incapable of doing so. The moral, says R. A. Fisher, is that no research worker can afford to neglect reading for himself all original papers in his own field of inquiry. It should be said, in defence of the biologists of the time, that their minds were filled with Darwin's theory of evolution, then a subject of acute controversy, and there is a limit to the new ideas which the normal human mind can recognise and assimilate.

Gregor Mendel (1822–84) was born of peasant stock in northeast Moravia, now in Czechoslovakia. but then part of the Austro-Hungarian empire. As a boy he used to help his father in the orchard with the grafting of fruit trees; the slips were obtained through the village priest, from the garden of the lady of the manor, the Countess Waldburg. Mendel's parents, though poor, were ambitious for their son, and sent him to Troppau High School, twenty miles from their village. Here he was entered on 'half rations', was obliged to serve as a private tutor to support himself, and became seriously ill. He went on to the Olmütz Philosophical Institute, but found it impossible to study and earn a living at the same time. At the age of twenty-one he decided to join the Augustinian monastery at Brünn, then one of the chief intellectual centres of the region, and after a four-year novitiate he was ordained priest.

He was directed by his abbot to teaching, since his talents lay in that direction, and he became a supply teacher at the Brünn Modern School. In 1850 he took the examination for high school teachers in physics and natural history; he failed. Six years later, after four terms at the University of Vienna, he took the examination again, and once more the examiners failed to discern in him the ability they demanded from their qualified school teachers. It was after this second failure that he began his world-famous experiments.

Mendel was much loved by his pupils. One of his acquaintances recollected:

I still seem to see him as he walked back to the monastery through the Bäckergasse, a man of medium height, broad-shouldered, and already a little corpulent, with a big head and a high forehead, his blue eyes twinkling in the friendliest fashion through his gold-rimmed glasses. Almost always he was dressed, not in a priest's robe, but in the plain clothes proper for a member of the Augustinian order acting as schoolmaster—tall hat; frock-coat, usually rather too big for him; short trousers tucked into top-boots.[1]

Mendel delighted to take his pupils round the monastery garden to show them his plants, fruit trees and beehives. Being sensitive to draughts, he hung an Eolian harp in the garden and, as soon as it sounded, he put on his hat.

His most important experiments were conducted with plants of the edible pea. He obtained a considerable number of varieties from seedsmen and, after a two years' trial to find which of them bred true when self-fertilised, he chose twenty-two varieties for his investigations. His object was to find the effect of cross-fertilisation on seven pairs of contrasting characters, such as tallness and dwarfness, yellow and green seeds, and round and wrinkled seeds.

In one series of experiments he crossed peas about 6–7 ft. high with peas $\frac{3}{4}$–$1\frac{1}{2}$ ft. high. He carefully removed the pollen-bearing organs from, say, the tall plants, to prevent self-fertilisation, and dusted their stigmas with pollen from the dwarf plants, afterwards covering the flowers with little paper or calico bags to stop insects from depositing pollen from other plants.

The first hybrid generation, F_1, were all tall—as tall as the tall parents (Fig. 51). Mendel then self-fertilised this first generation and obtained a second generation, F_2, some of whom were tall and some dwarf. He counted the two varieties of plants, and found that the tall ones were about three times as numerous as the dwarf. His actual figures were as follows: out of 1064 plants, 787 were tall and 277 dwarf, giving a ratio of 2·84 : 1. He realised that in a sample there is a probable error in the ratio depending on the size of the sample.

The fact that the hybrid peas are either tall or dwarf, and not of intermediate height, indicates that the 'factors' controlling height

[1] H. Iltis, *Life of Mendel.*

do not blend, but retain their individuality from generation to generation. This conception, that heredity is particulate, is the foundation stone on which western genetics, or neo-Mendelism, has been built. It is necessary to speak of western genetics to distinguish it from Soviet genetics, which we shall consider later.

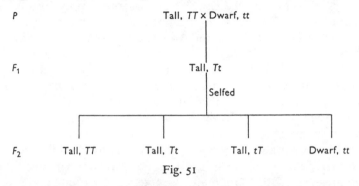

P   Tall, TT × Dwarf, tt

F_1 Tall, Tt

Selfed

F_2 Tall, TT Tall, Tt Tall, tT Dwarf, tt

Fig. 51

Mendel found that it made no difference which of the plants acted as the male parent and which as the female. The contributions of the two parents were equivalent. When fertilisation occurs the germ cells from the two parents combine and hence the combined or fertilised cell must carry factors determining height from both parents. Thus every plant must carry pairs of corresponding factors. In Fig. 51 the tall-determining factor is represented by T and the dwarf by t.

If the germ cells of each parent carried pairs of factors, then parents with factors TT and tt respectively would give rise to offspring with factors $TTtt$; likewise the next generation would have eight factors and so on. The germ cells, however, carry only one of the pair of height-determining factors, so that all offspring have only two of these factors.

This fundamental conception is known as Mendel's *law of segregation*, and it may be stated generally as follows: there are pairs of corresponding hereditary factors, which retain their identity from one generation to another; the factors in each pair are separated, or segregated, during the mechanism of reproduction.

The first generation, F_1, obtains factors T and t from its parents and, since the plants are all tall, T is said to be *dominant* and t *recessive*. The phenomenon of dominance had been noted before Mendel's time, but its significance was not understood.

The second generation, F_2, obtained by self-fertilisation of F_1, springs from the union of male and female germ cells, which may carry the factors T or t. Mendel's numerical results showed that the union occurs according to the laws of chance. Male germ cells carrying factors T or t, and female germ cells carrying factors T or t, will give rise with equal probability to offspring having factors TT, Tt, tT or tt. The first three offspring are tall, because T is dominant, and the fourth is dwarf. Hence the Mendelian ratio 3 : 1 is given a simple and satisfying explanation.

This second theoretical step is known as Mendel's *law of independent assortment*, and it may be stated as follows: single factors from each sex combine independently and at random with corresponding factors from the other sex, forming new pairs of factors in the offspring.

Mendel obtained precisely similar results for the other pairs of contrasted characters, such as round and wrinkled seeds, and yellow and green seeds, the dominant characters being round and yellow.

Mendel also performed experiments with plants having two pairs of differentiating characters. For example, he crossed plants having round, yellow seeds with plants having wrinkled, green seeds. His numerical results indicated that each factor was transmitted independently of the other. The ratio of each dominant character to its corresponding recessive character in the second (F_2) generation, that is of round to wrinkled and of yellow to green, was 3 : 1.

Mendel's experiments on the pea took place in the years 1856–63. He performed experiments with other plants including unfortunately, at Nägeli's suggestion, the hawkweed, whose reproduction is abnormal. In 1868 he was elected prelate or abbot of his monastery by his fellow monks, and thereafter much of his time was taken up with a struggle with the government about the taxation of monasteries. He retained his scientific interests to the end of his life, but his experiments virtually ceased. When he died in 1884 he was mourned as a kindly and conscientious prelate and no one had the least idea that he was a scientific investigator of genius.

His experiments displayed a remarkable sureness of touch and economy of effort. He chose the correct plant, the pea,[1] for his

---

[1] Peas, being self-fertilising, readily become a pure line which breeds true. Many organisms, including human beings, are complex hybrids from which the simple Mendelian results cannot be obtained.

main experiments, and selected certain limited, sharply-defined characters for investigation. He kept successive generations separate and exhibited his results in a simple mathematical form—to the bewilderment of contemporary botanists. Earlier experiments on the hybridisation of peas and other plants, in which some of his fundamental ideas, such as dominance, were adumbrated, enabled him to foresee clearly the lines his investigations should take.

The problem facing biologists at the beginning of the twentieth century was whether Mendel's laws and the resulting 3 : 1 ratio were of general application. Many experiments were performed, not only with plants, but also with snails, insects, fishes, amphibians, birds and mammals. The Mendelian laws were found to be universally valid.

Problems of inheritance, seemingly inexplicable on a blending theory, had a natural and convincing explanation on the Mendelian theory. Reversion, or the resemblance of an individual to a remote ancestor, was explained in terms of recessive factors, preserved hidden but unchanged from generation to generation, and showing their effect only when paired together, instead of with a dominant factor. Prepotency, or the resemblance of an individual to one parent more than to the other, was also explicable in terms of dominant and recessive factors.

*   *   *

Neo-Mendelism, which is the only body of biological theory comparable, in quantitative precision, with the theories of physics and chemistry, sprang from the marriage of Mendelism and of the microscopic observation of the cell, or cytology. Several important scientific advances have been made by the union of two quite dissimilar fields of observation and theory; an example we have already discussed is the union of thermodynamics and the kinetic theory of gases.

The cell is one of the basic concepts of biology and, like so many other fundamental scientific ideas, it dates back to the seventeenth century. The term cell was first used in 1665 by Robert Hooke to denote the honeycomb-like structures in cork. These are, in fact, the walls of dead cells which have lost their content. The primitive microscopes of the seventeenth century enabled men like Leeuwenhoek to observe unicellular creatures (not, of course, recognised as such) but the instruments were not powerful or perfect enough to allow a detailed examination of cell structure.

Notable improvements in the microscope, made in the late eighteen-twenties, led, in the following decade, to the foundation of the cell theory, that every living organism is made of cells, and that the various functions of the organism are the result of cell activities. After much confusion it was established that a new cell can come only from an existing cell; it was found, if we may

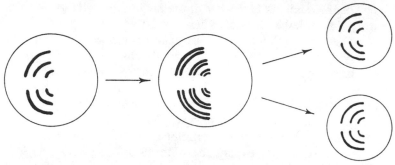

Fig. 52. Mitosis, or the division of body cells. The circles represent only the nuclei of the cells and the dark lines the chromosomes. The nuclei normally possess pairs of chromosomes, but just prior to division (middle nucleus) each chromosome splits into two.

repeat an old biological joke, that cells could multiply only by division.

The cell theory was not a theoretical necessity when it was put forward, and Darwin was uncertain whether or not to accept it. Its second phase of development coincided with further improvements in the microscope in the eighteen-seventies. Partly because Germany was the centre of the optical industry, most of the advances in the study of the cell took place in that country.

The cells of both animals and plants are made up of a jelly-like substance, termed protoplasm, and their activities are controlled by a small portion called the nucleus. The nucleus takes up readily certain dyes, which reveal its structure, and the pioneering investigators tried out many dyes in an empirical and haphazard manner, so that it was said that a cytologist had to be a dyer.

It was found that when cells divide their nuclei resolve themselves into pairs of rod-shaped objects, now called *chromosomes* (Plate XIX(a)), and that each chromosome splits longitudinally into two, one going to each of the new cells (Fig. 52). This is the mechanism of the formation of new body cells, and it is called *mitosis*. It ensures that every cell in the body has a complete set of chromosomes.

The cells in the body comprising, perhaps, millions, and including many specialised types forming the different organs and tissues, all originate from successive divisions of the fertilised egg cell. The fusion of the male germ cell or spermatozoon, which is nearly all nucleus, with the nucleus of the female germ cell or egg cell (Fig. 53), was first observed under the microscope in 1875

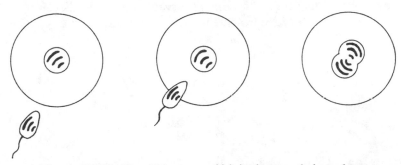

Fig. 53. Fertilisation. The sperm, which is almost entirely nucleus, unites with the nucleus of the egg cell.

by Hertwig in the case of the sea urchin. Heredity must be transmitted through the germ cells and it was suggested by Hertwig and Strassburger that the chromosomes were its material vehicle.

The subdivisions of the fertilised egg cell give rise mainly to body cells, but also to new germ cells. The direct lineage of the germ cells was stressed by August Weismann (1834–1913) in what he termed the continuity of the germ-plasm. He pointed out that there was no mechanism for the effects of any alterations in the body cells to be communicated to the germ cells. The development of the germ cells and of the body cells occur independently. Hence, as has been pointed out, a man is not the father of his children, but the uncle. Weismann was a convinced Darwinian, and he used his theory to show the improbability of Lamarckism; he refused to accept that characters acquired by the use and disuse of bodily organs could possibly be inherited.

Weismann had his own peculiar terminology; he suggested, in effect, that in the formation of the germ cells the nuclei underwent a halving of the chromosomes. If this were not so, assuming that the chromosomes are the carriers of heredity and that heredity is particulate, the hereditary factors of the offspring, being the sum of those of the parents, would be doubled in every successive generation. Weismann's theory was put forward before Mendel's work was known. It was influential in its stress on particulate heredity, and on the connection which must exist between heredity and the cell.

Weismann's ideas concerning the formation of the germ cells were fundamentally correct. The process, known as *meiosis* (meaning reduction), has been observed under the microscope, and its essentials are represented diagrammatically in Fig. 54. The

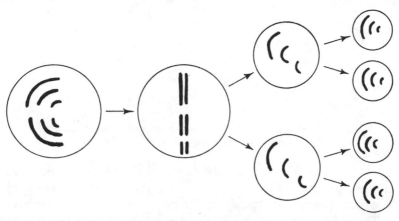

Fig. 54. Meiosis or the formation of the germ cells. The circles represent nuclei and the dark lines chromosomes. Before division corresponding chromosomes come together. On the first division the number of chromosomes is halved.

nucleus of the mother cell contains pairs of chromosomes, of the same total number and kind as every body cell. Corresponding chromosomes come together, the cell divides, and only one chromosome from each pair goes into the new cells. The chromosome number has thus been halved. The cells divide again, the chromosomes splitting longitudinally, giving rise to four germ cells, each containing only half as many chromosomes as the original mother cell. The whole process is elaborate, delicate and mysterious, and this brief and somewhat bald account is intended to convey only what, for geneticists, are its salient and significant features.

The close parallel between chromosomes and Mendelian factors soon became apparent: both occur in pairs, both are segregated in the germ cells and both are once more paired after fertilisation.

* * *

The possibility that each chromosome might represent a single Mendelian factor did not bear close examination. The edible pea has only seven pairs of chromosomes, and human beings have only twenty-three pairs, whereas their Mendelian factors are very much greater than this.

Another simple postulate, that each chromosome might carry many Mendelian factors, was developed by T. H. Morgan and his school in the U.S.A. Morgan called the Mendelian factors *genes* and he put forward the conception of material genes strung along the chromosomes, rather like beads on a string.

Morgan's investigations were conducted with the fruit fly *Drosophila*, which is ideal for breeding experiments. If one puts a male and female fly in a bottle with some banana they mate, the female lays eggs on the banana, and in about twelve days there may be three or four hundred mature offspring. The rapid rate of breeding has enabled millions of *Drosophila* to be studied.

One fly in about three hundred thousand may show a conspicuous mutation: its eyes may change from red to white, or its wings from grey to yellow, or its wings may become vestigial. Mutations are hereditary, and from a mutant fly a new race may be bred.

Morgan observed several hundred mutant characters in *Drosophila*, and he found that they were linked into four groups. Characters in the same group tended to remain together in the offspring, and not to be independently assorted as would be expected from Mendel's second law. Since *Drosophila* has four pairs of chromosomes Morgan concluded that each of the four groups of factors were carried by a single pair of chromosomes. The phenomenon of *linkage*, as it is called, had already been observed by Bateson and Punnett in their experiments on the breeding of sweet peas, but they had been unable to provide a convincing explanation.

The normal, wild *Drosophila* has red eyes and grey wings, and these two characters are linked together on one chromosome. A mutant race of *Drosophila* was bred having white eyes and yellow

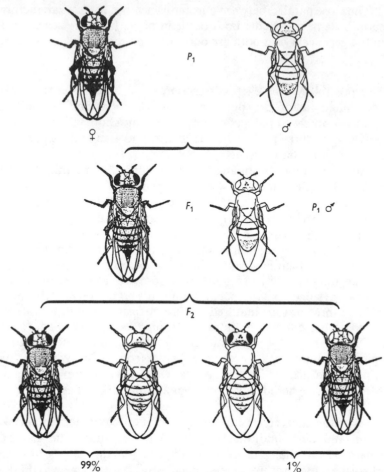

Fig. 55. A male *Drosophila* ♂, having two recessive, linked, mutant characters, yellow wings and white eyes, mated with a female ♀ having wild-type, linked characters, grey wings and red eyes. Their sons and daughters have grey wings and red eyes. If one of the daughters is mated with a male having yellow wings and white eyes there are four kinds of offspring: 99 % have either grey wings and red eyes or yellow wings and white eyes; 1 % have yellow wings and red eyes or grey wings and white eyes. The 1 % are cross-overs and represent interchanges between linkage groups. (From T. H. Morgan: *The Theory of the Gene*, by courtesy of Yale University Press.)

wings—characters which are likewise linked together. If one of these mutant males is mated with a normal female (Fig. 55), the offspring, F_1, are all normal because red eyes and grey wings are dominant characters, while white eyes and yellow wings are recessive. If one of the F_1 females is mated with a mutant male one would expect the offspring in the F_2 generation to consist of individuals having either red eyes and grey wings or white eyes and yellow wings, because of the linkage. In fact, 99 per cent of the offspring are of these two types; but the remaining 1 per cent consists of individuals in whom the linkage seems to have broken down because they have red eyes and yellow wings or white eyes and grey wings.

Morgan found the clue to this phenomenon in an observation of the abbé Jannsens, a Jesuit professor at the university of Louvain. Jannsens was investigating under the microscope the formation of the germ cells of yeast. He found that when the

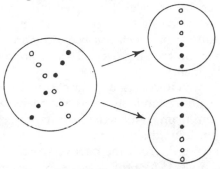

Fig. 56. Crossing over.

pairs of corresponding chromosomes came together, prior to the first division of the mother cell (Fig. 54), they formed cross-like figures (Fig. 56). Morgan realised that if the chromosomes split and exchanged parts of each other, the linkage between the genes would be broken. Thus two chromosomes, having genes *ABCDEF* and *abcdef*, might cross in the middle, exchange halves and give rise to chromosomes having genes *ABCdef* and *abcDEF*. The phenomenon is known as *crossing-over*.

Crossing-over need not occur in the middle of a chromosome; it can occur anywhere. It provides a method of preparing maps of the positions of the genes along the chromosomes. The fundamental idea underlying the method is that the probability of the linkage between two genes being broken by crossing-over is proportional to their distance apart on the chromosome. If two genes lie close together it is not very likely that the split in the chromosome will occur between them; if they lie at the ends of the chromosome the chance of a split between them is much more likely. By observing the frequency in the breakage of linkage between different pairs of characters, hundreds of genes have been mapped along the four pairs of chromosomes of *Drosophila*.

One of Morgan's co-workers, Hermann Muller, recorded the frequency of the natural mutations of *Drosophila*. He then made a most important discovery. He found that by irradiating the spermatozoa of the flies with X-rays he could increase the rate of mutation as much as one hundred and fifty times. Mutations can also be induced by chemical agents such as mustard gas and colchicine.

Individual genes have not been observed directly, but the detailed structure of the giant chromosomes in the salivary glands of *Drosophila* has been revealed by the electron microscope. The chromosomes are seen as a series of discs and bands, a change in the relative positions of which has been found to correlate with the phenomenon of crossing-over. This is impressive evidence in favour of the theory.

Genes are believed to consist of large chemical molecules whose structure could understandably be modified by X-rays and by similar radiation from radioactive substances. The cause of natural mutation is unknown; the normal amount of radiation from cosmic rays and from radioactive materials in the earth seems much too weak to provide an explanation. Since most mutations are deleterious and many are lethal, it is of vital importance that careful consideration shall be given to the possible effects on the heredity of the human race of the radiation from the radioactive materials forming the 'fall-out' of atomic and hydrogen bombs.

The original simplicity of the gene theory, based on the idea that each character of an organism is controlled by a single gene and that the body is a kind of mosaic, each part being paralleled by a gene in the chromosomes, has given place to the view that a single character depends on a large number of genes, and that a single gene may affect many characters. The effect of a gene is altered by the presence of other genes and all the genes combine to form what is termed the gene-complex. The effect of a major mutation of a single gene is usually modified by the combination of the other genes and therefore the gene-complex tends to evolve by small steps. A mutant gene may have a quite different effect in one gene-complex than in another, and hence have different effects in the offspring than in the parent. The need to elaborate the gene theory by the concept of the gene-complex has been criticised, particularly in Russia, as revealing a fundamental weakness. The theory has been compared to an inverted pyramid, whose foundation is inadequate to keep stable its large superstructure.

The chief Russian criticism of the gene theory, however, is the negligible part it allows to the environment in effecting changes in heredity. An hereditary change can occur only through a mutation of a gene or a reshuffling of the genes (or, rarely, through a change in the number of chromosomes). The genes cannot be affected by normal changes in the environment—assuming that fairly strong X-rays and mustard gas are not normal features of an environment.

The character of an organism is determined by its hereditary constitution and by the effects of the environment in which it develops. But the effects of the environment are not perpetuated in the offspring. Thus the baby of a woman who has acquired a deep sun-tan does not have a skin which is darker than normal.

Many experiments have been performed to test the effect of the environment in inducing hereditary changes, and they have all had negative or dubious results. For example, sixty generations of *Drosophila* (equivalent to 2000 years in human generations) have been bred in total darkness, without any effect on their powers of sight. During the course of evolution organisms have become stabilised and have not responded to considerable subsequent changes in climatic environment. The horse, for example, has exhibited no major evolutionary change during the past five million years.

Adherents of the gene theory believe that the heredity of an organism depends entirely on its parents and on its ancestors. They deny that hereditary changes can be induced by manipulating the environment. To the Russians this is an unacceptable doctrine.

* * *

The strange story of Soviet genetics is particularly illuminating because it raises fundamental issues concerning the nature and function of science as a whole.

Shortly after the Russian Revolution in 1917, as part of a vast expansion of scientific teaching and research, a chain of agricultural institutes was set up in the Soviet Union, including many genetical institutes, and these were closely linked with the new collective farms. The American geneticist, Muller, was invited to act as adviser in genetics and he took the first stocks of *Drosophila* to Russia in 1922. He spent several years there during the 1930's but left in 1937 because of his dissatisfaction with the political control of Soviet science. During this time, and indeed until 1948, nearly all university teaching in genetics was on orthodox, neo-Mendelian lines.

From about 1935 a conflict arose between neo-Mendelism and an alternative theory of genetics, known as Michurinism; the conflict ended in 1948 with the complete triumph of Michurinism, and resulted in the closing of several genetical institutes, the removal of a number of geneticists, and the prohibition of the teaching of neo-Mendelism in the Soviet Union.

Michurinism is really a form of neo-Lamarckism, and its fundamental tenet is that heredity can be altered by making suitable changes in the environment. Michurin (1855–1935) was a breeder of fruit trees and although, in the opinion of western geneticists, the theoretical basis of his doctrine was negligible, he was built up in Russia as a great and original thinker. His most prominent disciple was Lysenko.

Lysenko's chief aim was to improve Soviet agriculture, a desperately urgent task in view of the famines caused by the compulsory collectivisation of farms. He turned to practical value a discovery made earlier by Gassner, known as vernalisation. This is a process of treating the seeds of winter wheat, normally sown in the autumn, to make them crop in the summer when sown in the spring. The seeds are allowed to take up water, kept at a temperature just above freezing point for a few weeks, and then dried. As a result, when sown, they mature more quickly.

Lysenko claimed, as a result of his own experiments and those of his co-workers, that vernalisation could effect a permanent hereditary change. The claim was received with incredulity in the west, since vernalisation could not possibly affect the gene-complex, and it was suggested that insufficient care had been taken in selecting plants of genetic purity for the experiments; if pure lines were not used then natural selection might produce the effect which was ascribed to vernalisation. Moreover, in some of the experiments, proper controls and statistical tests had not been used.

Lysenko also claimed that hereditary changes in plants can be effected by grafting. Once again the strains used by Lysenko's followers were criticised as not sufficiently pure and, when the experiments were repeated in the U.S.A., they gave negative results.

On the basis of these experiments and guided by the doctrine of dialectical materialism, which is held to imply that there is a mutual interaction between an organism and its environment, Lysenko put forward the view that heredity is merely the concentrated influence of the environment during the organism's ancestral past, and that it is diffused throughout the body. He maintained that, by means of shock treatment, heredity can be de-stabilised or 'shaken' and that it is then possible to effect changes in it. Thus vernalisation gives rise to a shaken heredity and produces a permanent change in the response of the organism to its environment. Similarly hybridisation causes a shaken heredity, which accounts for the differences in the second hybrid generation (F_2 in Fig. 51). The theory does not explain, however, the 3 : 1 ratio, referred to contemptuously by Lysenko, with the other numerical Mendelian results, as the pea laws.

Lysenko was more convincing in his attack on neo-Mendelism than in his exposition of his own, somewhat vague theory. He denied that there was an organ of heredity and rejected the gene theory *in toto*. The concept of the gene was stigmatised as idealist, as opposed to a proper materialist concept, in the sense that the properties ascribed to the gene are metaphysical and incredible. The gene is a living and self-reproducing particle and yet it does not change or develop. It is like the unmoved mover of the scholastics. We shall see, in our discussion of positivism in the last chapter, that this need not be regarded as a serious criticism.

When the cell is in the 'resting' stage, the chromosomes disappear and they appear again only when the cell is about to divide. Supporters of the gene theory must assume that the linear arrangements of the genes are retained when the chromosomes disappear. There is an infrequent form of cell division, known as amitosis, in which chromosomes are not formed at all. Thus, argued Lysenko, although chromosomes may play an important part in heredity, the mechanism postulated by the gene theory is a naïve and over-simplified picture of a complex reality. The primary cause of hereditary change is the action of the environment on the whole organism, and it is ridiculous to explain it by random mutations or reshufflings of hypothetical genes.

The gene theory throws little light on embryology, and on the growth and development of an organism. If each cell of the body receives a complete and identical set of genes, how can the differentiation of tissues and organs be explained? The gene theory has been singularly unfruitful in stimulating practical experiments to improve domestic plants and animals. Most breeders in the west have continued along traditional and empirical lines with little assistance from academic geneticists.

Here, undoubtedly, was the main reason for the break of Lysenko and his followers with neo-Mendelism. They were preoccupied with the improvement of crops and animals and they were impatient of pure research which seemed to have no obvious application. They required a theory which would hold out a prospect of immediate practical results.

In the summer of 1948 at a meeting of the Lenin Academy of Agricultural Sciences, attended by about 700 scientific workers from every part of the Soviet Union, the relative merits of the two genetical theories were discussed at length. In the belief that the future of Soviet agriculture was at stake, it was decided that, since neo-Mendelism was of little practical value, Michurinism should be established as the guiding theory for future research. At the close of the discussion, Lysenko made it clear that the decision would receive the approval of the Central Committee of the Communist Party of the Soviet Union.

The decision to prohibit the teaching of neo-Mendelism in Russia was received in the west with profound disquiet. The international character of science had been repudiated and it was felt that the autonomy of science had been violated because a scientific theory had been condemned on other than scientific grounds. It was thought that Lysenko had been given official support largely because Marxism requires a belief that all problems can be solved by suitable control of the environment and that hereditary differences between individuals are of minor importance.

Many western geneticists, astonished at the cavalier dismissal of a well-established branch of biological science, came to the conclusion that Lysenko was a charlatan. Dr Harland met Lysenko in 1933 and remarked afterwards that discussing neo-

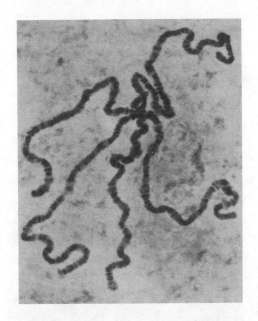

(a)

(b)

PLATE XIX. (a) Giant chromosomes of the fruit fly *Drosophila*. (b) The artificial disintegration of nitrogen, a photograph taken by P. M. S. Blackett using a Wilson cloud chamber. The almost parallel tracks are those of α-particles in nitrogen. An α-particle has entered a nitrogen atom, which has ejected a proton (the long track going to the left) and become oxygen (the short track veering to the right).

PLATE XX. Sir J. J. Thomson and Lord Rutherford.

Mendelism with him was like discussing the differential calculus with a man ignorant of arithmetic. On the other hand H. G. Morton, in his book *Soviet Genetics*, claimed that Lysenko was a scientist of the first rank.

In the Soviet Union all scientific research is controlled by the State, which provides all the necessary money. The State has, therefore, the power and indeed the right to close down any line of research of which it disapproves. This is a fundamental weakness of a society in which all power is vested in the State. Greater freedom and variety are possible in western society, where some power is distributed among individuals and individual organisations.

Marxists believe that the atmosphere of capitalism colours and distorts the aims and the very theories of western science. The argument, of course, cuts both ways; communism must likewise have an influence on Soviet science. Taking a long-term view, this need not necessarily be as wholly beneficial as Marxists maintain. Indeed, if science is merely an organ of communism, it must stand or fall with communism, just as Nazi racial 'science' shared the fate of Hitler and National Socialism.

* * *

The rise of Mendelism was at first thought to be fatal to Darwin's theory of natural selection. Evolution was considered to proceed by large variations, comparable with the Mendelian factors, and to be directed by some innate urge of the organism, rather than to proceed by small, random variations, sifted and selected according to their value in the adaptation of the organism to its environment. Bateson, in his presidential address to the British Association in Melbourne, Australia, in 1914, categorised Darwin's theory as of historical interest only and relegated it to the status of that of Lucretius. For twenty years he had been an outspoken critic of natural selection and by this date he had convinced his fellow biologists that the theory must be discarded.

The development of the gene theory, however, and the vast experimental work it inspired, showed that large variations are unusual and abnormal. The abrupt mutations of the evening primrose, which led de Vries to put forward his theory of discontinuous evolution, are the result of a peculiar increase in the number of chromosomes. It began to be realised that large variations are likely to be harmful because they tend to destroy the adjustment of the organism to its environment. Furthermore, the fossil record reveals gradual transitions, rather than abrupt changes in the evolution of organisms.

The re-establishment of natural selection took a mathematical form and was the work mainly of R. A. Fisher, J. B. S. Haldane and the American, Sewall Wright. Fisher, in his book *The Genetical Theory of Evolution* (1930) rejected such hypothetical agencies as an innate urge of an organism, capable of modifying the frequency or the direction of mutations. He accepted the occurrence of chance small variations and insisted on the inevitable effect of natural selection in perpetuating those that were favourable and in weeding out those that were harmful. He restored to natural selection the major role of controlling the direction of evolution. Taking as a fundamental theorem that the rate of increase of fitness of a population is proportional to its genetic variance, and using the experimental data of the rates at which mutations occur, he was able to account for the known rate of evolution.

The concept of the gene-complex and the idea that the effects of the mutation of a gene are reduced by the modifying effect of the remainder of the genes, confirmed the view that most variations are small. The innumerable possibilities of reshuffling of the genes led to the conviction that, even if no further gene mutations were to occur, there is a vast potential reserve of future variations in existing organisms. The strength of the gene theory is that it allows, on the one hand, for great genetical variation and, on the other, for great genetical stability.

Several instances of the actual operation of natural selection have been observed and perhaps the most striking is the case of the Peppered Moth. The moth is normally light in colour but it mutates into a dark (melanic) form. The lighter form is less conspicuous on lichen-covered trees in a rural wood and hence it is less likely to be eaten by birds than the darker form. On tree trunks darkened by soot, however, the darker form is less conspicuous, and has a stronger chance of survival. The spread of the darker form in the industrial regions of northern England appears to be wholly due to natural selection.

Natural selection is mechanistic but it simulates a purposive process because its effect is to produce more and more perfect adaptation of an organism to its environment. Too perfect an adaptation to an environment, however, may lead to extinction when the environment changes. The environment plays a decisive role in fixing, as it were, the pseudo-goal of natural selection.

This materialistic theory of evolution, now generally accepted in the west, is felt by many biologists to be only part of the truth and to be ultimately unsatisfying. In the words of Sir Alister Hardy, 'It is nonsense to consider the works of a Shakespeare, a Beethoven, or a Rembrandt, as the product of a machine.'[1] Professor Hardy suggests that there may be at work some evolutionary force at present unrecognised, such as telepathy. If there can be direct extra-sensory contact of one human mind with another there may be a similar telepathic force binding members of other species together in a common pattern of behaviour. Changes in behaviour effected by telepathy may result in changes in the gene-complex. There may, indeed, be a psychic animal world as yet unexplored. Such speculations, however, lie outside the present confines of biology.

[1] A. C. Hardy, *Telepathy and Evolutionary Theory*, J. Soc. Psychical Res. vol. xxxv (1950).

CHAPTER 20

THE STRUCTURE OF THE ATOM

THE CONCEPTION of atoms as tiny billiard balls, suggested by Dalton's symbolism and forming the basis of the impressive successes of the mathematical theory of gases, became patently inadequate towards the end of the nineteenth century.

Various attempts to classify the elements, in terms of similar properties, had been made from the time of Lavoisier. It was obvious, for example, that fluorine, chlorine, bromine and iodine formed a natural family. But significant progress was not made until Cannizzaro, in 1858, showed how atomic weights could be measured without any element of doubt.

Between 1863 and 1866 Newlands, a London industrial chemist, pointed out that, when the elements are arranged in the order of their atomic weights, they group themselves in octaves; the eighth, fifteenth, etc. elements have properties similar to the first; the ninth, sixteenth, etc. elements have properties similar to the second; and so on. The rule was only rough, however, and there were misfits. When it was discussed by the Chemical Society, one member asked Newlands whether he had considered arranging the elements in the order of the initial letters of their names.

Chemists became convinced of the periodic variation in chemical properties of the elements by papers published in 1869 by the German chemist Lothar Meyer and, more particularly, by Mendeléef, professor of chemistry at St Petersburg (now Leningrad). Mendeléef commanded attention because he used the periodic classification to make bold predictions. He said that there were gaps in his table, and these would be filled by hitherto undiscovered elements whose properties he described. Within twenty years the missing elements, gallium, scandium and germanium were isolated.

The periodic system of classification implied that atoms had a structure which varied periodically with increasing atomic weight. In his address as President of the British Association in

1894, Lord Salisbury selected the periodic law as one of the most promising pointers towards fruitful research.

There was also evidence that the structure of atoms was essentially electrical. Faraday's discovery in 1832–3 of the laws governing the splitting up of chemical compounds by an electric current, a phenomenon known as electrolysis, led him to the conclusion that 'the atoms of bodies which are equivalent to each other in their ordinary chemical action have equal quantities of electricity naturally associated with them'. In common with Davy, Berzelius and others, he regarded electricity as providing the force of chemical combination.

Faraday was not convinced of the existence of atoms and normally preferred to speak of the equivalent weights of elements, which are the weights combining with each other. He therefore did not stress an obvious deduction from the laws of electrolysis, that electricity itself must be atomic in character. The atomic unit of electricity was considered likely to be the charge carried by a single atom of hydrogen in electrolysis. It was estimated roughly by Johnstone Stoney in 1874 and given the name of electron.

Each kind of atom was known to emit its own characteristic blend of coloured light. When light was recognised as consisting of electromagnetic waves, the necessary inference was that atoms must contain some kind of electrical vibrators to generate the light and it was realised that Johnstone Stoney's electron might well be this vibrator.

The evidence then existing for the electrical structure of the atom seems plain enough today, but it was by no means so plain at the end of the nineteenth century, as Lord Salisbury's words to the British Association in 1894 can testify: 'What the atom of each element is, whether it is a movement, or a thing, or a vortex, or a point having inertia, whether there is any limit to its divisibility, and, if so, how that limit is imposed, whether the long list of elements is final, or whether any of them have a common origin, all these questions remain surrounded by a darkness as profound as ever.'

*   *   *

The phenomenon which led to the direct discovery of the electrical structure of the atom was the conduction of electricity through rarefied gases. A gas such as air, at normal atmospheric pressure, will not conduct electricity except as a short spark. But if the pressure of the gas is reduced to about $\frac{1}{100}$ of atmospheric pressure, a silent discharge passes in the form of a luminous glow, such as can be seen in the long discharge tubes now used for street and interior lighting. As the pressure of the gas is still further reduced, the luminous glow begins to retreat away from the cathode (i.e. the metal electrode at one end of the tube connected to the negative pole of the source of electrical supply); the glow begins to break up into striations and a new glow appears in the neighbourhood of the cathode. A whole series of complex and beautiful effects are observed.

At a pressure of about $\frac{1}{100,000}$ of an atmosphere, all glows have disappeared, the gas is dark, and the glass of the tube fluoresces with a faint, green light. In 1869 Hittorf found that a body, placed in this dark discharge, cast a shadow on the wall of the tube remote from the cathode. Invisible rays appeared to be coming from the cathode and in 1876 Goldstein gave them the name cathode rays (Kathodenstrahlen).

Sir William Crookes took the view that the cathode rays consisted of tiny particles and the fact that they could be deflected by a magnet, together with the direction of their deflection, indicated that they had a negative electric charge. He expressed the belief that these particles might be a primordial state of matter of which all atoms were composed.

The German investigators were unanimously opposed to the view of Crookes that the cathode rays were particles; they considered them to be æther waves, rather like the electromagnetic waves discovered by Hertz in 1887. This belief was strengthened by the further discovery of Hertz that the cathode rays could penetrate gold leaf.

If cathode rays consist of negatively charged particles they should be deflected by an electric field. Hertz tried to detect such an effect without any success. This, to his mind, was decisive evidence against the negative particle theory.

There was a feeling among German physicists that corpuscular theories were an archaic survival from the seventeenth and eighteenth centuries, and were alien to the science of their day. They overlooked the great advantage of a corpuscular theory, that it could be put into mathematical form and suggest a series of quantitative experimental tests.

*   *   *

The establishment of the nature of the cathode rays, and the realisation of its far-reaching implications for the structure of the atom, was mainly the work of Sir J. J. Thomson and his team of co-workers at the Cavendish laboratory in Cambridge, the first important research school in Britain. The laboratory was founded in 1874 and had as its earliest directors Clerk Maxwell and Lord Rayleigh. J. J. Thomson succeeded Rayleigh as Cavendish professor in 1884 at the age of 27. It was an unexpected appointment—apparently a bow at a venture on the part of the electors—and one college tutor was said to have remarked that things had come to a pretty pass when mere boys were made professors. During J. J. Thomson's tenure as professor, the Cavendish became the most famous laboratory in the world.

Thomson was not a deft experimenter, and stories are told of his laboratory assistant, Everett, shielding the fragile glass discharge tubes, which he had made, from unnecessary handling by their designer. Thomson's success rested on his ability to perceive with clarity a fundamental problem, to formulate an hypothesis, to conceive experimental tests and to marshal a concerted attack from all angles by the team working under him.

The great period of the Cavendish laboratory did not commence until eleven years after his appointment as its director. During those preparatory years he occupied his mind with the discharge of electricity through gases and wrote a book summarising the mass of diverse and bewildering phenomena associated with it.

The event which touched off the outburst of creative activity in the Cavendish laboratory was the discovery of X-rays by Röntgen in 1895. At this time Cambridge began to give degrees to research students for two years of work in the university; also the Commissioners of the Exhibition of 1851 changed their regulations to enable holders of their scholarships to spend two years at Cambridge instead of one year at their own university and one year at Cambridge or elsewhere. These scholarships were awarded to students from commonwealth universities and drew their funds from the profits of the Exhibition of 1851. The result was that Cambridge began to attract graduates from abroad, who formerly went to German universities, where research degrees had for years been available, and to attract also the cream of the physicists from the commonwealth universities. Rutherford arrived from New Zealand and Townsend from Dublin on the same day in 1895; Langevin came from France and Zeleny from the U.S.A.

Everything conspired towards what proved to be a veritable renaissance of British physics. A new sub-atomic territory stood waiting to be explored and a leader, with a mind prepared, found a team of research workers ready for his direction. Between 1896 and 1900 these workers published over one hundred papers. The Cavendish laboratory had then such meagre financial resources that the purchase of an instrument costing as much as £5 was preceded by long and careful deliberation and it acquired the nickname of the 'string and sealing-wax' laboratory. By contrast, a single machine built today for accelerating charged particles may cost thirty-five million pounds.

*   *   *

The discovery of X-rays by Röntgen, professor of physics at Würzburg, was made while he was experimenting on the discharge of electricity through gases at very low pressures. He discovered that a screen of barium platinocyanide, used normally for investigating the invisible ends of the spectrum, fluoresced brightly when placed near to the discharge tube. On interposing his hand between the tube and the screen, shadows of the bones were cast upon the screen. Photographic plates, wrapped in brown paper, became fogged when in the vicinity of the tube, and a silhouette of a key, resting on a box of plates, appeared on the plates after development.

It was clear that a very penetrating radiation was coming from the tube, and it appeared to originate at the fluorescent walls. The earlier investigators must often have produced X-rays without being aware of it, and Röntgen made the most of his good fortune by performing all the obvious experiments before announcing his discovery. It is said that even his wife did not share his secret before the public announcement was made.

The discovery created an immense sensation, not only in scientific circles, but among the general public. A misconception arose that the great penetrating power of X-rays could outrage modesty; as one writer flippantly put it:

> I hear they'll gaze
> Through cloak and gown—and even stays,
> These naughty, naughty Roentgen rays.

The rays could be produced quite simply by apparatus possessed by every physics laboratory, and were examined eagerly; as Rutherford put it, every professor of physics in Europe was on the warpath. Within a few weeks of the announcement of the discovery, X-rays began to be employed in hospitals for examining fractured bones, and for tracing foreign objects in the bodies of patients.

J. J. Thomson investigated the effect of X-rays on the electrical conductivity of a gas and found, to his immense satisfaction, that they caused a gas at ordinary atmospheric pressure to become a conductor of electricity. This opened up a new field of experiment. A gas in a conducting state could now be manipulated at ordinary pressure and an electric current could be passed through it, using an electrical pressure of only a few volts instead of the tens of thousands of volts which had hitherto been necessary.

The conductivity of a gas persists for a short time after the X-rays have been shut off. Thomson found that the conductivity could be destroyed by passing the gas through glass wool, a filtering process which suggested that the conductivity was due to particles. The conductivity could also be destroyed by passing the gas between charged plates, and this indicated that the particles were charged; the plates attracted the charged particles out of the gas.

It was known that positively and negatively charged particles, flowing in opposite directions, carry an electric current through a liquid solution. The charged particles are called ions and they are formed by the spontaneous splitting up of the molecules of the solute into atoms or groups of atoms, a process called ionisation. For example, a solution of sodium chloride, NaCl, contains the ions Na^+ and Cl^-.

A similar process of ionisation appeared to be occurring in a gas, under the influence of X-rays. Thomson and Rutherford found that the current through a gas, irradiated by X-rays of a constant intensity, between two metal plates, could not be increased beyond a certain maximum or 'saturation' value, however high the voltage applied across the plates. They assumed that the plates were then collecting the ions as fast as they were being produced by the X-rays.

An experiment of this kind, although it sounds so simple, presented considerable technical difficulties. Little was then known about how the X-rays were generated, and there was the problem of keeping them at constant intensity. The current was much too small to be measured with an ordinary galvanometer and it was necessary to employ a quadrant electrometer, a temperamental and exasperating instrument. Thomson said that he once went to a lecture during which it was used by its inventor, Lord Kelvin, and its deflection was invariably in the opposite direction to what Lord Kelvin said it should be.

Thomson became convinced that the cathode rays in a discharge tube were negative gaseous ions. He had no suspicion that they were other than negatively charged atoms until he measured their deflection in a magnetic field, an experiment already performed by Sir Arthur Schuster ten years earlier. The deflection was much greater than Thomson expected and suggested that the particles were lighter than atoms, although the evidence was by no means decisive. The large deflection was accounted for by Schuster, not by the small mass of the particles, but by their comparatively low velocity. Thomson believed that the sharp outlines of his beam of cathode rays indicated that the rays were travelling at a much higher velocity than Schuster imagined.

The obvious next step was to devise a method of measuring their velocity. The heat produced by the cathode rays was measured by causing them to strike an instrument (a thermopile) inside the discharge tube, and the total charge was also measured. By this means Thomson confirmed that the velocity was as high as he suspected. The experiment, together with the deflection of the rays by a magnetic field, enabled him to estimate the ratio of the charge e to the mass m of the cathode ray particles. His value for e/m was about 10^7 whereas the corresponding ratio for the lightest known atom, hydrogen, was 10^4. He was inclined to infer that the cathode ray particles were 1000 times lighter than the hydrogen atom.

Among the considerations which persuaded Thomson that the cathode ray particles were much smaller than atoms were the experiments of Hertz and Lenard showing that cathode rays can penetrate gold leaf or an aluminium window. No atom could do this; the particles must be much smaller. Also Thomson's own experiments showed that the deflection of cathode rays by a magnetic field was independent of the nature of the gas in the discharge tube. This implied that the cathode rays could not be charged atoms of the gas because otherwise a different deflection would be expected for each gas.

Thomson announced his results at a Friday Evening Discourse at the Royal Institution on 30 April 1897. Few of his hearers realised the importance of his conclusion, that particles existed which were lighter than the lightest known atom, and one of them said later that he thought the lecturer was pulling their legs. Thomson called the particles corpuscles and adhered to this name for some years; they are now called electrons.

Thomson's elation at this time is recorded by his biographer, Lord Rayleigh, son of the former Cavendish professor, then an undergraduate at Trinity. Thomson insisted on walking with him along King's Parade to his rooms in Whewell's Court, and explained that the cathode rays had turned out to be particles smaller than atoms. Later, when Rayleigh was working in the Cavendish, he remarked to Thomson how difficult he found research. 'Yes', replied Thomson, 'that is why so much credit is given for a scientific discovery.'

Hertz had been unable to produce a deflection of the cathode rays with an electric field, and Thomson found at first that he also could obtain no lasting effect, but he did detect a slight flicker in a narrow beam of cathode rays, passing between two metal plates, when a voltage was first applied between the plates. His experiments with gaseous ions gave him an insight into what was happening. The cathode rays were striking molecules of the gas in their path, ionising them, and the resulting positive and negative ions were attracted to the metal plates, thereby nullifying the electric field.

The cure was to remove most of the gas, but this was more easily said than done. The Töpler mercury pump employed by Thomson had to be worked up and down by hand. When the discharge passed, condensed gas on the walls of the tube and in the metal electrodes was released and the pump could not work fast enough to maintain the vacuum. However, by passing the discharge day after day and by continually pumping, a sufficiently good vacuum was obtained to enable a sustained deflection of the cathode rays to be produced.

Thomson now performed his most famous experiment in which he counterbalanced the effects on the cathode rays of magnetic and electric fields, applied simultaneously, so that the resulting deflection was zero. The result of the experiment was to confirm his previously obtained values of e/m and of the velocity of the cathode rays. The apparatus, shown in Fig. 57, is the forerunner of the cathode ray tube now manufactured by millions annually throughout the world for television and other purposes.

Thomson used to say that the trouble with experiments is that you spend long and laborious hours, perhaps weeks and months, getting the apparatus to work; once this is achieved the experiment is over far too quickly—perhaps in a matter of minutes.

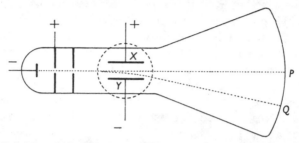

Fig. 57. J. J. Thomson's cathode ray tube. The dotted circle represents a pair of coils carrying a current which produces a magnetic field, causing the beam of cathode rays to be deflected from P to Q. This deflection was annulled by means of an electric field applied between the plates X and Y.

It was just possible that the value of e/m for the cathode ray particles resulted, not from m being $\frac{1}{1000}$ of the mass of a hydrogen atom, but from e being a thousand times the unit atomic charge calculated from electrolysis. Several experiments were therefore performed in the Cavendish laboratory to measure e direct. These experiments were carried out independently by Townsend, Thomson and H. A. Wilson, but the underlying method was substantially the same. A cloud was formed in a gas ionised by, say, X-rays, in such a way that each tiny water drop condensed on a negative ion; condensation occurs more readily on negative ions than on positive ions. The size of the drops could be calculated from the rate at which the cloud subsided under gravity; a small drop falls more slowly than a large one. The number of drops in the cloud was obtained by dividing the total mass of the cloud by the mass of each drop. The charge, e, on each drop, was equal to the total charge on the cloud divided by the number of drops. The value obtained for e was approximately equal to the charge carried by ions in liquids. No doubt remained that the high value of

e/m for the cathode ray particles was due to the very small value of m.

The existence of electrons was now established and their approximate charge and mass were known. Thomson proceeded to show that their occurrence was universal. Electrons were obtained not only in the discharge tube and by irradiating a gas with X-rays, but from metal surfaces subjected to ultra-violet radiation and from incandescent filaments. They appeared to be a constituent of all matter.

Thomson began to speculate on the structure of the atom. In its normal state the atom is uncharged and hence must contain equal quantities of positive and negative electricity. Thomson imagined an atom to consist of a sphere of positive electricity in which electrons were embedded— the so-called 'positive jelly model'.

To illustrate his model atom, Thomson recalled an experiment in which floating magnets arranged themselves in rings round a central magnet. The magnets were supported vertically in corks with, say, their north poles just above the surface of the water, and, in the middle of the water surface, there was a powerful magnetic south pole. The magnets under their mutual repulsion and the attraction of the central pole, formed an inner ring of five, surrounded by a second ring of fourteen and a third ring which became complete with twenty-six magnets. Thomson imagined the electrons in his atom to set themselves in similar rings or shells and suggested that this was the fundamental explanation of the periodic law of chemical properties. Atoms with complete shells of electrons were chemically inert; atoms having one electron too few in their outer shell had similar chemical properties, whatever the number of completed inner shells they contained; and so on.

Thomson's work has never received the recognition in Germany that it deserves, perhaps for the same reason that Wellington was never admitted to be a good general by Napoleon. The Germans had led the field in the investigation of the discharge of electricity in gases throughout most of the nineteenth century, and several of them were hot on the trail at the same time as Thomson. Wiechert measured e/m for the cathode rays and suggested, in a lecture in January 1897, that the particles had masses 2000 to 4000 times smaller than the hydrogen atom. Also in 1897 Kaufmann deflected cathode rays in a magnetic field and obtained an accurate value of e/m. But they did not carry their German contemporaries with them and they failed to realise the implications of their experiments.

*   *   *

In February 1896 a discovery even more pregnant and far-reaching than that of X-rays was made by Henri Becquerel in Paris. Becquerel saw the first X-ray photographs sent by Röntgen to the French Academy of Sciences on 20 January 1896. Since X-rays appear to come from the phosphorescent patch on the glass tube where the cathode rays strike it, Becquerel decided to investigate whether all phosphorescent substances emit similar rays. He discovered that salts of uranium affect a photographic plate wrapped in a light-tight envelope. In one of his experiments the uranium salt was placed on an aluminium medallion, resting on a photographic plate in an envelope, and was left for several days in a dark drawer. When the plate was developed there appeared upon it an image of the medallion. Thus uranium salts emit a penetrating radiation. The phenomenon was named radioactivity by Marie Curie.

Marie Curie (1867–1934) is the only individual who has received a Nobel prize on two occasions, her daughter being the only other woman to receive a Nobel prize for science. Her maiden name was Marya Skodovska and she was born in Warsaw. At the age of 23 she came to Paris, the city which symbolised freedom for the oppressed Poles of that time, and studied science at the Sorbonne, living, half-starved, in an attic. She met Pierre Curie, an able physicist having a poorly-paid post at the Paris School of Physics and Chemistry, and married him in 1895. They spent their honeymoon cycling through the lovely countryside of France.

In her search for a subject for a doctoral thesis, Marie boldly selected the newly-discovered phenomenon of radioactivity. The radiation from uranium, like X-rays, has the property of rendering a gas electrically conducting and hence causes the charge to leak away from an electroscope (Fig. 58). Utilising this property, Marie showed that the amount of the radiation was proportional to the weight of the uranium present; it did not depend at all on the weights or nature of the other elements with which the uranium might be combined. Radioactivity was therefore an atomic property. Marie began a systematic investigation of other metals and found that thorium also was radioactive.

She discovered that pitchblende, the ore from which uranium is extracted, was four times as radioactive as pure uranium. The ore must therefore contain an unknown radioactive element. To convince chemists of the existence of this element it was necessary to isolate it and to find its atomic weight. Pierre Curie now laid aside his own researches to assist Marie in this formidable task.

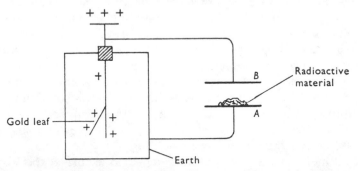

Fig. 58. The Curies' apparatus for testing the radioactivity of a material. The electroscope is given a charge (marked +) and the gold leaf is repelled into the position shown. When radioactive material is placed on plate A, the air between A and B becomes conducting. The charge on the electroscope leaks away and the gold leaf falls.

A ton of waste from the Bohemian uranium mines was presented to them free by the Austrian government but they had to pay for its transport. The peaty brown ore, mixed with pine needles, arrived in sacks on a heavy coal wagon at the large abandoned shed, with a leaky glass roof in the grounds of the School of Physics in Paris, where four years of arduous work was to take place.

The constituents of the pitchblende had to be separated by chemical means and each constituent tested for its activity by an electroscope, the inactive constituents being rejected and the active constituents subjected to further analysis. The sheer physical labour involved is evident from the fact that the Curies eventually analysed six tons of pitchblende and from this they extracted $\frac{1}{10}$ gram of pure radium. They handled about 40 lb. of matter at a time, working in an atmosphere of smoke and fumes from a bubbling cauldron. In the winter the shed in which they worked was cold and wet; in the summer, because of its glass roof, it was unbearably hot. 'And yet', wrote Marie, 'it was in this miserable old shed that the best and happiest years of our life were spent, entirely consecrated to work. I sometimes passed the whole day stirring a boiling mass, with an iron rod nearly as big as myself. In the evening I was broken with fatigue.'[1]

As they neared the end of their task, Pierre and Marie would go back to the shed in the evening to see the phials containing the radioactive liquors gleaming like glow-worms in the dusk.

In 1902 they announced that they had isolated a new element, to which they gave the felicitous name of radium, and that its atomic weight was 225. The following year they received a Nobel prize, jointly with Becquerel. The money was a godsend and one of their first acts was to engage a laboratory assistant at their own expense. When, shortly afterwards, it was decided to manufacture radium in America, because of its use in the treatment of cancer, the Curies decided not to patent their process of extraction, as it would be 'contrary to the spirit of science'.

In June 1903 Pierre Curie was invited to lecture at the Royal Institution in London and later Pierre and Marie, under the benign care of Lord Kelvin, then aged 79, attended a banquet. Pierre was fascinated by the diamonds and jewels worn by the women; he found himself calculating how many scientific laboratories could be bought with them.

In April 1906 Pierre slipped on a wet road in Paris, and the rear wheels of a heavy horse dray passed over his head. The death of her husband was, for Marie, almost a paralysing blow. Thereafter nothing desirable seemed left in her life apart from her scientific work. Even with her two daughters she was undemonstrative and could refer to their father only with the most painful effort. Usually dressed in black, she became a pale, taciturn, earnest, indomitable figure. Whenever Rutherford met her he always remarked how ill she looked. She died of pernicious anaemia; the cells in the marrow of her spine had been destroyed by the radiations with which she worked for over thirty years.

* * *

The unravelling of the complicated processes of radioactivity, and their explanation by a theory, supported and checked at every point by careful experiment, was primarily the work of Ernest Rutherford (1871–1937), later Lord Rutherford of Nelson. This

[1] Eve Curie, *Madame Curie.*

required a scientific imagination of a different order from that possessed by Marie Curie, as she herself recognised: 'I would advise England', she said to newspaper reporters in 1913, 'to watch Dr Rutherford; his work in radioactivity has surprised me greatly.'

Rutherford was an experimental physicist of supreme genius—of the same calibre as Faraday. His scientific ideas and experiments had a massive directness and simplicity. He had a boundless zest for the game of keeping ahead of the others working in his own field: 'I have to keep going', he wrote to his mother in New Zealand, 'as there are always people on my track.' He was never happier than when working in his laboratory. He once remarked to one of his research students, 'I *am* sorry for the poor fellows that haven't got *labs* to work in'. He had a sure instinct for experiments he could trust and would write in his laboratory notebooks, 'Very accurate experiment' or 'No good' or 'Good'. It is remarkable that he made so few mistakes and that he was so seldom tempted into blind alleys.

He began his investigations into radioactivity in 1897 in the Cavendish laboratory. He examined the radiations from uranium salts by means of their effect in ionising a gas, that is by their generation of positively and negatively charged gaseous particles or ions, thereby making the gas an electrical conductor. He discovered that the intensities of ionisation, measured by the rate of leak of a charged electroscope, near to the surface of the uranium salt was very much greater than that a few centimetres away. With his typical flair for seizing upon the implications of a small but significant fact, he deduced that the radiations were complex and consisted of at least two different types. The less penetrating he called α-radiation and the more penetrating β-radiation. He made a careful study of the absorption of α-radiation by different substances and found, for example, that the radiation could be completely absorbed by a thickness of a few centimetres of air or a stout sheet of paper. This work gave him a thorough grasp of the properties of the α-radiation, which was to serve as the tool or probe in much of his work on the structure of the atom. It was not long before the β-radiation was shown to consist of J. J. Thomson's corpuscles or electrons, and a third radiation, γ-rays, similar to X-rays, was found to be emitted during some radioactive changes (Fig. 59).

In September 1898 Rutherford left Cambridge and took up an appointment as research professor in physics at McGill University, Montreal. Here he was asked to suggest a topic for research for R. B. Owens and he selected the radiations from thorium which, like uranium and radium, is radioactive. Owens soon found that the radiations behaved in a most capricious manner and seemed to be peculiarly sensitive to draughts. Something was being emitted by the thorium salt which could be blown about by the air. It seemed to have the properties of a radioactive gas, and Rutherford called it an emanation. The emanation was led away from the thorium salt into a closed vessel and its rate of decline of activity, in causing the discharge of an electroscope, was measured. Its activity fell to half value in about a minute.

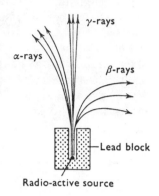

Fig. 59. The three types of rays emitted by a radioactive substance. The α- and β-rays are being deflected by a magnetic field.

Bodies in contact with the emanation were found to become temporarily radioactive. At first the phenomenon was called excited or induced radioactivity and J. J. Thomson suggested that it might be caused by the condensation of the emanation. But Rutherford showed that this could not be true because the rate of decay of its activity to half value was $11\frac{1}{2}$ hr., as compared with 1 min. for the emanation. He ascribed it to an active deposit from the emanation.

Similar phenomena were discovered to be associated with the radioactivity of radium. The Curies possessed the only radium salts in the world at this time and these salts were found to emit radium emanation, which also gave rise to excited radioactivity.

In the summer of 1900 Rutherford went to New Zealand to marry Mary Newton, whom he had met in his college days at Christchurch, and on his return to Montreal he was joined by Frederick Soddy, a young Oxford chemist. He and Soddy set to work to discover the chemical nature of thorium emanation. Their chief difficulty was the minute quantity they had to work with, less than one hundred thousandth of a cubic centimetre, but the emanation always signalled its presence unmistakably by its electrical effect. It appeared to resist the attacks of all of the most powerful chemical reagents. Only one family of gases behaved in this manner—the inert gases, discovered a few years earlier. When Sir William Macdonald, the benefactor of McGill, presented the university with a liquid air machine, the emanation was shown to be a genuine gas because it condensed in a metal tube immersed in liquid air. Sir William Macdonald was an eccentric millionaire, who made his money out of tobacco, but so great was his aversion to smoking that, whenever he was seen approaching the laboratory, Rutherford had to rush to open the windows.

The thorium emanation, apparently an inert gas, was found to be given off, not directly from thorium, but via an intermediate substance, called by Rutherford and Soddy thorium X. This could be separated from thorium by simple chemical means and had a half period of four days.

Thorium, thorium X and thorium emanation all gave off α-radiation. Rutherford succeeded in the difficult task of deflecting the α-radiation by a magnetic field and concluded that it consisted of positively charged particles with masses about the same as that of the lightest atoms hydrogen or helium. The fact that helium is found in radioactive minerals pointed to the probability of α-particles being charged helium atoms. The speed with which the α-particles are emitted is of the order of 10,000 miles per second, and they have immense energy. Rutherford saw in this energy the source of the heat which is generated continuously by radioactive substances, in apparent defiance of the principle of the conservation of energy. Radium generates enough heat to raise its own mass of water from freezing- to boiling-point every hour.

The mechanism of radioactivity was now becoming clear in Rutherford's mind. The atoms of thorium spontaneously disintegrate, hurling out a part of themselves in the form of α-particles, and they are transformed into a new kind of atom, thorium X. Thorium X is similarly transformed into thorium emanation, and thorium emanation into thorium active deposit.

The idea of atoms spontaneously exploding was so startling and revolutionary that Rutherford tried it out on his colleagues and some of them expressed disquiet that its publication might bring discredit on McGill university. It was put forward, in the joint names of Rutherford and Soddy, in a paper published in the *Philosophical Magazine* in September 1902.

The theory met with widespread incredulity and criticism. Professor H. E. Armstrong asked sarcastically why it was that radioactive atoms were seized with 'an incurable suicidal mania'. Lord Kelvin suggested that the vast amount of energy given out by radioactive substances had its source, not in atomic disintegration, but in the absorption of ætherial waves. As one of the creators of nineteenth-century physical science, he found it impossible to entertain the idea of disintegrating atoms.

To these criticisms Rutherford replied: 'The value of any working theory depends upon the number of experimental facts it serves to correlate, and upon its power of suggesting new lines of work. In these respects the disintegration theory, whether or not it may ultimately be proved correct, has already been justified by its results.'

* * *

Rutherford's theory of atomic disintegration was an important step towards a knowledge of the structure of atoms. When radioactive atoms throw out an α-particle, that is a charged helium atom, their atomic weight is reduced and they move down the periodic table into a new family of elements with different chemical properties.

Rutherford left McGill in 1907 and became professor of physics at Manchester university. In those days Europe was the centre of scientific activity; Canada and the U.S.A. were on the periphery and Rutherford was conscious of his isolation in Montreal. He was therefore glad to be back in England and was fortunate to find in Manchester a new, well-equipped laboratory, a skilful colleague in Geiger, and an outstanding laboratory assistant in William Kay.

Rutherford now began to devote himself to a thorough investigation of α-particles. He and Geiger developed an electrical method of detecting single α-particles—the forerunner of the Geiger counter, and an amazing technical feat at that time. But Rutherford came to prefer the simpler method, discovered by Crookes, of counting α-particles by viewing through a low-power microscope the tiny scintillations or flashing points of light which they made on striking a zinc sulphide screen.

An α-particle is a high-speed projectile of enormous energy in comparison with its size, and normally it proceeds in a straight line, ploughing through the outside of any molecules of air in its path. But the fact that it is sometimes slightly deflected shows that it must, at times, approach sufficiently near to the centre of a molecule or atom to be within the influence of extremely powerful electric forces.

Rutherford perceived that the scattering of α-particles was worth further investigation, and he allotted this task to Geiger and Marsden. They shot α-particles from a preparation of radium at a thin sheet of gold foil and determined the directions of scatter by the scintillation method. They found that the majority of the α-particles were scattered through only a few degrees, but about 1 in 10,000 actually bounced back from the gold foil. The model of the atom then generally accepted was a diffuse sphere of positive electricity in which electrons were embedded. Rutherford said that the bouncing back of an α-particle from such an atom was 'as incredible as if you had fired a fifteen inch shell at a sheet of tissue paper and it came back and hit you'.

He pondered over these results for a year and then one day, just before Christmas in 1910, as Geiger relates, 'he came into my room obviously in the best of tempers, and told me that now he knew what the atom looked like and what the strong scattering signi-fied'. The atom had a tiny massive, positively charged nucleus, with negative electrons circling round it—a kind of miniature solar system. It was possible to work out mathematically, assuming an inverse square law of repulsion between nucleus and α-particle, the probability of scattering through various angles; for example, eight times as many particles should be scattered between 60° and 120° as between 120° and 180°. During the next twelve months, by accurate experiments, Geiger and Marsden confirmed that the scattering did, in fact, conform with this theory. This was a beautiful example of a revolutionary discovery made by simple but ineluctable reasoning, based on experimental facts.

The fundamental structure of the atom was now established. Rutherford, with a broad grin, said of his critics: 'Some of them would give a thousand pounds to disprove it.'

* * *

The next step in the investigation of the structure of the atom was to determine how the positive charge on the nucleus increases from element to element in the periodic table. The scattering experiments of Geiger and Marsden gave a charge of roughly + 100 for the nucleus of the gold atom (the unit of charge being taken as the charge of the electron). It seemed possible, therefore, that the nuclear charge might be the ordinal number of the atoms in the periodic table, for example hydrogen 1, helium 2, lithium 3, and so on up to uranium 92. But the idea was speculative and unsupported by accurate experimental data. In order to provide such data H. G. Moseley (1887–1915), who was working under Rutherford at Manchester in 1913, decided to examine the X-ray spectra of as many elements as possible.

Until 1912 the true nature of X-rays was unknown. It seemed probable that they were waves of very short wavelength, but they could not, like light, be diffracted by the most closely-ruled gratings. A grating usually consists of a piece of glass on which are ruled many thousands of lines to the inch and this causes light to spread, that is to be diffracted. In 1912 von Laue, of the University of Munich, suggested that crystals, with their regularly spaced atoms, might behave as gratings and cause X-rays to be diffracted. The suggestion was fruitful. X-rays were shown to consist of waves, and methods were developed for measuring their wavelengths.

X-rays are produced wherever fast-moving electrons hit a substance. They include, besides a general radiation covering a whole range of wavelengths, like the spectrum of white light, certain more powerful beams giving rise to spectral lines on a photographic plate, characteristic of the substance emitting them. Moseley mounted a series of elements on a trolley in an X-ray tube and moved them so that they were struck in turn by a stream of electrons. The resulting X-rays were examined by a crystal. He discovered that there was a regular shift of the spectral lines from element to element. Each element, beginning with hydrogen, can be given a number, 1, 2, 3, etc., which is proportional to the square root of the frequency of one of the spectral lines. This number is now called the atomic number and it is equal to the positive charge on the nucleus of the atom and also to the number of planetary electrons. Hydrogen has a nucleus of charge $+1$ and a single revolving electron; helium has a nucleus of charge $+2$ and two revolving electrons; lithium has a nucleus of charge $+3$ and three revolving electrons; and so on.

Moseley pursued his research with daemonic energy, as though he had a premonition that his life would be brief. When his apparatus was functioning well he would work throughout the night and be seen leaving the laboratory by those who arrived early in the morning. His death in Gallipoli in 1915, from the bullet of a Turkish sniper, was a tragic object lesson to the British government. The national resources of scientific manpower were husbanded more carefully in the Second World War.

Moseley's discovery enabled the elements to be classified in a new and more satisfactory manner. Atomic number is more significant than atomic weight because chemical properties are determined by the number of planetary electrons in an atom. If the elements are arranged in ascending order of their atomic weights certain anomalies occur in the periodic classification: potassium precedes argon instead of following it, and the same is true of nickel and cobalt. If, however, the elements are arranged in the order of their atomic number, these anomalies disappear. Moseley found gaps in the table which have since been filled by hafnium (72) and rhenium (75).

The power of Moseley's X-ray technique was strikingly illustrated in 1914 when Georges Urbain, who had devoted his whole life to the study of the rare earths, brought preparations to Moseley for analysis. Moseley found what elements they contained and estimated their relative proportion, thereby confirming in a few days the work of twenty years by chemical methods. Urbain wrote later that Moseley's law gave the final exposition of the classification of the elements: '[sa loi] terminait un des plus beaux chapitres de l'histoire des sciences'.

*   *   *

The work of Moseley represented the end of a chapter, but by no means the end of the story. Although the structure of the atom was represented with satisfying simplicity by atomic number which is equal to the charge on the nucleus and to the number of planetary electrons, there was still no explanation of atomic weights. The first ten elements are as shown in the following table.

If all atoms are composed of the same unit particles, it might be expected that their weights would be multiples of that of the lightest atom, hydrogen, and hence that the atomic weights would be whole numbers. The departure from whole numbers is more marked with the heavier elements than with the lighter.

The clue to the solution of this problem was first obtained by

J. J. Thomson in the years 1910–11, when he was investigating the positive rays, which travel in the opposite direction to the cathode rays in the discharge tube. The positive rays can be separated from the cathode rays by boring a hole in the cathode, through which they stream. Thomson succeeded in measuring their deflections in magnetic and electric fields and hence determined their masses. The masses differ for different gases and the positive rays were found to consist of atoms which had lost one or more planetary electrons.

Atomic weights

Atomic number	Element	Atomic weight
1	Hydrogen	1·0080
2	Helium	4·003
3	Lithium	6·940
4	Beryllium	9·02
5	Boron	10·82
6	Carbon	12·010
7	Nitrogen	14·008
8	Oxygen	16·000
9	Fluorine	19·00
10	Neon	20·183

Thomson's experiments showed that neon, the tenth element in the above table, of atomic weight 20·183, consists of two kinds of atoms, of masses 20 and 22. The existence of atoms having identical chemical properties, but different masses, was realised by Soddy from his experiments on radioactivity and he called them isotopes, a term derived from the Greek, and indicating that the atoms had the same place in the periodic table. Soddy deduced that lead, which is the end point of the uranium radioactive series, should have an atomic weight of 206 while the lead from the thorium series should have an atomic weight of 208; his prediction was confirmed by experiment. Since ordinary lead has an atomic weight of 207·2, it must consist of a mixture of isotopes.

J. J. Thomson's method of measuring the masses of atoms was elaborated by Aston and, when his apparatus was perfected in 1919, scarcely a week passed without the announcement of new isotopes. The most complex element is tin, with no fewer than ten isotopes, and the ninety-odd elements consist of about 275 stable isotopes.

At the time of Aston's researches atoms were regarded as composed of two fundamental particles having a strange asymmetry: the electron with a negative electric charge, and the proton with an equal positive charge, but with a mass 1840 times as great. To these a third particle was added—the neutron, with zero charge and a mass equal to the mass of a proton.

The hydrogen atom of mass 1 has a nucleus consisting of a single proton, round which revolves a single electron. A comparatively rare isotope, heavy hydrogen of mass 2, has a nucleus consisting of a proton and a neutron, round which revolves a single electron.

The second lightest atom, that of helium, has a nucleus of charge 2 and mass 4, consisting of two protons and two neutrons, and it has two revolving electrons.

In a similar manner the structure of the atoms of all the elements can be explained. The heaviest naturally occurring element, uranium, has a nucleus of charge 92 and 92 revolving electrons. It has several isotopes, the commonest having masses of 238 (92 protons and 146 neutrons), and of 235 (92 protons and 143 neutrons).

The masses of all atomic nuclei are not exactly what would be expected from adding up the masses of the protons and neutrons they contain. This has proved to be of great theoretical importance. Astronomers account for the radiation of the sun and stars by the building up of more complex elements from hydrogen, with a slight annihilation of mass and a consequent release of an enormous quantity of energy. The atomic bomb and the nuclear reactor also derive their energy from the annihilation of mass.

* * *

Once it was clear that the nucleus of an atom was made up of particles, Rutherford began to speculate on the possibility of knocking out one or more of these particles, and hence of transmuting one element into another. He naturally decided to employ the α-particle, the most powerful atomic projectile then known. He bombarded nitrogen with α-particles from radium and detected, by their scintillations, particles coming from the nitrogen which had a longer range than the α-particles. He showed that these long-range particles were probably hydrogen nuclei or protons. The experiments, which were performed during the last two years of the First World War, were very tiring and trying to the eyes, since faint scintillations had to be observed on a screen in a darkened room.

His conclusion, that a nitrogen atom, struck by an α-particle, disintegrates and is converted into oxygen, may be represented by the following equation:

$$_2\mathrm{He}^4 + {}_7\mathrm{N}^{14} = {}_8\mathrm{O}^{17} + {}_1\mathrm{H}^1$$

$\quad\quad$ α-particle $\quad$ Nitrogen $\quad$ Oxygen $\quad$ Proton

The subscripts represent charges and the superscripts masses. The sums of both charges and masses on each side of the equation must balance.

The disintegration of a nitrogen atom, on being struck by an α-particle, was later actually photographed by Blackett (Plate XIX(b)), using an apparatus which has proved of immense value in atomic research, the cloud chamber invented by C. T. R. Wilson. The path of a particle in the cloud chamber is rendered visible by a trail of condensation, rather as the path of a high-flying aeroplane is sometimes indicated by vapour trails in the sky.

Rutherford succeeded Sir J. J. Thomson as Cavendish Professor of Experimental Physics at Cambridge in 1919. He and Chadwick continued the experiments on artificial disintegration and showed that about a dozen of the lighter elements can be transmuted, like nitrogen, by bombardment with α-particles. But the α-particle, which in Rutherford's hands had led to such fundamental discoveries, was beginning to reach the limits of its usefulness. Today large and expensive machines are employed to provide more copious supplies of swift particles for the breaking of atoms.

Physicists now have such a remarkable understanding of the structure of the atom that they can not only split existing atoms, but synthesise new ones. There are two gaps in the periodic table below lead which nature has not filled. The elements which fit these gaps have been prepared artificially in radioactive form. They are unstable and soon transform themselves into different elements; they have atomic numbers 43 (technetium) and 61 (promethium).

A number of elements heavier than uranium, the heaviest naturally-occurring element, have also been created in small quantities.

* $\quad$ * $\quad$ *

The explanation of the way in which atoms emit light was provided by Niels Bohr and will be discussed in chapter 22. The work of Bohr and his followers has relegated Rutherford's nuclear atom to the same category as the æther and lines of force. It is merely an inadequate model, but a model which has given rise to the nuclear reactor and to the atomic bomb.

CHAPTER 21

THE THEORY OF RELATIVITY

NEWTON FOUND it necessary, when presenting the foundations of classical mechanics in his *Principia*, to define absolute time, absolute space and absolute motion. His definitions were as follows:

Absolute, true, and mathematical time, of itself, and from its own nature, flows equably without relation to anything external.

Absolute space, in its own nature, without relation to anything external remains always similar and immovable.

Absolute motion is the translation of a body from one absolute space into another.

Newton was not entirely comfortable about these definitions because he realised that what we observe in nature is the relative motion of two bodies rather than the absolute motion of a single body. He admitted that absolute space is unobservable: 'The parts of that immovable space, in which those motions are performed, do by no means come under the observation of our senses.'

His concept of absolute space was attacked on theological grounds by the philosopher Berkeley. Berkeley maintained that if there is such a thing as absolute space then 'there is something beside God which is eternal, uncreated, infinite, indivisible, immutable'. This is of scientific interest because it prompted Newton, in a Scholium to the second edition of the *Principia*, to justify absolute space and absolute time in terms of the existence of God: '[God] is not eternity and infinity, but eternal and infinite; he is not duration or space, but he endures and is present. He endures for ever, and is everywhere present; and by existing always and everywhere, he constitutes duration and space.'

Newton's concept of absolute motion did not rest entirely upon theological considerations for it was supported by his rotating bucket experiment. Imagine a bucket, containing water, to be hung from a twisted cord and to be allowed to spin. At first there is relative motion between the water and the bucket but soon the water acquires the spinning motion of the bucket and it then, under the action of centrifugal force, becomes heaped up at the side of the bucket. 'This ascent of the water', wrote Newton, 'shows its endeavour to recede from the axis of motion, and the true and absolute circular motion of the water, which is here directly contrary to the relative, becomes known, and may be measured by this endeavour.' This was an apparent illustration of the absolute motion of a single body, as opposed to the relative motion of two bodies.

The conclusions drawn from the rotating bucket experiment were criticised towards the end of the nineteenth century by Ernst Mach, whose ideas provided an invaluable stimulus to Einstein in the formulation of the theory of special relativity. Mach did not accept that the experiment proved the existence of absolute space. He maintained that the centrifugal force was produced, not by absolute rotation, but by relative rotation with respect to the earth and other celestial bodies. He claimed also that uniform linear motion cannot be related to space itself, but only to the centre of all the masses in the universe.

A possible method of testing the existence of absolute space became obvious when it was accepted that space is filled with a medium, the æther, through which light and all electromagnetic radiations are propagated. A celebrated experiment was performed in 1887 by Michelson and Morley in a cellar in Cleveland, Ohio, to detect the motion of the earth through the æther. It can be shown by simple calculation that the time taken to swim across a stream and back is less than to swim an equal distance downstream and up again, the reason being that, in the latter case, the current impedes the swimmer for a longer time than it assists him. In the Michelson–Morley experiment the swimmer was a beam of light and the stream was the æther which, if stagnant in space, should appear to be flowing past the earth as the latter moves in its orbit round the sun. Light was split into two beams and sent backwards and forwards between mirrors in two directions at right angles. A very slight difference in the times of travel of the two beams could have been detected because, when the beams were recombined, any lack of fit between the waves in the two beams would give rise to interference. The apparatus was mounted on a massive stone floating in mercury, so that it could be rotated into any desired direction. In no case was a difference in time between two perpendicular paths detected, showing that there is no drift of the æther with respect to the earth.

Michelson drew the conclusion that the æther must be carried with the earth, rather as the atmosphere is carried, and he suggested that a relative motion between earth and æther might be detected at a higher altitude. Experiments were performed on the summits of the Jungfrau and of the Rigi in Switzerland, but no æther drift was detected. In 1921, Miller claimed, as a result of experiments performed at the Mount Wilson Observatory, California, at an altitude of 6000 ft., that he had detected a relative velocity between the earth and the æther of 2 miles per second, but by this time physicists were convinced that no relative velocity existed and his result was not accepted. Further experiments have not confirmed his findings.

In 1892, Fitzgerald put forward an explanation of the negative result of the Michelson–Morley experiment, on the assumption that the æther remains at rest, and is not carried with the earth. He suggested that, just as a ship pushing through a calm sea must suffer a definite, if small, contraction because of the pressure on its bow, so a body moving through the æther must suffer a similar contraction. But this contraction must be the same for all bodies, whatever the nature of the substance of which they are composed. The contraction cannot be measured because measuring rods shrink in the same proportion.

Fitzgerald's contraction became more convincing when it was supported by the mathematical theory of the Dutchman Lorentz, who was the pre-eminent theoretical physicist at the turn of the century. Lorentz assumed that the cohesion of all bodies is due to the forces between electric charges in the molecules, and this suggested that the contraction of all bodies might be the same. He considered what changes must be made in Maxwell's equations for the propagation of light, when the source emitting that light is moving. He found that it was necessary to transform all lengths, in the direction of motion of the source, from l_0 to l, where $l = l_0 \sqrt{(1 - v^2/c^2)}$, v being the velocity of the source and c the velocity of light. This expression is known as the Lorentz transformation. Lorentz found that it was necessary to transform the time in a similar manner, but he regarded the new time as artificial, with no physical significance.

It was at this point that Einstein gave a completely different interpretation, using the Lorentz transformation as the leitmotiv of his theory.

* * *

Albert Einstein (1879–1955) was born in Ulm, Southern Germany, of Jewish parents. He went to school in Munich until the age of sixteen when his father, who was a not very successful, small business man, moved to Milan.

After a few months in Milan, Einstein was sent to study at the Polytechnic in Zurich which, at the time, had an international reputation. With customary Jewish family solidarity, his fees were paid by relatives. There he studied the works of the masters of theoretical physics, Kirchhoff, Boltzmann, Maxwell and Hertz and learnt how a mathematical structure is built up. He read the philosophers David Hume and Ernst Mach, whose critical scepticism gave him the background and confidence needed to propound, a few years later, views of time and space which were contrary to those on which science had been founded.

While in Zurich he renounced his German nationality and became a Swiss citizen. He never had strong national roots; he was essentially detached, a citizen of Europe or of the world, rather than a Jew, a German, a Swiss or, at the end of his life, an American.

In 1902 he took a post at the Swiss Patent Office at Berne as a technical expert, third class. He found the work comparatively undemanding; it provided a modest living and left free most of his mental vitality for his scientific work. He called it his cobbler's job and retained it for seven scientifically productive years. He was able to pursue his researches in physics during office hours on pieces of paper, which he hid in a drawer whenever anyone came into his room.

Max von Laue, later a Nobel prizewinner for physics, recollected a visit which he paid to Einstein at this time; it is reminiscent of Kafka.

After a written appointment I looked him up in the Patent Office. In the general waiting room an official said to me: 'Follow the corridor and Einstein will come out and meet you.' I followed his instructions, but the young man who came to meet me made so unexpected an impression on me that I did not believe he could possibly be the father of the relativity theory. So I let him pass and only when he returned from the waiting room did we finally become acquainted. I cannot remember the actual details of what we discussed, but I do remember that the cigar he offered me was so unpleasant that I 'accidentally' dropped it into the river from the Aare bridge.'[1]

* * *

Einstein's special theory of relativity was published in 1905. It is called the special theory because it is confined to systems moving in relative uniform motion; it was later extended in the general theory of relativity to cover all types of motion.

Newton, in the *Principia*, gave a simple example of relative uniform motion. Suppose that a ship is moving with a velocity v with respect to the sea, and a sailor walks along the deck with a velocity w with respect to the ship in the direction of the ship's motion, then the sailor's velocity with respect to the sea is $v+w$.

Newton's principle of the addition of velocities breaks down in the case of light. Suppose that the ship is moving with velocity v directly towards a lighthouse, and the lighthouse sends towards the ship a beam of light, which travels with velocity c. To the lighthouse-keeper the light appears to be approaching the ship with a velocity $c+v$. But an observer on the ship, as the Michelson–Morley experiment showed, cannot detect any motion of the ship with respect to the æther, and to him the velocity of the light appears to be c. The velocity of light is always c, whatever the motion of the source of light or of the observer.

This is the paradox of the Michelson–Morley experiment which Fitzgerald and Lorentz attempted to resolve by a physical contraction of bodies moving through a stationary æther. Einstein denied the existence of a stationary æther, which might make possible the detection of absolute rest or of absolute velocity. All motion is relative, he maintained, and to speak of absolute rest and absolute velocity, or of absolute space and absolute time, is meaningless. Two systems, moving relatively to each other, have different spaces and different times.

The key to the special theory of relativity is Einstein's analysis of the concept of simultaneity. He imagined two observers, one in a railway train, moving at velocity v on a straight track and the other stationary at the side of the track. He supposed that as the man on the train passes the man on the track two lightning flashes occur, one a certain distance ahead, and the other an equal

[1] Carl Selig, *Albert Einstein.*

PLATE XXI. Marie Curie.

PLATE XXII. Nernst, Einstein, Planck, Millikan and von Laue, taken in Berlin in 1931.

distance behind them. The beams of light from the two flashes reach the man on the track at the same instant, because they have to travel equal distances, and he concludes that the flashes were simultaneous. The man on the train, however, finds that the light from the flash ahead reaches him earlier than the light from the one behind, because the former has a shorter distance to travel than the latter as a result of the train's motion, and hence he concludes that the flashes were not simultaneous.

It is not possible to argue that the flashes were really simultaneous, in view of the privileged position of rest of the man on the track, for this man is moving in space with the velocity of the earth. Nor would the man on the train accept the suggestion that the conflict of opinion about simultaneity is caused by the fact that the light from the flash ahead approaches him at a relative velocity of $c + v$ and that from behind at $c - v$, because when he measures these velocities he finds that they are both c.

Einstein introduced the idea that no observer can say that his measurements are any more fundamental or correct than those of any other observer and he assumed that the simultaneity of two events is relative to motion and has significance only with reference to a specific observer.

Einstein's analysis of concepts in terms of real or imaginary experiments has had a great influence on the development of modern physics. He did not accept the concept of absolute simultaneity because, when he subjected it to the test of imaginary experiment, he found it to be meaningless. If the velocity of light were infinite, then there would be such a thing as absolute simultaneity. But because light, which must be used to test the simultaneity of distant events, takes a finite time to travel, simultaneity is relative to the motion of the observer.

Physical concepts defined in terms of experiments or operations have been termed by Bridgman operational concepts. He has suggested that many of the questions asked about social and philosophical subjects will be found to be meaningless when examined from the point of view of operations.

Let us return to Einstein's two observers, one in a moving train and the other standing beside the railway track. We have already seen that they disagree about simultaneity. Let us suppose that each observer wishes to measure the length of the train. The man in the train can use a measuring rod. The man on the track cannot do this, but he can arrange for the positions of the ends of the train with respect to the track to be noted simultaneously—say by means of a row of assistants having synchronised clocks—and then he can measure the distance between these two points on the track by means of a measuring rod. The man in the train does not agree that the positions of the ends of the train were noted simultaneously and it is not surprising that his measurement of the length of the train turns out to be different from that of the man on the track.

Einstein, using the condition that the velocity of light must be the same for both observers, showed that the measurement of the man on the track is shorter than that of the man on the train in the ratio $\sqrt{(1 - v^2/c^2)}$, where v is the velocity of the train relative to the track and c is the velocity of light. Since c is 186,000 miles per second and v may be of the order of 60 miles per hour or $\frac{1}{60}$ mile per second, it is obvious that the difference in the two measurements is inappreciable. The difference only becomes significant when the relative velocity of a body is very high and comparable with the velocity of light.

All lengths in the train appear to the man on the track to be shortened in the ratio $\sqrt{(1 - v^2/c^2)}$, and the rates of clocks on the train appear to him to be slower in this ratio. The relations are reciprocal. To the man in the train, distances on the track appear to be shortened and clocks on the track appear to tick more slowly than his own.

A body moving at 90 per cent of the velocity of light relative to an observer would appear to that observer to have shrunk to half its length in the direction of motion. Similarly time on the moving body would appear to be going at only half the rate of the time of the observer.

The expression $\sqrt{(1 - v^2/c^2)}$ becomes meaningless if v is greater than c because the square root of a negative quantity is imaginary. Hence Einstein assumed that speeds greater than that of light are impossible. At the speed of light a body would appear to have shrunk to nothing and a clock on it to have stopped.

The special theory of relativity is based on two fundamental postulates:

(1) The velocity of light *in vacuo* is invariable, whatever the relative motion of the source and the observer.

(2) The laws of nature are valid for all uniformly moving systems.

The second postulate is an extension of a principle known to Galileo and Newton and it was first put forward by Henri Poincaré in 1904. On the deck of a uniformly moving ship it is possible to play games with balls and with other moving objects exactly as on land. In fact it is impossible, by experiments conducted inside the ship, to discover whether or not it is moving, assuming that the sea is absolutely calm. This can be generalised in the statement that the laws of mechanics are valid for all uniformly moving systems. Einstein expressed the equations defining the propagation of electromagnetic radiation in such a way that they are invariant with respect to uniform motion. Not merely does this give aesthetic satisfaction, but it accounts for the uniformity of natural laws in all terrestrial experiments, despite the motion of the earth in space.

The mass of a body increases as its velocity increases, and at the speed of light, its mass is infinite. If m_0 is the mass of a body at rest, and m is its mass when moving with velocity v, then

$$m = \frac{m_0}{\sqrt{(1 - v^2/c^2)}}.$$

Einstein's prediction of this increase of mass was confirmed by experiments on electrons, travelling with velocities comparable with that of light, in the discharge tube.

The significance of Einstein's conceptions of relative space and relative time became clearer when, three years after the publication of his Special Theory of Relativity, Minkowski showed that although there is no absolute interval in space, nor in time, between two events, there is an absolute interval in space-time.

This space-time interval is $\sqrt{(s^2 - ct^2)}$, where s is the spatial distance between two events, t is the time interval, and c is the velocity of light. The values of s and t may vary for observers moving relatively to each other, but the expression $\sqrt{(s^2 - ct^2)}$ is invariant.

* * *

Einstein's fame slowly grew; he became a professor at Zurich and in 1913 he was visited by Planck and Nernst with the offer of the post of director of the new Kaiser Wilhelm Physical Institute in Berlin, where he would be able to devote nearly all his time to research. He accepted this offer on the understanding that he retained his Swiss nationality.

He never quite fitted into the Prussian atmosphere of Berlin. He had an essentially bohemian nature, was amused by the conventions of academic society, and was careless about his clothes. He would lecture, for example, wearing an open-necked shirt, a sports jacket and sandals—a rather more scandalous matter in the early days of this century than it would be today.

His principle of special relativity applied only to systems in relative uniform motion and the laws of nature, as then formulated, did not hold for accelerated systems. He felt that it was unsatisfactory to give preference to uniform motion and, during the ten years after the publication of his special theory, he was continually occupied with the problem of generalising it. In October 1911 he wrote to his friend Laub: 'I work very hard but without much success. I have to discard again nearly everything that comes into my mind.' And on another occasion he remarked: 'Now I know why there are so many people who love chopping wood. In this activity one immediately sees the results.'

In November 1915 he wrote to Sommerfeld: 'During the last month I experienced one of the most exciting and most exacting times of my life, true enough, also one of the most successful.' At last he was able to put his equations of general relativity into a satisfactory form, and in 1916 he published them.

In an accelerated system forces come into play which are similar to the effects of gravitation. When the brakes of a moving train are suddenly applied, luggage may fall off the rack and a plumbline hanging in the train would be deflected from the vertical as though it were under the influence of a horizontal gravitational field in addition to the vertical gravitational field of the earth. Similarly, it is impossible to devise an instrument which will indicate the direction of the vertical in an aeroplane, because the effects of the acceleration of the aeroplane are indistinguishable from those of gravity. Einstein's General Theory is based on his Principle of Equivalence: 'A gravitational field of force at any point of space is in every way equivalent to an artificial field of force resulting from acceleration, so that no experiment can possibly distinguish between them.'

Einstein illustrated the principle by considering a freely-falling lift. Observers in the lift would find that bodies do not fall relative to the lift, but remain suspended in space, because both bodies and lift are falling with the acceleration due to gravity. The observers would find that their weight had disappeared and that, with a slight spring from the floor, they could rise gently to the ceiling. The phenomena are the same as would occur if the lift were in outer space, away from the influence of the earth's gravitational field.

Suppose that the lift were transported to outer space and suddenly accelerated by means of a cable attached to the roof. The observers would find that the floor suddenly pressed upwards on their feet, as if their weight had been restored, and unattached bodies would fall to the floor of the lift. The acceleration of the lift would produce exactly the same effects as a gravitational field directed downwards from the ceiling to the floor.

Suppose that a bullet were fired from outside through the lift as the latter accelerated 'upwards' in outer space. The bullet would travel in a straight line through space but, to the observers in the lift, because of their upward acceleration, it would appear to fall in its trajectory, as a bullet does on the earth, and the hole made by the bullet as it left the lift would be nearer to the floor than the hole it made on entry. In a similar way a beam of light flashed across the lift would follow a curved path. By the Principle of Equivalence, bending of light should likewise occur in a gravitational field and Einstein predicted that light from a star, passing through the intense gravitational field near to the surface of the sun, should be slightly deflected.

The path taken by light is that of shortest time, and, if it is curved, space also is curved; in such space the theorems of Euclidean geometry are not valid. A familiar example of a non-Euclidean, two-dimensional space is the surface of a sphere. The shortest distance between two points on it, called a geodesic, is part of a great circle and the sum of the angles of a triangle drawn on it is not two right angles, nor is it possible to draw a line through a given point parallel to another line. If the sphere has a large radius, a small portion of its surface is approximately Euclidean.

Einstein replaced the idea of gravitational forces by the conception of space which is made non-Euclidean by the presence of matter. In the Newtonian universe, space is regarded as Euclidean and a geodesic is a straight line. A planet which moves in a curved path is considered to be pulled out of its natural, rectilinear path by a gravitational force. According to Einstein, the path of a planet is a geodesic, which is curved because space is curved, and no force need be postulated. In a similar way, balls running on a warped or ruckled billiard table would swerve in their path. Eddington, when he sliced a ball at golf, used to remark that space seemed unusually curved in that part of the course!

The four-dimensional geometry of the universe, as represented by Einstein's structural field theory, is determined by the distribution of masses in it. Einstein was once asked by reporters in New York to explain the essence of relativity in a few sentences. He replied: 'If you will not take the answer too seriously and consider it only as a kind of a joke, then I can explain it as follows. It was formerly believed that if all material things disappeared out of the universe, time and space would be left. According to the relativity theory, however, time and space disappear together with the things.'

The general theory of relativity proved successful in explaining the precession of the orbit of Mercury, where Newton's theory of gravitation had failed. The ellipse, in which Mercury curves round the sun, slowly precesses (Fig. 60) with a period of 3,000,000 years. Hitherto astronomers could explain the precession only by the influence of an unknown planet, for which they had ready a name, Vulcan, but which had never been found. The precession of the orbits of the other planets is inappreciable because they are considerably farther from the sun than Mercury.

Fig. 60

Einstein suggested a further means of testing his theory. Time should be affected by a gravitational field and a clock on the sun should run with a slightly slower rhythm than on the earth. The light from atoms on the sun should be of slightly lower frequency than from corresponding atoms on the earth, and there should be a slight shift of the solar spectral lines towards the red end of the spectrum. The effect in the case of the solar spectrum is too small to be measurable but it has been detected in the spectrum of the light from the companion of Sirius, a white dwarf of enormous density.

* * *

The most convincing test of the general theory of relativity is the bending of light by the sun, and on 29 May 1919 a total solar eclipse occurred which was suitable for the test, the sun then being in the middle of a group of bright stars called the Hyades. These stars could be photographed during the eclipse, when their light just grazed the rim of the sun, and also when they were away from the sun's neighbourhood. A slight change in their relative positions during the eclipse would indicate a deflection of light by the sun, and enable this deflection to be calculated. At the instigation of Eddington the Royal Society and the Royal Astronomical Society sent out two expeditions, one to Sobral in Northern Brazil, and the other to the isle of Principe in the Gulf of Guinea, West Africa. Eddington himself was in charge of the second expedition.

On 6 November 1919 the Royal Society met in London and it was announced that the results of the expeditions confirmed Einstein's theory. The measured mean deflection of light by the sun was 1·64 seconds of arc, as compared with Einstein's prediction of 1·75 seconds of arc.

It was from this dramatic meeting that Einstein's world fame began. *The Times* sent a special correspondent to Berlin and asked him to contribute an article. 'It is thoroughly in keeping with the great and proud traditions of scientific work in your country', he wrote, 'that eminent scientists should have spent much time and trouble and your scientific institutions have spared no expense, to test the implications of a theory that was perfected and published during the war in the land of your enemies.' Then followed a typical Einstein joke: 'By an application of the theory of relativity to the taste of the reader, today in Germany I am called a German man of science, and in England I am represented as a Swiss Jew. If I come to be regarded as a *bête noire* the description will be reversed, and I shall become a Swiss Jew to the Germans, and a German to the English.'

In Germany the theory of relativity was violently attacked. Mortified and bewildered by defeat in war, the Germans looked for scapegoats and they found them in pacifists, socialists and the Jews. Einstein had refused at the beginning of the war to sign the Manifesto of the Ninety-two German Intellectuals which, in answer to allied propaganda, asserted that the representatives of German science and culture supported the war. His pacifist views were well known. At meetings of the Prussian Academy of Sciences the chairs on either side of him were left conspicuously empty and he was virtually ostracised. The theory of relativity was described by certain members of The Society of German

Scientists and Physicians as 'a fundamentally erroneous and logically untenable fiction', and it was referred to in the newspapers as 'the greatest hoax in history'.

When the Nazis came to power, Lenard and Stark, both Nobel prizewinners, maintained that there are two types of physical science: an Aryan, empirical science which is true and fruitful, and a Jewish, theoretical science which is subversive and nonsensical. Some German physicists, realising the damage that Lenard was doing to German science, made desperate researches into his ancestry, hoping to find some non-Aryan forbears, but without success. Many Jewish scientists fled from Germany into exile, to the enrichment of the countries in which they settled.

Ineffable folly of this kind is apt to prove expensive. During the First World War German science was, on the whole, superior to that of the allies. But in the Second World War it was outclassed, especially in operational research, in radar, and above all in the release of atomic energy.

Einstein was a freethinker and became fully conscious of his Jewish origin only as a result of anti-Semitism in Germany. He was not inhibited by the divided loyalties of a man like Planck, and he made full use of his world reputation to bring discredit upon a regime which showed such contempt for human rights. Inevitably he was forced to leave Germany and his property was confiscated. In 1933 he accepted a post at the Institute for Advanced Study in Princeton, U.S.A. which he retained until the end of his life.

Einstein's great fame, lasting for nearly forty years, was a source of embarrassment to him, and he used to refer to 'my mythical namesake who makes my life a singular burden'. He was continually subjected to requests, invitations, interrogations and challenges, which he bore with great patience. For example, in 1919 he received a telegram from New York, 'Do you believe in God stop prepaid reply fifty words'. He replied: 'I believe in Spinoza's God who reveals himself in the harmony of all being, not in a God who concerns himself with the fate and actions of men.'

It has been suggested that the fundamental malaise of our civilisation today is the loss of a belief in absolute values and that this loss can be attributed, in some measure, to Einstein's theory of relativity. But the relativity of moral concepts and ethics, which anthropological research has revealed in different civilisations and types of society, bears only the most superficial resemblance to Einstein's theory. Just after the First World War, Lord Haldane told the Archbishop of Canterbury, Dr Randall Davidson, that Einstein's theory would have a profound influence on theology, and that he should become acquainted with it. Some time later it was reported to the President of the Royal Society, Sir J. J. Thomson, that 'The Archbishop, who is the most conscientious of men, has procured several books on the subject and has been trying to read them, and they have driven him to what, it is not too much to say, is a state of intellectual desperation'. On meeting Einstein at dinner and asking him what would be the effect of relativity on religion, the Archbishop was deeply relieved to receive the reply that it would have no effect at all since it is a purely scientific theory. Nevertheless relativity, like evolution, has had a deep influence on our modern climate of thought.

The remarkable fact that two such different theories as Newton's Theory of Gravitation and the General Theory of Relativity could account for the same facts led Einstein to the belief that scientific theories are free creations of the human mind, rather than unique inductions from experience. We shall discuss this in the last chapter.

CHAPTER 22

THE QUANTUM THEORY

THE QUANTUM THEORY was put forward in 1900 by Max Planck to account for the radiation from hot bodies. When a body such as a poker is heated, it begins to emit invisible heat radiation known as infra-red radiation; it then becomes red-hot, giving off red light as well as infra-red radiation; its colour gradually changes until finally it may become white-hot, when it is emitting light of nearly all the colours of the spectrum. If the poker is coated with a matt, black substance such as lamp-black, it gives off more radiation than if it is highly polished, and Kirchhoff showed that the radiation from a so-called perfect black body, easily realisable in practice, is the maximum possible and depends only upon the temperature of the body and not upon the material of which it is made.

Planck saw that the explanation of the distribution of energy among the different colours or wavelengths in black body radiation might throw light on the fundamental nature of energy; he said that the radiation 'represents something absolute, and since I always regarded the search for the absolute as the loftiest goal of all scientific activity, I eagerly set to work'.

Physical theory, as it then existed, now called classical physics to distinguish it from the modern physics to which the quantum theory gave rise, predicted, as Lord Rayleigh proved, that the radiation from a hot body should be confined to infinitely short wavelengths.

But the maximum energy at low temperatures is found to be emitted in the comparatively long wavelength region of the infra-red, and the maximum gradually shifts, as the temperature of the body is raised, into the shorter wavelength region of the visible spectrum. Planck found that he could derive a mathematical formula, which fitted the experimental facts, only by making what seemed at the time the fantastic assumption that a body cannot radiate energy continuously, but only intermittently in discrete quantities, which he termed quanta. It was as though a hose-pipe could not deliver a continuous jet of water, but only a spray of independent spurts.

The energy of each quantum is $h\nu$ where h is a universal constant, called by Planck the elementary quantum of action, and ν is the frequency of the radiation which, on the wave theory of continuous emission, is the number of waves emitted per second. The paradox of the quantum theory is immediately apparent: the prescription of the size of a quantum or particle of energy must be made in terms of a concept, frequency, which can apply only to continuous emission in the form of waves. The wave theory of radiation was regarded at this time as one of the most firmly established of all physical theories.

Planck's work was stimulated by the experimental determination of the distribution of energy in black body radiation by two teams of workers, Lummer and Pringsheim, and Rubens and Kurlbaum, in the Physikalisch-Technische Reichsanstalt in Berlin, the national physical laboratory of Germany. Planck submitted his radiation formula to the Berlin Physical Society on 19 October 1900 and the next morning Rubens stated that, during the night, he had compared his experimental readings with those predicted by the formula and had found satisfactory agreement. Since that day, as experimental measurements have improved, their agreement with Planck's formula has become more perfect.

On 14 December 1900 Planck gave a lecture to the Berlin Physical Society in which he explained the significance of the elementary quantum of action h, for which he deduced, as we now know, a remarkably accurate value from the experiments on black body radiation. The value is unimaginably small, 6.56×10^{-34} joule-sec. But although such a little thing, the elementary quantum of action has immense consequences. Nature, in its ultimate working, cannot be continuous; it must operate in jerks like the minute hand of a watch. The old Latin tag 'Natura non facit saltus'—that nature never makes a jump—is apparently the opposite of the truth; nature cannot do anything but jump.

Max Born has written of Planck: 'His was, by nature, a conservative mind; he had nothing of the revolutionary and was thoroughly sceptical about speculations. Yet his belief in the compelling force of logical reasoning from facts was so strong that he did not flinch from announcing the most revolutionary idea which has ever shaken physics.'[1]

Planck tried for years to fit the quantum of action into classical theory but without success. His theory was not taken seriously at first and later it was said that he had not realised the significance of his discovery. But his son Erwin recollected that, during a walk in the Grunewald in Berlin in 1900, his father said to him: 'Today I have made a discovery as important as that of Newton.'

Planck came of a family of lawyers and scholars and was a man of high integrity, the finest type of Prussian. He was a theoretical physicist and he did his work, not in a laboratory, but standing at his tall, old-fashioned desk in his study. He was mainly responsible for attracting Einstein to Berlin, and the two of them made Berlin, in the years just before and after the First World War, the greatest centre of theoretical physics in the world. He and Einstein often played music together, Planck at the piano, and Einstein with his violin.

Planck's long life, 1858–1947, covered Germany's rise to power and its tragedy in the First and Second World Wars. As the doyen of German scientists and as President of the Kaiser Wilhelm Society for the Advancement of Science, the highest scientific post in Germany, Planck went to plead with Hitler for his Jewish colleague, Fritz Haber, whose discovery of the method of fixing nitrogen from the air for making explosives was of great assistance to Germany in the First World War. Hitler flew into one of his violent rages and ranted against the Jews, while Planck listened in silence. Although he made no attempt at overt opposition to the Nazis, Planck's attitude was well known. Goebbels wrote in his diary: 'It was a great mistake that we failed to win science over to support the new state. That men such as Planck are reserved, to put it mildly, is the fault of Rust and is irremediable.'[2]

Planck suffered too from personal tragedies; of his four children, his eldest son fell at Verdun in 1916, his two daughters died during the First World War, and his second son was executed by the Nazis for suspected complicity in the plot against Hitler's life. His house in Berlin was destroyed in an air raid, and his fine library, the fruits of a long lifetime's collecting, was completely lost.

[1] Obituary Notices of the Royal Society.
[2] The hapless Rust, whom Goebbels selects as the scapegoat of the Nazi party, was Minister of Education.

In the last twenty years of his life he was concerned to combat the scientific and philosophical trends, such as the abandonment of causality, to which the quantum theory gave support. He may have been right; but he may, on the other hand, have been an unwitting illustration of a rather bitter and cynical remark which he made in his *Scientific Autobiography* in connection with the rejection of his early researches in thermodynamics: 'A new scientific truth does not triumph by convincing its opponents and making them see the light, but rather because its opponents eventually die, and a new generation grows up that is familiar with it.'

*　　*　　*

Certain characteristic phenomena, exhibited by light and all invisible electromagnetic radiation, are readily explicable in terms of waves, but seem quite incomprehensible on a quantum theory of radiation. Two overlapping beams of light from the same source annul each other in places, producing darkness, and reinforce each other in other places producing increased brightness, a phenomenon known as interference (Plate XI and p. 154). Again, a beam of light spreads on passing through a narrow aperture or round a narrow obstacle, and gives rise to what is called a diffraction pattern (Plate XII and p. 151).

Planck assumed, at first, that light and all invisible electromagnetic radiation are absorbed, as well as emitted, in the form of quanta. But reflecting on the difficulties which were raised by such a view, he revised it and suggested that light behaved as quanta during emission only. It was then possible to regard the light as a group of waves and to account for such phenomena as interference and diffraction.

In 1905 Einstein put forward an uncompromising corpuscular theory of light in which light was regarded as quanta both during emission and during absorption. The year 1905 was an *annus mirabilis* for Einstein, then 26 years old; in it he published three contributions to physics of major importance, his theory of light, the special theory of relativity and the theory of the Brownian movement.

Einstein's theory of light was put forward to account for the photoelectric effect, which is the ejection of electrons from certain metals by light. The experiments of Lenard elicited the remarkable fact that the velocity of these electrons is rigorously independent of the intensity of the light. If the light falling on the metal is intense, more electrons are emitted than if the light is weak, but the velocities of the electrons are the same in each case. According to the wave theory, light spreads from a source in waves of ever increasing radius and the energy of the waves is spread more and more thinly with increasing distance from the source. It was, therefore, as incomprehensible that distant and nearby electrons should acquire the same energy from waves of light as that a submarine, miles away from a depth charge, should sustain the same damage as one very near to the explosion. Two submarines at different distances can be damaged equally by means of torpedoes, and it was that type of explanation that Einstein applied to the photoelectric effect. He suggested that an electron is emitted from a metal when it receives a direct hit from a quantum of light, now known as a photon.

Ten years later, the American physicist Millikan made an experimental verification of Einstein's mathematical theory. He illuminated the clean surfaces of various metals with different coloured lights and measured the velocity of the electrons emitted in each case. He called his apparatus a vacuum barber shop, because he had to shave each metal in a vacuum to obtain a completely clean surface before irradiating it with light. From his readings he was able to calculate a value for h which agreed very well with that of Planck, derived from black body radiation. But Millikan remarked, at the close of his paper, that he still regarded the theory as quite untenable.

Physicists were in the uncomfortable position of requiring two apparently irreconcilable theories of light. Some phenomena such as interference and diffraction were intelligible only in terms of a wave theory; other phenomena such as radiation from hot bodies and the photoelectric effect could be explained only in terms of a quantum or corpuscular theory.

* * *

The power and range of the quantum theory were strikingly demonstrated in 1913 when it was applied by Niels Bohr to Rutherford's nuclear atom. Thereafter it could no longer be regarded as a disparate appendage to the wave theory; as its stature grew, it came to dominate all further advances in theoretical physics.

Bohr, a young Danish physicist, came to England in 1912 and worked for a time in the Cavendish laboratory. There he acquired the reputation of asking searching and awkward questions about current ideas and theories, with the result than men walked away hurriedly when they saw him approaching. He moved to Manchester and was working with Rutherford at about the time that the nuclear theory of the atom was announced. Rutherford's atom consisted of a massive, positively charged nucleus round which electrons revolved, rather like a miniature solar system. Bohr directed his attention to the way in which such an atom could radiate light.

The lightest atom is that of hydrogen and this can be most simply represented by a single electron revolving in a circle round a single proton; the proton has a positive charge, equal in magnitude to the negative charge of the electron, and a mass 1840 times as great. According to the laws of classical physics, however, such an atom should be unstable; the electron should spiral towards the nucleus, radiating continuously waves of decreasing wavelength.

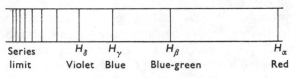

Series limit	H_δ	H_γ	H_β	H_α
	Violet	Blue	Blue-green	Red

Fig. 61. Hydrogen spectrum.

Light from hydrogen rendered luminous in an electric discharge tube, when passed through a prism, produces a spectrum, not of a continuous band of rainbow colours like that of white light, but of a few bright, narrow, spectral lines (Fig. 61). Thus a hydrogen atom radiates light of only a few particular wavelengths. These can be represented by a simple mathematical formula, discovered by Balmer, a Swiss schoolmaster, in 1885.

Bohr made the extraordinary assumption that the electron in the hydrogen atom can revolve only in orbits of particular radii and that when in these orbits it does not radiate. It radiates only when it jumps from one orbit to another and the radiation is a quantum, $h\nu$, whose frequency depends on the orbits between which the jump is made (Fig. 62).

Bohr's mathematical theory gave values for the wavelength of the light emitted by hydrogen with an accuracy which was quite astonishing. It predicted hitherto unobserved radiations from hydrogen in the ultra-violet, and in the infra-red; these were looked for and found. It showed that certain faint lines, hitherto ascribed to hydrogen and puzzling because they did not conform to the mathematical formula expected, were really emitted by ionised helium, each atom of which had lost one of its electrons and hence had a single electron revolving round a nucleus with a double positive charge.

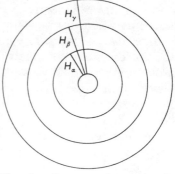

Fig. 62. Quantum jumps in the hydrogen atom giving rise to the spectral lines H_α, H_β, H_γ.

The impression made by the theory is typified by the reaction of Einstein, described in a letter from Hevesy to Rutherford: 'The big eyes of Einstein looked bigger still, and he told me "Then it is one of the greatest discoveries".'

Bohr proceded to account in a general way, and with fair success, for the spectra of elements whose atoms are more complex than that of hydrogen. The mathematical difficulties involved in an exact solution of the motion of two or more electrons revolving round a nucleus are insuperable. In the case of an atom like that of sodium, which has eleven electrons but a particularly simple spectrum, Bohr assumed that only the outermost electron is responsible for the emission of visible light; the remaining ten electrons form two completed inner rings.

Bohr's fundamental contribution to the quantum theory, at this stage, was to identify the emission of a quantum with the jump of an electron, called a quantum jump, and to show that the atom could exist only in certain stationary states, called quantised states. The existence of stationary or quantised states was confirmed experimentally in 1914 by Franck and Hertz. They found that mercury vapour could be made to emit radiation, corresponding to a particular spectral line of mercury, when bombarded with a beam of electrons having just sufficient energy to knock the outermost electron of the mercury atoms from one stationary state to another; the radiation was emitted as the disturbed electron jumped back.

The Rutherford–Bohr model of the atom, with its quantum jumping, was not fully satisfactory, however, even in the case of the simplest atom, hydrogen. The orbits and velocities of the electron, calculated according to classical theory, were unobservable details which were really irrelevant. The frequency of revolution of the electron bore no direct relation, contrary to the expectation of classical theory, to the frequency of the radiation emitted. Bohr made masterly use of what he termed the correspondence principle, which is that the laws of quantum theory converge towards, and in limiting cases become identical with, those of classical theory.

In 1925 Heisenberg abandoned completely the idea of a model atom and developed a theory of quantum mechanics which dealt only with observable quantities such as spectral frequencies and intensities. The theory was very abstruse, and took the form of tables of numbers, called matrices, whose manipulation had been worked out by mathematicians in the theory of algebraic equations. It gave results which corresponded more satisfactorily with the experimental data than the results of the Rutherford–Bohr model atom.

Heisenberg was working at the Copenhagen Institute for Theoretical Physics, established by Bohr in 1920. This became the leading centre of research into quantum theory in the world, and to it came many physicists who achieved an international reputation, including Dirac, Born, Jordan and Frisch. During the occupation of Denmark in the Second World War, the Institute was eventually closed by the Nazis after Bohr, for years an active member of the Danish underground resistance movement, had escaped to Sweden in a fishing boat. Bohr was brought from Sweden to England in the bomb bay of a Mosquito plane, in order to assist in the designing of the first atomic bomb. Because of his very long head, the earphones he was given for the flight did not reach his ears; he did not hear the instruction to put on his oxygen equipment, and became unconscious for most of the journey.

* * *

The greatest achievement of the Copenhagen school was the reconciliation of the quantum and wave theories, in what is known as wave mechanics. This was based on the work of de Broglie and Schrödinger.

By analogy with the dual nature of radiation, as both quanta and waves, de Broglie put forward in 1924 the view that electrons, and in general all material particles, can behave as waves. To an earlier generation of physicists such a suggestion would have been regarded as preposterous. But Davisson in America in 1927 and G. P. Thomson in England in 1928 obtained diffraction patterns with beams of electrons, very similar to the patterns obtained with light, showing that electrons behave like waves.

In 1926 Schrödinger applied de Broglie's theory of electron waves to the hydrogen atom and obtained an equation which is the basis of wave mechanics. It provided a very satisfactory explanation of Bohr's stationary states by showing that the orbit of an electron must contain an integral number of electron wavelengths, just as a taut violin string can vibrate only in an integral number of loops. Schrödinger was able to prove that his wave mechanics was equivalent to Heisenberg's quantum mechanics, despite the very different mode of approach of the two theories.

At the invitation of Bohr, Schrödinger visited Copenhagen in September 1926 to lecture on wave mechanics. Schrödinger's interpretation of the physical meaning of his mathematical theory was that the hydrogen atom is comparable to the circular haloes one sees round a distant lamp, when viewed through a misty window. The waves associated with the revolving electron formed a diffraction halo round the nucleus of the atom, consisting of spherical rings of negative electrical charge, whose density varied in a wave-like manner outwards from the nucleus. Rings of maximum density corresponded to Bohr's orbits.

Schrödinger believed that he could escape from the dilemma of the dual, wave and corpuscular, nature of matter and of radiation by regarding material particles and quanta as 'wave packets' or intense concentrations of waves in a small space. He thought that he could explain the radiation from the hydrogen atom in terms of classical wave theory and dispense with quantum jumps. When Bohr maintained that quantum jumps were still essential, Schrödinger is said to have retorted: 'If we are going to stick to this damned quantum jumping, then I regret that I had anything to do with quantum theory.'

Schrödinger's mathematical theory, when applied to an atom having two revolving electrons, requires that the electron waves shall be in 6-dimensional space, and for n electrons the space must be 3n-dimensional. Thus the electron waves cannot be waves in ordinary space. Max Born, one of Bohr's collaborators, suggested that they represent the probability of an electron's presence, and this proved to be an idea of decisive importance.

The apparent density of the rings in Schrödinger's picture of the atom represent, not a density of electric charge, but the probability of the electron being there, the electron being most probably in one of Bohr's orbits. Similarly, a beam of light can be regarded as swarms of quanta whose probable distribution is represented by waves.

That the quantum theory is concerned, not with certainties, but with probabilities is still further emphasised by a principle put forward by Heisenberg, known as the Principle of Indeterminacy (or Uncertainty) and one of the most important keys to an understanding and interpretation of the theory. The principle asserts that it is impossible to determine simultaneously and accurately both the position and the velocity of a small particle like an electron.

Heisenberg illustrated his principle in terms of the disturbance caused by the means of observation. In order to observe an electron we must use radiation, and if we wish to locate the position of the electron with accuracy we must use radiation of short wavelength. Hence the frequency, v, of the radiation must be high and the energy of the quantum hv will be large. When the quantum hits the electron, the velocity of the latter will be disturbed. If, on the other hand, we use radiation of long wavelength, having quanta of low energy, we cannot locate the position of the electron with accuracy although we do not seriously disturb its velocity. An accurate measurement of velocity means a blurring of position and a precise knowledge of position entails an imprecise knowledge of velocity.

The principle of indeterminacy may be expressed mathematically as follows:

$$\Delta x . \Delta v = h/m,$$

where Δx and Δv are the uncertainties of position and velocity of a particle of mass m. The product of the two uncertainties is proportional to h, the elementary quantum of action. If h were zero there would be no uncertainty; making h equal to zero transforms the equations of wave mechanics into the equations of classical mechanics. Thus the root of the uncertainty is the existence of the elementary quantum of action. It is not possible to calculate separately the extent to which the means of observation creates uncertainty in the position and velocity of a particle, so that the means of observation disturb an object in an unpredictable way.

In order to predict in classical mechanics we must know accurately the positions and velocities of the particles in the system at some instant. Laplace maintained that a super-intelligence, having a knowledge of the positions and velocities of all the particles in the universe, could predict the whole of the future. But, according to the Principle of Indeterminacy, such a knowledge in the realm of atomic events is impossible and the utmost we can predict is the probability of occurrences. In the case of larger scale phenomena the relative size of the elementary quantum of action is so small that our predictions of probability have virtual certainty.

The fact that we cannot attribute to an atomic object such as an electron a precise position and a precise velocity simultaneously is an aspect of its dual nature as both particle and wave. When its position is precise we observe it as a particle; when its position is imprecise we observe it as a wave. Bohr put forward a theory of complementarity, in which he maintained that we must of necessity use the two complementary concepts of particle and wave, both of which are equally valid, to describe an electron completely. The two concepts are mutually exclusive and never conflict. In order to observe an electron as a particle we must use an experimental arrangement which is incapable of detecting waves; when we observe its wave aspect our apparatus cannot detect it as a particle.

The theory of complementarity introduces the important idea that a single concept may have only a limited and partial relevance to a particular atomic event. The realm of the atom is foreign to our experience and is not amenable in a normal way to our language. To describe a single entity like an electron we require the two concepts, based on our everyday experience, of the particle and the wave. As de Broglie has put it, for a description of the electron 'we could hold, contrary to Descartes, that nothing is more misleading than a clear and distinct idea'.

Bohr has suggested that the concept of complementarity may find applications in other sciences such as biology and psychology.

The determinist methods of classical physics and chemistry, which are used to study life, may be incompatible with the existence of life, and there may be a complementary relation between living and lifeless matter. The apparent conflict between our knowledge of free will and the theory that our conduct is rigidly determined by our physical constitution and by our experience may be resolved by some form of complementarity.

*　　*　　*

The consequences of indeterminacy and of complementarity were clarified by many discussions at Copenhagen between Bohr and his co-workers. An opportunity for wider discussion occurred in October 1927 at the Solway Congress on Physics in Brussels and here Bohr's ideas were subjected to the formidable criticism and opposition of Einstein. Several imaginary experiments were suggested by Einstein which might be incapable of coherent explanation in terms of Bohr's theory of complementarity or by which the principle of indeterminacy might be infringed.

Imagine a semi-silvered mirror which reflects half the light which falls upon it and transmits the other half. In the case of a beam of light consisting of millions of photons (the name given to a quantum of light), we can say that half the photons are reflected, and half are transmitted; this is a statistical explanation. But what happens when a single photon strikes the mirror? Will it appear as an indivisible particle or as a group of waves? Bohr's answer is that it depends on the means we use to observe it. If we attempt to observe it as a particle, for example, by means of the photoelectric effect, there is an equal probability that it will be reflected or transmitted as a photon. If we attempt to observe it as waves, and combine the reflected and transmitted beams, we obtain an interference pattern. Thus the photon behaves simultaneously as both a particle and as waves—quite unlike any object of our normal, everyday experience. As a particle it appears to be *either* reflected or transmitted; but as waves it appears to be *both* reflected and transmitted.

Again, imagine a single electron passing through a very tiny hole in a screen. By virtue of its wave nature it will spread and give rise to a diffraction pattern. This diffraction pattern represents the probability of finding the electron particle at a particular place. At the maxima of the pattern there is a strong probability of finding the electron, and at the minima a small probability. When we find the electron at a particular place, the observation changes the probability at that place to certainty, and the probability distribution contracts to a point. Einstein claimed that this was contrary to our accepted notions of space and time. Bohr argued that it was an example of the complementary properties of the electron.

In classical physics causality was accepted as an act of faith and it was believed that, given the requisite data, any physical event could be predicted with certainty. According to the quantum theory, however, an atomic event cannot, in principle, be predicted as certain but only as probable. Hence causality, as applied to atomic events, cannot, in principle, be tested and it is therefore a concept without physical meaning (see pp. 343-4).

Einstein expressed concern at the Solway Congress about the abandonment of the principle of causality, and neither he nor Planck ever accepted the Copenhagen interpretation of quantum theory. The majority of physicists did so and, while they still regarded Einstein with boundless admiration and affection, some of them looked upon him regretfully, in this field, as a Lost Leader:

> Then let him receive the new knowledge and wait us,
> Pardoned in heaven, the first by the throne.

CHAPTER 23

COSMOGONY

DURING THE PRESENT century the construction of large tele-scopes and the replacement of visual observation by the camera and by the photoelectric cell have so enormously increased the depth to which astronomers can probe into space as to make possible an investigation of the universe as a whole, and to provide a firm basis for cosmogony, which is the study of how the universe was created.

Two opposing theories of creation now await the decision of further observation, and this decision may cause a profound re-orientation of human thought, comparable, perhaps, with that caused by the Copernican theory or by evolution. One theory is that the universe was created thousands of millions of years ago, but at a certain point of time, and that it must ultimately die when no further energy is available. The other theory is that the universe is continuously created, at the same rate now as in the past, that it had no beginning in time, and will have no end.

In this chapter we shall trace the development of that part of astronomy which led to the formation of the two theories.

* * *

The Greeks regarded the universe as a finite system enclosed by the sphere of stars, which revolved every twenty-four hours about the earth as centre. They thought that the universe could not be very large because otherwise the velocities of the stars, in their diurnal circuit, would be incredibly great. When Copernicus postulated that the stars were at rest, and accounted for their apparent revolution by a rotation of the earth in the opposite sense, it was no longer necessary to regard the universe as comparatively small.

Copernicus realised that the stars must be very distant from the earth because they did not appear to close up or spread out as the earth moved in its orbit round the sun. He still regarded the stars as confined to a sphere, and did not suggest that they were scattered through space. This step, taken by Thomas Digges in 1576 and by Giordano Bruno (1548–1600), enlarged the universe from a finite to an infinite sphere. The mental adjustment to an infinite universe took place in the sixteenth and seventeenth centuries, so that the almost inconceivable distances of modern astronomy have lost their power to stupefy. Pascal wrote of the astronomy of his time: 'The eternal silence of these infinite spaces frightens me.'

The faint, broad belt of light extending right round the sky, known as the Milky Way, was shown by Galileo's telescope to consist of innumerable faint stars and in 1750 Thomas Wright suggested that it was a disc-like stellar system with the solar system at its centre. When we look at the Milky Way we are looking along the plane of the disc, and hence we see many stars, but when we look in a different direction we are looking out of the disc and see comparatively few stars.

Wright further suggested that the faint cloud-like patches in the sky, known as nebulae, were similar stellar systems of comparable size at enormous distances from us. This idea of island universes, or galaxies as they are now called, was elaborated by the philosopher Kant in 1755. The assumption behind these speculations was the Principle of Uniformity, that one sample of the universe is like any other sample.

Many data about our galaxy and about the nebulae were compiled by Sir William Herschel (1738–1822), one of the greatest observational astronomers, and a pioneer in the examination of the stellar universe as a system. Herschel constructed his own telescopes and kept making larger and larger ones until he had an instrument with a speculum mirror of 40 in. aperture; the larger the aperture the greater the number of stars that can be seen. His object was to resolve the stars in the nebulae, but he found that with each increase in size of his telescope, for every additional nebula in which he could resolve the stars, ten more nebulae appeared in which the stars were not resolvable.

His observations confirmed that our galaxy is shaped roughly like a disc or a grindstone. He believed, correctly, that some nebulae, such as that in the constellation of Orion, are truly gaseous and form part of our galaxy. He thought that other nebulae lay outside our galaxy and were comparable in size with it. But his views on this difficult problem were conjectural and uncertain. He was unable to measure the distance from the earth of even the nearest stars and a century was to elapse before methods were devised of measuring the distance from the earth of nebulae.

*   *   *

In 1838 Bessel, the director of the Königsberg Observatory, made the first measurement of the distance of a star from the earth. He had reason to believe that a faint star in the constellation of the Swan, 61 Cygni, must be comparatively near to the earth because, on comparing his readings of its position with those taken a century earlier at Greenwich, he found that it had moved slightly, and more than any other star, with reference to the background of stars. His method of measuring its distance, indicated in Fig. 63,

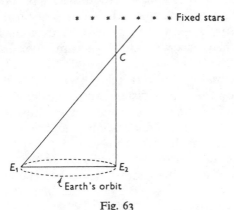

Fig. 63

was similar to that used by a surveyor for measuring terrestrial distances. His baseline was $E_1 E_2$, the diameter of the earth's orbit, which was known to be 186 million miles. He observed the positions of the star C, at intervals of six months, when the earth was at E_1 and E_2 respectively, and found that these positions, relative to the rest of the stars, differed by about the width of a penny at a distance of five miles. The angle $E_2 C E_1$ was 0·31 seconds of arc, and from it the distance of the star was calculated to be about eleven light years, each light year being about six million million miles, which is the distance travelled by light, at a speed of 186,000 miles per second, in one year.

In 1839 Henderson announced that Alpha Centauri, a star in the Southern Hemisphere and now known to be the star nearest to the earth, is about four light years distant. Since then the distances of several thousands of the nearer stars have been measured in this way.

For stars more than about 300 light years away, the parallactic angle ($E_1 \hat{C} E_2$ in Fig. 63) is as small as the probable error and hence the method is inapplicable. The distances of these stars can be determined by comparing their apparent brightnesses with that of a star whose distance is known. Assuming, as did Newton, Huygens and Herschel, for lack of other evidence, that all stars have the same intrinsic brightness, that is that they would all look equally bright if at the same distance away, then a star, whose apparent brightness is one quarter that of another star, must be twice as far away, since apparent brightness is proportional to the inverse square of the distance. It is now known that stars vary in intrinsic brightness, but it is possible, from an analysis of a star's light, when dispersed into a spectrum, to estimate what the intrinsic brightness is.

As a simple analogy, suppose that we wish to estimate, on a dark clear night, how far away a distant lamp is. Suppose that we compare this lamp with a nearer lamp, which we know to be 100 yards away, and find that it is only one-ninth as bright. If it were the same type of lamp as the nearer one we could say that its distance away must be 300 yards. But suppose that we know that it is twice as powerful as the nearer lamp, that is that its intrinsic brightness is twice as great. We conclude that its distance away is not 300 yards, but $\sqrt{2} \times 300 = 420$ yards.

When stars are very remote and their light is consequently very faint, it is difficult to analyse their light and to determine their intrinsic brightness. A discovery made in 1912 by Miss Leavitt of the Harvard College Observatory provided a new method of determining the intrinsic brightness of a certain class of stars, even at very great distances. These stars are known as Cepheid variables; they are extremely bright and they fluctuate regularly in brightness, like lighthouses in very slow motion, with a period which, for some stars, may be less than a day, and for others, as much as a fortnight. Miss Leavitt found that the periods of the Cepheid variables in the Large Magellanic Cloud were proportional to their apparent brightnesses and hence, since they were all the same distance away, to their intrinsic brightnesses. Nearer Cepheid variables, at a known distance, were used to establish the exact relation between period and intrinsic brightness. Thus a Cepheid variable's period gives its intrinsic brightness and, by measuring its apparent brightness, its distance can be calculated.

This powerful new method enabled Harlow Shapley, professor of astronomy at Harvard, to measure the distances of stars on the confines of the Milky Way and hence to estimate that the diameter of our galaxy is about 300,000 light years. It has since been discovered that there are vast clouds of hydrogen gas and fine dust in what were thought to be the empty spaces of the galaxy and an allowance for the obscuration they cause has reduced Shapley's calculated diameter to 100,000 light years.

In 1923 Edwin Hubble (1889–1953) turned the great 100-inch telescope of Mount Wilson on to the nebulae in Andromeda (Plate XXIII(a)) and discovered there a Cepheid variable. By the end of 1924 he had found thirty-six variable stars in this nebula, of which twelve were Cepheids. These enabled him to estimate that the nebula is about a million light years distant and to prove that it lies outside, and far beyond our own galaxy. Because of its great distance away its size must be enormous and its diameter was estimated to be about the same as that of our galaxy.

In this way was solved the long-standing problem of the nebulae. Direct observation with the 100-inch telescope proved that most nebulae are not part of our galaxy, but are galaxies themselves, and that the speculations of Wright, Kant and Herschel were fundamentally correct.

The nebula in Andromeda is the nearest to our galaxy, and is the only nebula that can be seen with the naked eye. Nebulae occur in clusters or groups, sufficiently near to each other to be bound by mutual gravitational forces; the group to which the Milky Way belongs consists of nearly twenty nebulae, including the nebula in Andromeda. The estimated number of observable galaxies runs into hundreds of millions.

Determinations of the distances of several hundred nebulae have now been made, making use both of Cepheid variables and of other bright stars, whose intrinsic brightnesses are known. But, beyond about 10 million light years, individual stars in the nebulae are not resolvable, even by the 200-inch telescope at Mount Palomar, which is the largest in the world.

For greater distances than this the apparent brightness of the nebula as a whole is compared with the apparent brightness of similar nebulae whose distances are known. This method is applicable for distances up to several hundred million light years.

At still greater distances the individual nebulae are too dim for quantitative measurements but the apparent brightness of a cluster of nebulae can be compared with that of a nearer cluster whose distance is known. The greatest distance yet measured is that of the cluster in Hydra, about two thousand million light years. Individual distances are not likely to be very accurate, but reliance can be placed on a statistical average of many results.

The measurement of these staggering distances from data which are merely points or patches of light in the sky, most of them invisible to the naked eye, is truly remarkable. The basic distance, from which are calculated the distances of the nearer stars, is the diameter of the earth's orbit. From the distances of the nearer stars are obtained, step by step, the distances of the nebulae, to the very faintest at the limit of observation of the 200-inch telescope. The chief nebular measurement is that of apparent brightness and this is made photoelectrically. The light is focused by the telescope on to a sensitive metal surface and causes electrons to be ejected; each electron, in a multiplier tube, ejects six more electrons, these six eject thirty-six, and so on until an amplification of 10 million is achieved.

*   *   *

The light from nebulae, when dispersed into a spectrum, exhibits what is called a 'red-shift'; the spectral lines are all shifted slightly towards the red end of the spectrum (Plate XXIV). This indicates that the nebulae are receding from us at very high speed. The universe appears to be expanding.

The effect of the motion of a source on the waves it emits is named after Doppler, its discoverer, and may be observed most directly in the case of sound waves. The pitch of the whistle of a railway engine appears lower than its true pitch when the engine recedes from us at speed, because the sound waves are more spread out, that is the wavelength is increased. The same thing happens in the case of a source of light and, from the increase in wavelength of the light, it is possible to calculate the recession velocity of the source.

The first determinations of the recession velocities of nebulae were made by V. M. Slipher of the Lowell Observatory, U.S.A. from 1912 onwards. In 1929 Hubble compared Slipher's determinations of the recession velocities of nebulae with his own determinations of distances and he discovered a simple relation, now called Hubble's law, that the velocity is proportional to the distance. Thus one nebula, which is twice as far away from us as another, moves away from us twice as fast.

Hubble's law can be stated in the form that the distance away divided by the velocity of every nebula is the same. Since distance divided by velocity is the time required to cover the distance, this ratio gives the age of the expanding universe, assuming the velocity of each nebula to have remained uniform. The recession velocity of the cluster of nebulae in Hydra, distant about 2000 million light years, is about one-fifth of the velocity of light, and this gives a value of $2000/\frac{1}{5} = 10,000$ million years for the age of the expanding universe. The accepted value at the moment, based on the straight-line graph of the distances and velocities of all the nebulae studied, is about 9000 million years.

When Hubble made the first estimate of the age of the expanding universe the value he obtained was too small; it was 1800 million years. The reason was that his nebular distances were too small. In 1946 Baade, of the Mount Wilson and Mount Palomar Observatories, discovered that there are two types of Cepheid variables. Those used by Hubble to measure the distance of the nebula in Andromeda, and hence of more distant nebulae, belong to what Baade called star population I, and are considerably brighter than those belonging to star population II, which were used by Shapley to measure the dimensions of the Milky Way. A re-evaluation of all nebular distances is now in progress.

Hubble's estimate of 1800 million years for the age of the expanding universe was less than the age of the earth, as calculated from the radioactive contents of the rocks. He was therefore inclined to doubt whether the red-shift in the spectra of the nebulae really indicated an expansion of the universe. It was conceivable that the light 'grew tired' during the millions of years of its journey through space; if it lost energy, a reduction in frequency and hence a shift of the spectral lines towards the red would occur. But such a proposal could not be explained or justified on the known laws of physics. Now that the age of the expanding universe has been raised to 9000 million years, Hubble's difficulty has disappeared; astronomers and physicists accept the evidence as indicating that the universe is expanding.

* * *

In the study of the expanding universe observation and theory have proceeded side by side, each stimulating the other. Although the observations were made in the U.S.A., the theory originated in Europe.

Modern cosmological theory dates from Einstein's paper of 1917, in which he applied the theory of general relativity to the universe as a whole. According to this theory, space is not flat nor Euclidean, but it is curved, and its curvature depends upon the amount of matter it contains. Einstein conceived that the matter in the universe is sufficient to cause space to curve back upon itself, so that the universe is a finite but unbounded sphere. He had no idea that the universe might be expanding and he was able to account for a static universe only by introducing a mysterious cosmological constant into his equations.

Einstein's static universe is unstable; a slight disturbance would cause it to expand or to contract. When observation showed that the universe is expanding, this was hailed as a triumph for the relativity theory.

If we imagine the directions of the velocities of the nebulae to be reversed, so that the universe contracts instead of expands, the nebulae will eventually come together simultaneously in a concentrated mass. This suggests a simple picture of the origin of the universe; in the beginning, all matter was jammed together, there was a gigantic explosion, and the expansion started.

The most influential and successful of the propounders of this kind of evolutionary model, in mathematical terms, is the abbé Lemaître. He postulated that, before the explosion, there was a conglomeration of matter of enormous density. The nebulae did not form in the first phase of the expansion, which lasted for thousands of millions of years. After the initial impetus was exhausted the universe settled down to the nearly static condition described by Einstein and it was then that clusters of nebulae began to form from the primeval matter. The process gave rise to instability and a force of cosmical repulsion overcame the force of gravitation, causing the present expansion of the universe.

334

Lemaître's theory puts the initial explosion of the universe at between 20,000 and 60,000 million years ago.

Gamow has worked out an alternative picture of how the universe may have evolved from a super-dense state to its present condition. He imagines that the packed mass of incredibly dense matter, before the expansion began, consisted of free neutrons, which are uncharged particles of matter. Many of these changed into positively charged protons and negatively charged electrons and, during the first twenty or thirty minutes of the expansion, the temperature was so high that they combined to form atoms. The chemical elements, in their present relative abundances, were formed in the first half hour. Gamow's theory differs from that of Lemaître in that it postulates that the present expansion of the universe is still part of the initial explosion, so that the explosion occurred only about 9000 million years ago. Hubble's law is an obvious consequence; a piece thrown out by the explosion at twice the speed of another piece would have travelled twice as far.

An entirely different theory of the origin of the universe was suggested by Bondi and Gold in 1948 and, a year later from a different point of view, by Hoyle. One of the fundamental assumptions of astronomy, which we have called the Principle of Uniformity and which Bondi and Gold called the Cosmological Principle, is that the universe is similar throughout; the universe presents the same aspect from every point except for local irregularities. Bondi and Gold, as the basis of their theory, took a step further. They assumed that the universe is not only uniform in space but also in time; that its appearance has not changed fundamentally during the whole of its history and that it will not change in the future. This assumption they called the Perfect Cosmological Principle.

Evolutionary theories of the universe require the nebulae to be thinning out so that, as time proceeds, fewer and fewer will be visible from the earth. This is contrary to the Perfect Cosmological Principle, and Bondi and Gold assumed that new nebulae are continuously being formed to replace those that have disappeared.

We may compare the nebulae on the evolutionary theory to cross-country runners who, bunched together at the beginning of the race, become more and more spread out, because of their different speeds, and disappear one by one over the horizon. On the continuous creation theory, the runners never become more spread out because new runners rise from the ground to keep the distribution constant; the race has no beginning and the supply of runners is endless.

The way in which the new nebulae are thought to be formed is by the condensation of new matter, probably hydrogen atoms, which are continuously created throughout space. The necessary rate of creation in the whole observable universe is 10^{32} tons per second, but, so vast is space, a new hydrogen atom would appear in a normal-sized room only once in every ten thousand years, so that continuous creation could not possibly be detected terrestrially.

The creation of new matter does not violate the principle of the conservation of energy and mass because the new matter provides exact compensation for that which disappears beyond the limit of observation. But it does falsify the deduction, made in the nineteenth century from the second law of thermodynamics, that the universe is running down and that eventually it will suffer a 'heatdeath'. The universe is assumed to be in a steady state; the availability of its energy remains constant.

The theory of continuous creation has been strongly attacked by Dingle: 'It has no other basis than the fancy of a few mathematicians who think how nice it would be if the world were made that way.' Dingle takes particular exception to the term 'perfect cosmological principle'. A spade should be called a spade, he says, and not a perfect agricultural principle. Hoyle escapes this last castigation because he developed this theory from a modification of the field equations of general relativity, preferring to proceed from well-established laws rather than from a new principle. It is, however, rash of Dingle to condemn strange new ideas in science in quite so forthright a manner. A feature of twentieth-century science has been prodigal speculation, inexorably pruned by facts.

It should be possible, perhaps in the next decade, to decide between the evolutionary theory and the theory of continuous creation by observational tests. 'For a theory to be of any value it must be vulnerable', said Gold; in other words, useful theories must make predictions to be tested, and the two theories of creation can do this.

The evolutionary theory postulates that the total mass of the universe is constant and therefore, as a result of the expansion, the spatial density of the nebulae must be smaller now than it was thousands of millions of years ago. When we look at the most distant observable nebulae we are seeing them as they were 2000

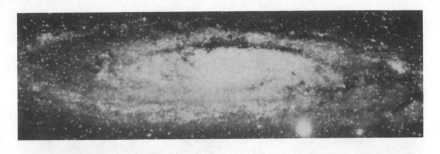

(a)

(b)

(c)

PLATE XXIII. (a) The Great Nebula M31 in Andromeda. (b) The Nebula M51 in Canes Venatici, showing the spiral form (which indicates that it is rotating). (c) The Nebula (NGC 4565) in Berenice's Hair, an end-on view.

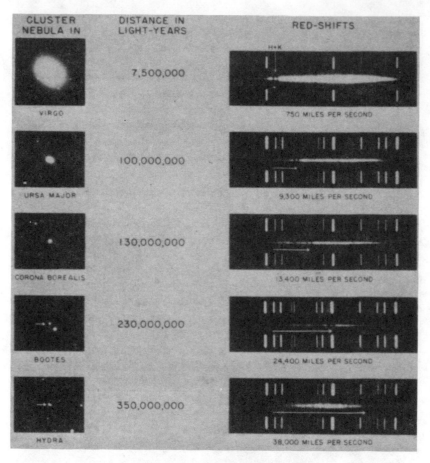

CLUSTER NEBULA IN	DISTANCE IN LIGHT-YEARS	RED-SHIFTS
VIRGO	7,500,000	H+K 750 MILES PER SECOND
URSA MAJOR	100,000,000	9,300 MILES PER SECOND
CORONA BOREALIS	130,000,000	13,400 MILES PER SECOND
BOOTES	230,000,000	24,400 MILES PER SECOND
HYDRA	350,000,000	38,000 MILES PER SECOND

PLATE XXIV. Red-shifts of the nebulae. At the left are the apparent sizes of the nebulae; since the true sizes are comparable, the apparent sizes decrease in proportion to distance. On the right are the red-shifts. The lines above and below the cigar-shaped nebular spectra are comparison spectra. The red-shifts of the dark, absorption H and K lines of calcium are indicated by horizontal arrows.

million years ago. If we could make an estimate of the number of nebulae in a known volume at this range, or preferably at treble the range, it should be significantly greater than the number of nearer nebulae in the same volume. The theory of continuous creation, on the other hand, requires the spatial density of the nebulae to be the same in all parts of the universe.

An unfortunate result of the expansion of the universe is to impose upon us an observational limit. The recession causes a weakening of the light from the nebulae, a phenomenon which is quite different and distinct from the red-shift effect. If we imagine light as consisting of quanta, or small particles of energy, the number of quanta received per second from the nebulae is reduced by the recession. When the nebulae are receding at the velocity of light, the intensity of the light received from them is zero.

It seems likely, because of the unsteadiness of the atmosphere, that the range of the largest optical telescope, the 200-inch instrument at Mount Palomar, is near to the practicable limit. The telescope can reach its maximum penetration of 2000 million light years only on a limited number of nights in the year. The golden era of the big optical telescopes is coming to a close.

Radio telescopes, in use from the 1940's and still in process of development, are able to range well beyond the optical limit, and provide one of the alternative means of observation. These telescopes may be able to find whether there is hydrogen in the vast spaces between the nebulae, which is one of the important tests of the continuous creation theory. Another possible means of observation, free from the handicaps imposed by the earth's atmosphere, is the artificial satellite.

According to the evolutionary theory the appearance of the most distant nebulae, representing their condition thousands of millions of years ago, should be more youthful than that of the nearer nebulae. If, on the other hand, matter is continuously created, the population of nebulae, both nearest to us and furthest away, should include individuals of widely different ages, rather like a human population of children and adults. No method has yet been devised for estimating the age of a nebula, but the shapes and brightnesses of nebulae vary, and this may indicate differences in age.

Other tests to decide between the two theories may also be devised, and the immediate future of cosmogony should be particularly intriguing.

CHAPTER 24

THE TWENTIETH CENTURY

IN THE PRECEDING five chapters we have described some of the most significant advances of science in the twentieth century. In this chapter we shall discuss first the modern view of the nature of physical science, brought about mainly by the theory of relativity and the quantum theory, but having its roots in the last quarter of the nineteenth century. Secondly, we shall give a brief account of the official philosophy of science in the Soviet Union, dialectical materialism. Thirdly we shall consider science as a social function, and how far it is desirable that scientific research should be planned by the State. Finally we shall examine the shift in the centre of gravity of science from Western Europe to the U.S.A. and the possibility of a further shift to the U.S.S.R.

* * *

Science was regarded by Newton and his successors in the eighteenth and nineteenth centuries as the discovery of the laws of nature, imposed by the Creator or immanent in nature. It was believed, for example, that it is the essential property of matter to attract other matter according to the inverse square law. The law was there in nature to be discovered and, once discovered, it represented final truth. Science was believed to be the investigation of objective physical reality.

From about 1600 to 1900 physical science was mechanistic, endeavouring to explain all phenomena in terms of simple mechanical models. Phenomena were reduced, in the final analysis, to the motion of particles and Newton's laws of motion were the core of physics. Newton stated his belief that God created matter in small, hard, massy particles or atoms, and that these were the ultimate stuff of the universe. The first real breach in the mechanistic picture, not widely recognised as such at the time, was Maxwell's theory of the electromagnetic field.

Towards the end of the nineteenth century the Viennese

physicist Ernst Mach (1838–1916) attacked the prevailing view of science and his thought has had a dominating influence in the twentieth century. His philosophy of science is known as positivism, a word coined by Comte to represent a philosophy confined to positive facts and observable phenomena, ignoring all metaphysical explanation of ultimate causes. A metaphysical explanation is one based on intellectual intuition and having no basis in sense observation.

Mach criticised Newton's basic concepts of absolute space, absolute time and absolute motion, showing them to be metaphysical since they are unverifiable by observation. He restated the laws of motion in terms of actual experiments so as to purge them of metaphysical assumptions and to relate them directly to observation.

The fundamental tenet of positivism is that all genuine knowledge is based on our sense perceptions and a proposition has meaning only if it is verifiable by sense observation. At the same time that Mach was developing positivism the American C. S. Peirce (1839–1914) was founding pragmatism, which is based on a similar idea. Peirce came to philosophy from physics and, like Mach, anticipated the modern scientific practice of defining all scientific concepts in terms of operations or imaginary experiments. He held that the meaning of the statement 'This is hard' must be given in terms of the operation, 'If this substance were rubbed with other substances it would not be scratched'. Peirce's theory of meaning was extended by William James (1842–1910), mainly with the object of mediating between science and religion, into a pragmatic theory of truth. James maintained that the way we find whether or not a belief is true is to see whether its practical consequences are satisfactory. Put briefly, a belief is true if it works.

Mach maintained that the laws of science are no more than economical descriptions or summaries of our sense observations. The law of gravitation, for example, merely describes how all bodies appear to attract each other. The value of the law, apart from any aesthetic satisfaction which it may give us, is that it enables us to predict the paths of the planets; it does not tell us anything of the inherent nature of matter.

What had been accepted as the ultimate physical reality, atoms, Mach rejected. He regarded atoms as auxiliary concepts, useful in summarising and classifying our sense observations, but having no existence in the world of sense. Explanation by reduction to the movement of atoms, he argued, was no more understandable than the phenomena to be explained.

Mach's views were bitterly attacked in Germany and they are heresy today in Soviet Russia. Planck objected that if it became generally accepted by men of science that they were investigating only their sense perceptions, and that their findings told them nothing about the objective external world, their desire to continue scientific activity would weaken and die. Lenin felt that Mach's doctrine was dangerous because it left the investigation of reality open to religion, which he regarded, like Marx, as mainly an insidious means of fooling the working classes.

Mach failed to realise the significance in science of what he called auxiliary concepts and of the role of mathematics. His conception of science as an economical description of observed facts, with the object of prediction, was an inadequate account of the elaborate mathematical and deductive structures of physical theory. The simplest example of these structures is geometry. The theorem of geometry that the sum of the three angles of a triangle is equal to two right angles is not merely a concise description of a property of all triangles; it is a logical deduction from a few axioms in which the word triangle is not even mentioned. Kant had regarded Euclidean geometry as an outstanding example of the power of pure reason, independent of sense observation, to give an account of the external world, and he had explained it as a result of the way in which our minds apprehend space. After his death, however, non-Euclidean geometries were discovered, based on different axioms, but equally the product of pure reason, in which the sum of the three angles of a triangle is not equal to two right angles.

A system of geometry is a logical structure deduced from arbitrary axioms, whose conclusions are logically certain, but do not necessarily correspond with experience. Only if the conclusions correspond with experience can the geometry be regarded as a valid physical theory and the conclusions are then just as uncertain, when applied to experience, as those of any other empirical theory. Euclidean geometry describes space in which the gravitational field is negligible; space in which the gravitational field is not negligible is best described by Riemannian geometry, although Euclidean geometry can be used so long as due allowance is made for the effect of the gravitational field.

Henri Poincaré (1854–1912) pointed out that one geometry is no more true than another, but merely more convenient, and he extended the idea to all scientific theory. Physical theories, he held, are not unique inductions from the observations, but free creations of the mind which must be checked for their validity by testing their conclusions against experience. He compared the fundamental principles of theoretical physics, such as the law of inertia and the principle of the conservation of energy, to the axioms of geometry, maintaining that they are merely conventions, defining scientific terms, like the axiom of Euclid that two straight lines cannot enclose a space. 'Principles are conventions and definitions in disguise', he wrote. And since a scientific theory is a logical structure deduced from the principles, 'the aim of science is not things themselves, as the dogmatists in their simplicity imagine, but the relations between things; outside these relations there is no reality knowable.'[1]

The positivism of Mach and the conventionalism of Poincaré are fused together in the modern positivistic view. Physical science is twofold; it consists of (1) a collection of sense observations, (2) theoretical mathematical structures based on a minimum of fundamental principles, whose validities are established by checking their conclusions against the sense observations.

The connection between the world of sense and the world of scientific principles is established by means of the verifiability criterion of meaning. All physical concepts are defined operationally in terms of imaginary experiments; any scientific concept which cannot be defined in this way is scientifically useless since it cannot be related to sense experience. Thus theoretical science begins and ends with observation; it begins with observation in defining its concepts and ends with observation in proving its validity.

It was Einstein who convinced physicists that their concepts must be defined operationally. He showed, for example, that the concept of the absolute simultaneity of two events is metaphysical because it cannot be verified in principle by experiment. If a suitable experiment were carried out, two events which appeared simultaneous to one observer would appear not to be simultaneous

---

[1] Henri Poincaré, *Science and Hypothesis* (1905), pp. 138 and xxlv.

to another observer moving relatively to the first. The essential features of the Special Theory of Relativity were the removal of non-observables from physics and the idea that physics is concerned solely with the relations between what we observe.

Einstein's General Theory of Relativity was also influential in establishing the modern view of science, because it provided a striking confirmation of Poincaré's contention that a physical theory is a free creation of the mind rather than a unique induction from the observations. The theory accounted for the same observations as Newton's Gravitational Theory in an entirely different manner. Einstein compared the two theories to two sets of clothes made for the same body.

Thus a scientific theory represents only partial or relative truth. It serves to correlate our sense observations, and to represent them in the form of a deductive structure, by means of which predictions about physical events can be made. One theory is more true than another, from a pragmatic point of view, if it is superior as a means of making predictions and has a wider range of applicability.

A useful and revealing analogy is the comparison of a scientific theory to a map.[1] A map is a free creation of the mind in the sense that the cartographer is at liberty to invent or choose the method of projection and the conventional signs. In optics three theories are in current use, the ray theory, the wave theory and the quantum theory. The ray theory may be compared to a motor map giving roads and towns but no contours, the wave theory to a relief map, and the quantum theory to a large-scale map showing individual farmhouses and fields. The motor map is simplest and it is more convenient for motoring but it is not so complete a representation of the countryside. No map can depict the countryside as it actually appears to our eyes. Similarly the wave theory and the quantum theory account for more optical phenomena than the ray theory, but their representation of the world of light and colour is still only partial and symbolic.

Eddington went a stage further than Poincaré and Einstein. He maintained that the world of scientific concepts (theoretical physics) as opposed to the world of sense (the facts of observation) is wholly subjective and can be foreseen from what he called epistemological considerations, that is to say, the grounds and

[1] See Stephen Toulmin, *The Philosophy of Science* (Hutchinson's University Library), where the map analogy is discussed in some detail.

philosophy of knowledge. He believed that the laws of nature are based, fundamentally, on chance, and that our discovery of them merely reveals the laws of our thinking. He claimed to have calculated from *a priori* principles several numerical constants of nature, for example that the total number of protons and electrons in the universe is $2 \times 136 \times 2^{256}$. It is as though one could calculate the total number of contour lines on a complete set of maps of the world before making any survey of the ground. There seems to be doubt whether Eddington's mathematics is sound and his thesis is still *sub judice*. If the thesis were true it would clearly have profound significance.

A logical or mathematical structure tells us nothing more than is contained or implied in its fundamental principles. From the axioms of Euclid an all-powerful mind would perceive at once the truth of all the theorems of Euclid's geometry. Hence theoretical science is summarised in its fundamental principles.

One of the most fundamental of the principles of classical science, the law of causality, namely that one event is necessarily connected with another, was shown by Hume to be metaphysical because it cannot be proved or disproved by experience. Kant believed that our idea of causality is inevitable because of the way our minds work. The modern positivist holds that causality is an arbitrary convention, and that the concepts of science are so defined as to preserve it. When causality cannot be reconciled with the other concepts of science it must be rejected.

This view of causality was forced upon physicists by the quantum theory. The only test of causality is the ability to predict one phenomenon from another. Heisenberg's Principle of Uncertainty implies that exact prediction is impossible in principle because we can never know exactly both the position and the velocity of an atomic particle such as an electron. All that we can predict are probabilities and not certainties.

Some physicists like Planck were reluctant to abandon strict causality, maintaining that it should be regarded as present in nature rather than as a free creation of the mind which can be abandoned in the quantum theory. He held that a particle can be conceived as having exact position and exact velocity; the difficulty is that our means of observation are inadequate to enable us to measure both simultaneously. But according to the verifiability criterion of meaning, since we cannot observe in principle a particle having exact position and exact velocity, a particle with these properties is meaningless and does not exist. Since we cannot predict in principle the future behaviour of an atomic particle, causality is metaphysical or scientifically meaningless.

Mach's view that atomic particles do not exist is no longer tenable. Several methods of inferring their existence beyond doubt have been discovered; they can make a flash on a screen, or a track in a cloud chamber or on a photographic plate. Anything which can be related to sense observation is held, by definition, to exist. An atom, like other objects such as a table, is a logical construct from sense data.

Since all scientific observations are sense observations, it is not possible by any experiment to obtain information about an external reality which may be behind the world of sense. According to the verifiability criterion of meaning, therefore, such an external reality is metaphysical and hence, for the positivist, it does not exist.

It is at this point, if we wish to go further, that we must pass from the realm of science to philosophy or metaphysics. Everything turns on whether we are prepared to accept that there is reliable knowledge outside sense observation, and hence to reject the verifiability criterion of meaning, apart from its essential use in the realm of science. The realist takes the commonsense view that the external world does exist and holds that the problem is not its existence, but how we know that it exists. Whereas the positivist takes sense experience as primary and bases existence upon it, the realist reverses the process, taking existence as primary. The realist argues that the verifiability criterion of meaning, and hence of existence, is itself metaphysical because it cannot be checked by reference to experience.

The realist is not prepared to state what is the ultimate nature of the external world. The materialist, on the other hand, believes that it is material and the idealist that it is mental.

The school of philosophy known as logical positivism, founded in Vienna and now widely accepted in the western world, sidesteps both materialism and idealism by putting the ancient problem of whether the world is matter or mind in the form: Is matter or mind a more convenient concept to group events in order to represent the world?

Materialism and idealism are then seen, like the theories of science, as partial truth with a limited range of applicability. Logical positivists accept as knowledge only that which is verifiable and hence that which is based on, or related to, sense perception. They regard scientific knowledge as the only true knowledge, taking the task of philosophy to be a critique and analysis of scientific propositions. They contrast their scientific philosophy with the speculative philosophy of, say, Plato or Descartes, who believed that reason, unaided by sense, could reveal truth.

Science deliberately excludes values and purpose, although its ultimate justification must be in terms of value. Scientific assertions are factual, as opposed to the normative assertions of ethics and aesthetics. A normative assertion is one that provides a norm or pattern to which our judgements and actions should comply: for example, that this is good and that bad, this beautiful and that ugly. The logical positivist maintains that when you say that a certain thing or act is good, you are not making a verifiable statement about it; you are merely expressing personal feelings. The world contains only as much value and purpose as you yourself put into it.

On the other hand some modern philosophers, like G. E. Moore, have maintained that values are inherent in what is evaluated. When you perceive that something is good it is rather like seeing that an object is blue, except that goodness is a 'non-material' quality. Values are irreducible and unanalysable; they are grasped intuitively and emotively. Moore argued that if moral judgements were merely expressions of personal feelings, it would not be possible to debate and disagree about them. It is not to be expected that knowledge of value can be as reliable as factual knowledge because people do not agree in their emotional reactions as they do in their perceptual observations.

Values and purpose must therefore be sought and appraised outside science, although no adequate discipline has yet been developed for dealing with them.

* * *

Positivism, the fashionable philosophy of science in the western world, is proscribed in the Soviet Union where belief in dialectical materialism is obligatory. Dialectical materialism is a complete metaphysical system. Some of its adherents in the west display a touch of mystical fervour, quoting its major prophets, Marx, Engels and Lenin with as much reverence and trust as a Plymouth brother quotes his Bible. Behind the philosophy there is a political aim, the conservation and extension of communism.

The fundamental tenets of dialectical materialism are: (i) that the world is material, everything mental and spiritual being the product of material processes; (ii) that matter is an objective reality, its laws being discoverable by science; and (iii) that there is a dialectical process in the universe, which accounts for the emergence of novelty in evolution and human history.

Lenin wrote scornfully of the idea that science is concerned only with our sense perceptions and that we cannot know the real nature of the external world, contending that Mach's doctrine leads to the absurdity of solipsism, which is the belief that we can know only the contents of our own minds. He held that our sense perceptions are images or reflections of matter and that matter is primary because it existed before consciousness evolved.

The plausibility of materialism has grown less with the development of physics in the twentieth century. The fundamental property of matter is its mass or inertia. The mass of a body depends on its motion, increasing as the velocity increases. We cannot tell whether or not a body has absolute motion; we can observe only the relative motion of two bodies. The mass of a body depends, therefore, on our own choice, and our minds cannot be merely mirrors of an external world.[1]

Again, mass can be annihilated giving rise to an equivalent amount of energy. Atoms and sub-atomic particles under certain conditions are best regarded as waves. We cannot, in principle, find both the position and velocity of a sub-atomic particle. The modern physicist, therefore, has abandoned all pretensions to be describing the fundamental nature of matter; he is content with correlating physical phenomena.

Dialectical materialism differs from the mechanical materialism of the eighteenth-century *philosophes* and of some nineteenth-century scientists by reason of its dialectical element, which Marx took over from Hegel, applying it to the material world instead of to the Hegelian world of ideas. The basic idea of dialectics is that all change in the universe occurs in the following manner: any

[1] See H. Dingle, 'The Scientific Outlook in 1851 and 1951', *British Journal for the Philosophy of Science*, II (1951), p. 86.

process must give rise to a counter-process, and these unite to produce something new. In the words of Hegel, thesis leads to antithesis, and they combine into a synthesis. The supreme example, for all Marxists, of the dialectical process is the capitalist thesis, the proletariat antithesis, and the communist synthesis.

It is not easy to find convincing examples of dialectical change in the history of physical science. Engels maintained that the dialectical process is to be found in nature itself and not merely in the study of it. However, examples of the former are not very plausible and we will consider the situation which arose when, at the beginning of the twentieth century, the wave theory and the quantum theory were in conflict. This was what the Marxists call the struggle or interpenetration of opposites. The conflict was resolved in the dialectical synthesis of quantum mechanics and of the theory of complementarity. It was realised that certain concepts, such as the concept of causality, had a limited validity— called by Marxists a dialectic negation.

The physicists who were responsible for developing quantum mechanics and the theory of complementarity received no assistance from the doctrine of dialectical materialism. Indeed, the laws of dialectical change, if taken into account by physicists in their work, would serve only to hamper, and there is evidence that Russian physicists ignore them, just as in the west the research of very few physicists is influenced by positivism. Russian biologists, however, have been compelled to adjust their genetical theories to comply with the tenets of dialectical materialism.

To the biologist, positivism makes little appeal. His branch of science is largely descriptive and is less symbolised and deductive than physics. The only body of abstract biological theory is the gene theory. Faced with the mystery and complexity of evolution some biologists in the west have found dialectical materialism suggestive and stimulating; for the majority, however, it is too vague and elastic to be helpful.

* * *

Science is now so much an integral part of modern civilisation that it is no longer merely the private activity of individuals, like composing music or painting pictures. It is a social function. The health, water-supply, sanitation, power, communications and some of the food-supply of dense industrial populations are dependent in considerable measure on a supply of scientifically trained technicians. Any industrial nation which neglects scientific research, pure as well as applied, will find itself lagging behind its competitors within ten or twenty years.

Since science has such important social consequences, should it be socially controlled? In Great Britain during the 1930's a number of writers, notably J. D. Bernal, J. G. Crowther, Lancelot Hogben, and Benjamin Farrington, put forward the Marxist view that science advances in response to the economic needs of society and they urged that all scientific research should be planned by the State. The Association of Scientific Workers adopted the same policy. Alarmed by what they regarded as a threat to the freedom of science, a group of British scientists formed in 1940 the Society for Freedom in Science and this now has a considerable membership in the U.S.A. and in the rest of the world.

Research in Russia is planned by the Soviet Academy of Sciences which has some degree of autonomy but is a government organ. The speed with which Russian scientists produced atomic bombs and their priority in launching the first rocket satellites have been striking evidence of the efficiency of their planning.

Those who advocate planning usually blur the distinction between pure science and applied science. They hope that the argument for planning applied science, which is a strong one, will carry the argument for pure science with it.

The outstanding example in Great Britain of the State planning of applied research was the setting up by the government, after the Second World War, of the Atomic Energy Authority, with the main object of the development of atomic power stations. The formation of the Atomic Energy Authority followed the success of the Anglo-American wartime research project for the development of the atomic bomb.

Most research in applied science in western countries is planned, not by the State, but by large combines or cartels, such as Imperial Chemical Industries, the General Electric Company, the Standard Oil Company, Bell Telephone Laboratories, Du Pont, Schneider Creusot and I.G. Farben. The direction of industrial research laboratories is usually tolerant and enlightened because it is realised that the by-products of an investigation may prove to be more valuable than the purpose of the investigation itself. Two Americans working in industrial research laboratories won

Nobel prizes for research in pure science—Langmuir of the General Electric Company and Davisson of the Bell Telephone Laboratories.

There is much to be said for fostering applied science under a variety of conditions. Too strong a central directive and control can have disastrous results. Hitler's obstruction of technical developments, particularly in the field of rockets, may have lost him the war. Nor is resistance to new technical ideas and a failure of imaginative grasp confined to dictators. Two instances from the field of aeronautics may be cited: the statement in 1910 by the British Secretary of State for War, 'We do not consider that aeroplanes will be of any possible use for war purposes', and the discouragement by the Air Ministry of Whittle's development of the gas-turbine.

It may be argued that blatant errors like these would be avoided if the government were advised by a committee of eminent scientists. This may be so in applied science. But no committee of planners, however eminent, can forecast which lines of research in pure science will prove to be the most fruitful and valuable. The characteristic of scientific discovery has been its unpredictability. Much of the finest scientific investigation is inspired by curiosity and is uninfluenced by economic needs. The freedom of the individual research worker to follow his own bent is, in the long run, more efficient than planning because it is more likely to ensure that the unpredictable will emerge.

The highest scientific authority in Great Britain, the Royal Society, has a tradition of independence of political controversy and freedom from all government interference. Very large sums of money are provided annually by the government for research, and the danger of political control is minimised by not making the grants direct. The grants are allocated, without undue governmental conditions, by the University Grants Committee, by the Department of Scientific and Industrial Research, and by the Agricultural Research Council.

The fact that science today requires large financial aid from the State means that governments control the maintenance and expansion of scientific work. In the sense of determining how much science the State requires and can afford, government planning is unavoidable.

*   *   *

A useful index of how science has flourished during the present century in the different countries of the world is provided by the comparative numbers of Nobel prizes awarded to scientists of different nationalities.

Alfred Nobel was a Swedish industrialist and the discoverer of dynamite. He died in 1896; in his will, he directed that the interest on his large estate should be distributed annually in prizes for Physics, Chemistry, Physiology and Medicine, Literature and the Furtherance of Peace. The prizes were first given in 1901 and, although their monetary value has now declined, being about £15,000 for each subject, their prestige has increased. The award of a Nobel prize is regarded as the highest of scientific honours. The winners are selected by the Nobel Committee of the Royal Swedish Academy of Sciences, to whom suggestions are sent from outstanding scientists throughout the world.

It is the convention to credit a Nobel prize to the country in which the recipient is resident at the time of the award. The country which gains most from this is the U.S.A., to which many European scientists fled during the Nazi and Fascist regimes. In the table below the convention causes some distortion of the picture but the figures are broadly significant.

Number of Nobel prizes for science awarded in the periods 1901–39 and 1943–59

Nationality	1901–39	1943–59	Total
American	15	42	57
German	35	8	43
British	22	15	37
French	16	–	16
Dutch	8	1	9
Swedish	6	3	9
Austrian	6	1	7
Russian	2	4	6
Swiss	4	2	6
Danish	4	1	5
Italian	3	1	4
Canadian	2	1	3
Belgian	2	–	2

In addition there was one Nobel prize-winner in 1901–59 of each of the following nationalities: Argentinian, Czech, Finnish, Hungarian, Indian, Japanese, Portuguese and Spanish.

The years 1901–59 have been divided into two periods, 1901–39 and 1943–59 (no awards being made in the war years 1940–42).

The most striking feature of the table is the comparative decline in the second period of the countries of western Europe, which exhausted themselves materially and intellectually in the two world wars. Only America and Russia have gained more awards in the second period than in the first.

In the first period, Germany was pre-eminent; in the second period it declined to third place. The damage inflicted on German science by the 1914–18 war was accentuated by the racial and ideological policy of the Nazis. During the 1930's some two thousand eminent scientists left Germany and Austria and the number of students studying science in German universities declined by two-thirds.

Britain, of all the western European countries, has best maintained its position. France, with so glorious a history of scientific achievement, has suffered a sad, and no doubt temporary, eclipse.

Will American science maintain its superiority in the remaining decades of the twentieth century? A strong challenge is to be expected from Russia, now producing more scientists and engineers than any other country in the world. Russian science is being stimulated and encouraged by every possible means, including higher salaries for scientists, in relation to the rest of the community, and smaller taxation, than in the U.S.A. and Britain. Incentives to make able men work hard and to take up the most socially useful professions are more than repaid by the general rise in the prosperity of the whole community which results.

In contrast to the success of Russian scientific education, the U.S.A. is reaping the reward of a spate of educational theory advocating too much choice to the child as to what and how he should learn, while the educational systems of France and England are still handicapped by their powerful classical and literary traditions.

PART II

NOTE TO PART II

This selection of extracts, taken mainly from the classics of science, follows the same sequence as Part I. The extracts are numbered seriatim, in the order of the chapters in the first part, and they are arranged under these chapter headings.

They have been selected for the general reader rather than primarily for the science specialist. Their selection rested upon their liveliness and human interest as well as upon their scientific value. Hence only a minority of the extracts have been taken from scientific papers, which are apt to be too technical for the general reader; many have come from books, lectures and letters.

The extracts are usually quite short and, I hope, to the point. But occasionally I have made them rather longer, when the subject is one of general interest: for example the extracts from Lamarck and Darwin on evolution, from Hutton on his geological studies in the Scottish border country, and from Pasteur's dramatic account of his first treatment of victims bitten by rabid dogs.

Some extracts have been selected because they are amusing as well as instructive; these include the account of Archimedes from Plutarch's *Lives*, Swift's attack on astrology, the duologue from Diderot, Voltaire's *conte*, Goethe's conversation with Eckermann about his theory of colours and the article by Raphael Demos. Thomas Young's ponderous reply to the vindictive attacks made upon him in the *Edinburgh Review* can hardly be called an amusing document, but it has its human interest and it includes a notable statement of the principle of interference.

The extracts illustrating the chapters on the seventeenth, eighteenth, nineteenth and twentieth centuries are mainly philosophical. For the seventeenth century, brief extracts from Bacon's *Novum Organum*, Descartes' *Discours de la Méthode* and Hobbes' *Leviathan* are accompanied by a rather longer passage from *New Atlantis*. For the eighteenth century, Locke, Berkeley, Hume and Kant were inevitable choices; something astringent and sparkling from Diderot and Voltaire give the flavour of the Enlightenment; while Goethe represents the school of *Naturphilosophie*. The nineteenth century is represented by Comte, Tyndall, Huxley, Spencer and Haeckel. One could question whether any of them ever produced a classic but they had a wide influence and they illustrate vividly the impact of science on nineteenth-century thought. For twentieth-century philosophy of science, extracts from Mach, Poincaré, Einstein and Eddington were almost obligatory, while the rival philosophy of Russian science is represented by Engels and Lenin.

The sources of the extracts are listed at the end of the book and I wish to express my thanks to the various publishers who have given permission for their reproduction.

A.E.E.M.

CAMBRIDGE
April 1960

CHAPTER I

SCIENCE AND TECHNOLOGY IN ANCIENT AND MEDIEVAL TIMES

1. From ARISTOTLE'S *Physics*

[The following is the opening passage of Aristotle's *Physics* and it illustrates the great achievement of the Greeks in creating science as an abstract, logical system based on principles, replacing the animism of earlier civilizations.]

In all sciences that are concerned with principles or causes or elements, it is acquaintance with these that constitutes knowledge or understanding. For we conceive ourselves to know about a thing when we are acquainted with its ultimate causes and first principles and have got down to its elements. Obviously, then, in the study of Nature too, our first object must be to establish principles.

Now the path of investigation must lie from what is more immediately cognizable and clear to us, to what is clearer and more intimately cognizable in its own nature; for it is not the same thing to be directly accessible to our cognition and to be intrinsically intelligible. Hence, in advancing to that which is intrinsically more luminous and by its nature accessible to deeper knowledge, we must needs start from what is more immediately within our cognition, though in its own nature less fully accessible to understanding.

Now the things most obvious and immediately cognizable by us are concrete and particular, rather than abstract and general; whereas elements and principles are only accessible to us afterwards, as derived from the concrete later when we have analysed them. So we must advance from the concrete whole to the several constituents which it embraces; for it is the concrete whole that is the more readily cognizable by the senses. And by calling the concrete a 'whole' I mean that it embraces in a single complex a diversity of constituent elements, factors, or properties.

2. From PLUTARCH'S *Lives*

[This entertaining account of the machines devised by Archimedes, and their use in the defence of Syracuse, is particularly instructive in its last paragraph, which illustrates one of the reasons why the scientific revolution did not occur in hellenistic times.]

These machines he had designed and contrived, not as matters of any importance, but as mere amusements in geometry; in compliance with King Hiero's desire and request, some little time before, that he should reduce to practice some part of his admirable speculation in science.... Archimedes, however, in writing to King Hiero, whose friend and near relation he was, had stated that given the force, any given weight might be moved, and even boasted, we are told, relying on the strength of demonstration, that if there were another earth, by going into it he could remove this. Hiero being struck with amazement at this, and entreating him to make good this problem by actual experiment, and show some great weight moved by a small engine, he fixed accordingly upon a ship of burden out of the king's arsenal, which could not be drawn out of the dock without great labour and many men; and, loading her with many passengers and a full freight, sitting himself the while far off, with no great endeavour, but only holding the head of the pulley in his hand and drawing the cords by degrees, he drew the ship in a straight line, as smoothly and evenly as if she had been in the sea. The king, astonished at this, and convinced of the power of the art, prevailed upon Archimedes to make him engines accommodated to all the purposes, offensive and defensive, of a siege....

When, therefore, the Romans assaulted the walls in two places at once, fear and consternation stupefied the Syracusans, believing that nothing was able to resist that violence and those forces. But when Archimedes began to ply his engines, he at once shot against the land forces all sorts of missile weapons, and immense masses of stone that came down with incredible noise and violence; against which no man could stand; for they knocked down those upon whom they fell in heaps, breaking all their ranks and files. In the meantime huge poles thrust out from the walls over the ships sunk some by the great weights which they let down from on high upon them; others they lifted up into the air by an iron hand or beak like a crane's beak, and, when they had drawn them up by the prow, and set them on end upon the poop, they plunged them to the bottom of the sea; or else the ships, drawn by engines within, and whirled about, were dashed against steep rocks that stood jutting out under the walls, with great destruction of the soldiers that were aboard them. A ship was frequently lifted up to a great height in the air (a dreadful thing to behold), and was rolled to and fro, and kept swinging, until the mariners were all thrown out, when at length it was dashed against the rocks, or let fall. . . .

In fine, when such terror had seized upon the Romans, that, if they did but see a little rope or a piece of wood from the wall, instantly crying out, that there it was again, Archimedes was about to let fly some engine at them, they turned their backs and fled, Marcellus desisted from conflicts and assaults, putting all his hope in a long siege. Yet Archimedes possessed so high a spirit, so profound a soul, and such treasures of scientific knowledge, that though these inventions had now obtained him the renown of more than human sagacity, he yet would not deign to leave behind him any commentary or writing on such subjects; but, repudiating as sordid and ignoble the whole trade of engineering, and every sort of art that lends itself to mere use and profit, he placed his whole affection and ambition in those purer speculations where there can be no reference to the vulgar needs of life.

3. From *The Book of Beasts* (a twelfth-century bestiary)

[This extract illustrates the medieval view that everything in nature is symbolical of the spiritual life. Bestiaries were based on the compilation of an anonymous person called Physiologus, who lived between the second and fifth centuries A.D., probably in Egypt.]

Scientists say that Leo [the Lion] has three principal characteristics.

His first feature is that he loves to saunter on the tops of mountains. Then, if he should happen to be pursued by hunting men, the smell of the hunters reaches up to him, and he disguises his spoor behind him with his tail. Thus the sportsmen cannot track him.

It was in this way that our Saviour (i.e. the Spiritual Lion of the Tribe of Judah, the Rod of Jesse, the Lord of Lords, the Son of God) once hid the spoor of his love in the high places, until being sent by the Father, he came down into the womb of the Virgin Mary and saved the human race which had perished. Ignorant of the fact that his spoor could be concealed, the Devil (i.e. the hunter of humankind) dared to pursue him with temptations like a mere man. Even the angels themselves who were on high, not recognizing his spoor, said to those who were going up with him when he ascended to his reward: 'Who is this King of Glory?'

The Lion's second feature is, that when he sleeps, he seems to keep his eyes open.

In this very way, Our Lord also while sleeping in the body, was buried after being crucified—yet his Godhead was awake. As it is said in the *Song of Songs*, 'I am asleep and my heart is awake', or, in the Psalm, 'Behold, he that keepeth Israel shall neither slumber nor sleep'.

The third feature is this, that when a lioness gives birth to her cubs, she brings them forth dead and lays them up lifeless for three days—until their father, coming on the third day, breathes in their faces and makes them alive.

Just so did the Father Omnipotent raise Our Lord Jesus Christ from the dead on the third day. Quoth Jacob: 'He shall sleep like a lion, and the lion's whelp shall be raised.'

4. From JONATHAN SWIFT's *Predictions for the Year 1708 by Isaac Bickerstaff Esq.*

[After thousands of years of credulity about astrology educated opinion ceased to take it seriously as a result of the new scientific outlook on the universe of the seventeenth century.]

My first prediction is but a trifle, yet I will mention it to show how ignorant these sottish pretenders to astrology are in their own concerns; it relates to Partridge the Almanackmaker. I have consulted the star of his nativity, by my own rules, and find that he will infallibly die upon 29th March next, about eleven at night, of a raging fever; therefore advise him to consider of it and settle his affairs in time.

[When Mr Partridge protested that he had not died, Swift retorted that over a thousand gentlemen had looked at Partridge's almanack and exclaimed, NO MAN ALIVE EVER WRIT SUCH DAMNED STUFF AS THIS.]

CHAPTER 2

THE COPERNICAN THEORY

5. From NICOLAUS COPERNICUS' *On the Revolutions of the Celestial Orbs* (1543)

[Book I of *On the Revolutions* gives a brief general account of the heliocentric universe. In the following extract Copernicus first considers the diurnal rotation of the Earth and suggests that the Earth would not fly to pieces under the action of the centrifugal force caused by its rotation because its motion is natural and not forced. He then explains how the Earth, with the Moon, revolves round the Sun, together with the rest of the Planets, whose distances from the Sun must be in the order of their periods of revolution. Saturn, with a period of 30 years, is furthest away from the Sun, and Mercury, with a period of 80 days, is nearest. The Sun, in the centre, is described as the Ruler of the Universe.]

If then, says Ptolemy, Earth moves at least with a diurnal rotation, the result must be the reverse of that described above [i.e. that terrestial elements move in straight lines up or down]. For the motion must be of excessive rapidity, since in 24 hours it must impart a complete rotation of the Earth. Now things rotating very rapidly resist cohesion or, if united, are apt to disperse, unless firmly held together. Ptolemy therefore says that Earth would have been dissipated long ago, and (which is the height of absurdity) would have destroyed the Heavens themselves; and certainly all living creatures and other heavy bodies free to move could not have remained on its surface, but must have been shaken off. Neither could falling objects reach their appointed place vertically beneath, since in the meantime the Earth would have moved swiftly from under them. Moreover clouds and everything in the air would continually move westward.

The Insufficiency of these Arguments, and their Refutation

For these and like reasons, they say that Earth surely rests at the centre of the Universe. Now if one should say that the Earth *moves*, that is as much as to say that the motion is natural, not

361

forced; and things which happen according to nature produce the opposite effects to those due to force. Things subject to any force, gradual or sudden, must be disintegrated, and cannot long exist. But natural processes being adapted to their purpose work smoothly.

Idle therefore is the fear of Ptolemy that Earth and all thereon would be disintegrated by a natural rotation, a thing far different from an artificial act. . . .

Why then hesitate to grant Earth that power of motion natural to its shape, rather than suppose a gliding round of the whole Universe, whose limits are unknown and unknowable? And why not grant that the diurnal rotation is only apparent in the Heavens but real in the Earth? It is but as the saying of Aeneas in Virgil— 'We sail forth from the harbour, and lands and cities retire'. As the ship floats along in the calm, all external things seem to have the motion that is really that of the ship, while those within the ship feel that they and all its contents are at rest.

* * *

Since then there is no reason why the Earth should not possess the power of motion, we must consider whether in fact it has more motions than one, so as to be reckoned as a Planet.

That the Earth is not the centre of all revolutions is proved by the apparently irregular motions of the Planets and the variations in their distances from the Earth. These would be unintelligible if they moved in circles concentric with the Earth. Since, therefore, there are more centres than one, we may discuss whether the centre of the Universe is or is not the Earth's centre of gravity.

Now it seems to me gravity is but a natural inclination, bestowed on the parts of bodies by the Creator so as to combine the parts in the form of a sphere and thus contribute to their unity and integrity. And we may believe this property present even in the Sun, Moon and Planets, so that thereby they retain their spherical form notwithstanding their various paths. If, therefore, the Earth also has other motions, these must necessarily resemble the many outside motions having a yearly period. For if we transfer the motion of the Sun to the Earth, taking the Sun to be at rest, then morning and evening, risings and settings of Stars will be unaffected, while the stationary points, retrogressions, and progressions of the Planets are due not to their own proper motions, but to that of the Earth, which they reflect. Finally, we shall place the Sun himself at the centre of the Universe. All this is suggested by the systematic procession of events and the harmony of the whole Universe, if only we face the facts, as they say, 'with both eyes open'....

We therefore assert that the centre of the Earth, carrying the Moon's path, passes in a great orbit among the other Planets in an annual revolution round the Sun; that near the Sun is the centre of the Universe; and that whereas the Sun is at rest, any apparent motion of the Sun can be better explained by the motion of the Earth. Yet so great is the Universe that though the distance of the Earth from the Sun is not insignificant compared with the size of any other planetary path, in accordance with the ratios of their sizes, it is insignificant compared with the distance of the sphere of the Fixed Stars.

I think it is easier to believe this than to confuse the issue by assuming a vast number of spheres, which those who keep Earth at the centre must do. We thus rather follow Nature, who producing nothing vain or superfluous often prefers to endow one cause with many effects. Though these views are difficult, contrary to expectation, yet in the sequel we shall, God willing, make them abundantly clear at least to mathematicians.

Given the above view—and there is none more reasonable—that the periodic times are proportional to the sizes of the orbits, then the order of the spheres, beginning with the most distant, is as follows. Most distant of all is the Sphere of the Fixed Stars, containing all things, and being therefore itself immovable. It represents that to which the motion and position of all the other bodies must be referred. Some hold that it too changes in some way, but we shall assign another reason for this apparent change, as will appear in the account of the Earth's motion. Next is the planet Saturn, revolving in 30 years. Next comes Jupiter, moving in a 12 year circuit: then Mars, who goes round in 2 years. The fourth place is held by the annual revolution in which the Earth is contained, together with the orbit of the Moon as on an epicycle. Venus, whose period is 9 months, is in the fifth place, and sixth is Mercury, who goes round in the space of 80 days.

In the middle of all sits Sun enthroned. In this most beautiful temple could we place this luminary in any better position from which he can illuminate the whole at once? He is rightly called the Lamp, the Mind, the Ruler of the Universe; Hermes Trismegistus names him the Visible God, Sophocles' Electra calls him the All-Seeing. So the Sun sits as upon a royal throne ruling his children the planets which circle round him.

6. From GALILEO'S letter to Madame CHRISTINA OF LORRAINE, Grand Duchess of Tuscany, *Concerning the Use of Biblical Quotations in Matters of Science* (1615)

Some years ago, as Your Serene Highness well knows, I discovered in the heavens many things that had not been seen before our own age. The novelty of these things, as well as some consequences which followed from them in contradiction to the physical notions commonly held among academic philosophers, stirred up against me no small number of professors—as if I had placed these things in the sky with my own hands in order to upset nature and overturn the sciences....

Persisting in their original resolve to destroy me and everything mine by any means they can think of, these men are aware of my views in astronomy and philosophy. They know that as to the arrangement of the parts of the universe, I hold the sun to be situated motionless in the centre of the revolution of the celestial orbs while the earth rotates on its axis and revolves about the sun. They know also that I support this position not only by refuting the arguments of Ptolemy and Aristotle, but by producing many counter-arguments; in particular, some which relate to physical effects whose causes can perhaps be assigned in no other way. In addition there are astronomical arguments derived from many things in my new celestial discoveries that plainly confute the Ptolemaic system while admirably agreeing with and confirming the contrary hypothesis. Possibly because they are disturbed by the known truth of other propositions of mine which differ from those commonly held, and therefore mistrusting their defence so long as they confine themselves to the field of philosophy, these men have resolved to fabricate a shield for their fallacies out of the mantle of pretended religion and the authority of the Bible. These they apply, with little judgment, to the refutation of arguments that they do not understand and have not even listened to.

First they have endeavoured to spread the opinion that such propositions in general are contrary to the Bible and are consequently damnable and heretical....

With regard to this argument, I think in the first place that it is very pious to say and prudent to affirm that the holy Bible can never speak untruth—whenever its true meaning is understood. But I believe nobody will deny that it is often very abstruse, and may say things which are quite different from what its bare words signify. Hence in expounding the Bible if one were always to confine oneself to the unadorned grammatical meaning, one might fall into error. Not only contradictions and propositions far from true might thus be made to appear in the Bible, but even grave heresies and follies. Thus it would be necessary to assign to God feet, hands, and eyes, as well as corporeal and human affections, such as anger, repentance, hatred, and sometimes even the forgetting of things past and ignorance of those to come. These propositions uttered by the Holy Ghost were set down in that manner by the sacred scribes in order to accommodate them to the capacities of the common people, who are rude and unlearned. For the sake of those who deserve to be separated from the herd, it is necessary that wise expositors should produce the true senses of such passages, together with the special reasons for which they were set down in these words. This doctrine is so widespread and so definite with all theologians that it would be superfluous to adduce evidence for it.

7. From GALILEO'S *Dialogue concerning the Two Chief World Systems—Ptolemaic and Copernican* (1632)

[There are three interlocutors: Salviati, who represents the views of Galileo; Simplicio, who is a follower of Aristotle; and Sagredo who represents an intelligent layman, uncommitted to either side. The extract is taken from The Third Day, about three-quarters of the way through the Dialogue, and in it the arrangement of the planets round the Sun is discussed in some detail.]

Salviati. Now if it is true that the centre of the universe is that point around which all the orbs and world bodies (that is, the planets) move, it is quite certain that not the earth, but the sun, is to be found at the centre of the universe. Hence, as for this first general conception, the central place is the sun's, and the earth is to be found as far away from the centre as it is from the sun.

Simplicio. How do you deduce that it is not the earth, but the sun, which is at the centre of the revolutions of the planets?

Salviati. This is deduced from most obvious and therefore most powerfully convincing observations. The most palpable of these, which excludes the earth from the centre and places the sun there, is that we find all the planets closer to the earth at one time and further from it at another. The differences are so great that Venus, for example, is six times as distant from us at its farthest as at its closest, and Mars soars nearly eight times as high in the one state as in the other. You may thus see whether Aristotle was not some trifle deceived in believing that they were always equally distant from us.

Simplicio. But what are the signs that they move around the sun?

Salviati. This is reasoned out from finding the three outer planets—Mars, Jupiter, and Saturn—always quite close to the earth when they are in opposition to the sun, and very distant when they are in conjunction with it.[1] This approach and recession is of such moment that Mars when close looks sixty times as large as when it is most distant. Next, it is certain that Venus and Mercury must revolve around the sun, because of their never moving far away from it, and because of their being seen now beyond it and now on this side of it, as Venus's changes of shape conclusively prove. As to the moon, it is true that this can never separate from the earth in any way, for reasons that will be set forth more specifically as we proceed.

Sagredo. I have hopes of hearing still more remarkable things arising from this annual motion of the earth than were those which depended upon its diurnal rotation.

Salviati. You will not be disappointed, for as to the action of the diurnal motion upon celestial bodies, it was not and could not be anything different from what would appear if the universe were to rush speedily in the opposite direction. But this annual motion, mixing with the individual motions of all the planets, produces a great many oddities, which in the past have baffled all the greatest men in the world.

---

[1] Heavenly bodies in opposition differ in position by 180°, and in conjunction they have the same longitude, i.e. are passing each other.

Now returning to these first general conceptions, I repeat that the centre of the celestial rotation for the five planets, Saturn, Jupiter, Mars, Venus, and Mercury, is the sun; this will hold for the earth too, if we are successful in placing that in the heavens. Then as to the moon, it has a circular motion around the earth, from which as I have already said it cannot be separated; but this does not keep it from going around the sun along with the earth in its annual movement.

Simplicio. I am not yet convinced of this arrangement at all. Perhaps I should understand it better from the drawing of a diagram, which might make it easier to discuss.

Salviati. That shall be done. But for your greater satisfaction and your astonishment, too, I want you to draw it yourself. You will see that however firmly you may believe yourself not to understand it, you do so perfectly, and just by answering my questions you will describe it exactly. So take a sheet of paper and the compasses; let this page be the enormous expanse of the universe, in which you have to distribute and arrange its parts as reason shall direct you. And first, since you are sure without my telling you that the earth is located in this universe, mark some point at your pleasure where you intend this to be located, and designate it by means of some letter.

Simplicio. Let this be the place of the terrestrial globe, marked *A*.

Salviati. Very well. I know in the second place that you are aware that this earth is not inside the body of the sun, nor even contiguous to it, but is distant from it by a certain space. Therefore assign to the sun some other place of your choosing, as far from the earth as you like, and designate that also.

Simplicio. Here I have done it; let this be the sun's position, marked *O*.

Salviati. These two established, I want you to think about placing Venus in such a way that its position and movement can conform to what sensible experience shows us about it. Hence you must call to mind, either from past discussions or from your own observations, what you know happens with this star. Then assign it whatever place seems suitable for it to you.

Simplicio. I shall assume that those appearances are correct which you have related and which I have read also in the booklet of theses; that is, that this star never recedes from the sun beyond a certain definite interval of forty degrees or so; hence it not only never reaches opposition to the sun, but not even quadrature, nor so much as a sextile aspect.[1] Moreover, I shall assume that it displays itself to us about forty times as large at one time than at another; greater when, being retrograde, it is approaching evening conjunction with the sun, and very small when it is moving forward toward morning conjunction, and furthermore that when it appears very large, it reveals itself in a horned shape, and when it looks very small it appears perfectly round.

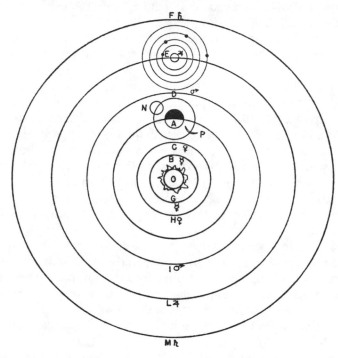

These appearances being correct, I say, I do not see how to escape affirming that this star revolves in a circle around the sun, in such a way that this circle cannot possibly be said to embrace and contain within itself the earth, nor to be beneath the sun (that is, between the sun and the earth), nor yet beyond the sun. Such a circle cannot embrace the earth because then Venus would

[1] Quadrature means that heavenly bodies differ in position by 90°, sextile aspect by 60°.

sometimes be in opposition to the sun; it cannot be beneath the sun, for then Venus would appear sickle-shaped at both conjunctions; and it cannot be beyond the sun, since then it would always look round and never horned. Therefore for its lodging I shall draw the circle *CH* around the sun, without having this include the earth.

Salviati. Venus provided for, it is fitting to consider Mercury, which, as you know, keeping itself always around the sun, recedes therefrom much less than Venus. Therefore consider what place you should assign to it.

Simplicio. There is no doubt that, imitating Venus as it does, the most appropriate place for it will be a smaller circle, within this one of Venus and also described about the sun. A reason for this, and especially for its proximity to the sun, is the vividness of Mercury's splendour surpassing that of Venus and all the other planets. Hence on this basis we may draw its circle here and mark it with the letters *BG*.

Salviati. Next, where shall we put Mars?

Simplicio. Mars, since it does come into opposition with the sun, must embrace the earth with its circle. And I see that it must also embrace the sun; for, coming into conjunction with the sun, if it did not pass beyond it but fell short of it, it would appear horned as Venus and the moon do. But it always looks round; therefore its circle must include the sun as well as the earth. And since I remember your having said that when it is in opposition to the sun it looks sixty times as large as when in conjunction, it seems to me that this phenomenon will be well provided for by a circle around the sun embracing the earth, which I draw here and mark *DI*. When Mars is at the point *D*, it is very near the earth and in opposition to the sun, but when it is at the point *I*, it is in conjunction with the sun and very distant from the earth.

And since the same appearances are observed with regard to Jupiter and Saturn (although with less variation in Jupiter than in Mars, and with still less in Saturn than in Jupiter), it seems clear to me that we can also accommodate these two planets very neatly with two circles, still around the sun. This first one, for Jupiter, I mark *EL*; the other, higher, for Saturn, is called *FM*.

Salviati. So far you have comported yourself uncommonly well. And since, as you see, the approach and recession of the three outer planets is measured by double the distance between the earth and the sun, this makes a greater variation in Mars than in Jupiter because the circle *DI* of Mars is smaller than the circle *EL* of Jupiter. Similarly, *EL* here is smaller than the circle *FM* of Saturn, so the variation is still less in Saturn than in Jupiter, and this corresponds exactly to the appearances. It now remains for you to think about a place for the moon.

Simplicio. Following the same method (which seems to me very convincing), since we see the moon come into conjunction and opposition with the sun, it must be admitted that its circle embraces the earth. But it must not embrace the sun also, or else when it was in conjunction it would not look horned but always round and full of light. Besides, it would never cause an eclipse of the sun for us, as it frequently does, by getting in between us and the sun. Thus one must assign to it a circle around the earth, which shall be this one, *NP*, in such a way that when at *P* it appears to us here on the earth *A* as in conjunction with the sun, which sometimes it will eclipse in this position. Placed at *N*, it is seen in opposition to the sun, and in that position it may fall under the earth's shadow and be eclipsed.

Salviati. Now what shall we do, Simplicio, with the fixed stars? Do we want to sprinkle them through the immense abyss of the universe, at various distances from any predetermined point, or place them on a spherical surface extending around a centre of their own so that each of them will be the same distance from that centre?

Simplicio. I had rather take a middle course, and assign to them an orb described around a definite centre and included between two spherical surfaces—a very distant concave one, and another closer and convex, between which are placed at various altitudes the innumerable host of stars. This might be called the universal sphere, containing within it the spheres of the planets which we have already designated.

Salviati. Well, Simplicio, what we have been doing all this while is arranging the world bodies according to the Copernican distribution, and this has now been done by your own hand.

8. From letters written by JOHANNES KEPLER to HERWART

[In the first extract Kepler states his intention to find a mechanical interpretation of the universe. In the second he explains that he favours the
Copernican theory for mechanical or physical reasons.]

Prague, 10 February 1605

My aim is to show that the heavenly machine is not a kind of
divine, live being, but a kind of clockwork (and he who believes
that a clock has a soul, attributes the maker's glory to the work),
insofar as nearly all the manifold motions are caused by a most
simple, magnetic, and material force, just as all motions of the
clock are caused by a simple weight. And I also show how these
physical causes are to be given numerical and geometrical
expression.

Prague, 28 March 1605

You ask me, Magnificence, about the hypotheses of Copernicus
and you seem to be pleased that I insist on my opinion. . . .

[One of my main ideas aimed against Tycho is] if the sun moves
round the earth, then it must, of necessity, along with the other
planets become sometimes faster, sometimes slower in its movements, and this without following fixed courses, since there are
none. But this is incredible. Furthermore, the sun which is so much
higher ranking than the unimportant earth would have to be moved
by the earth in the same way as the five other planets are put in
motion by the sun. That is completely absurd. Therefore it is
much more plausible that the earth together with the five planets
is put in motion by the sun and only the moon by the earth.

CHAPTER 3

THE MECHANICAL UNIVERSE

9. From ARISTOTLE'S *Physics*

[This extract explains why light bodies rise and heavy bodies fall, in their motion to their 'proper place'. It illustrates Aristotle's thesis that all movement is change from potential to actual existence, and that Nature is essentially teleological, that is, it has inherent properties tending to a goal or purpose. Galileo and Newton completely rejected this thesis in their creation of modern physics.]

If the question is still pressed why light and heavy things tend to their respective positions, the only answer is that they are natured so, and that what we mean by heavy and light as distinguished and defined is just this downward or upward tendency. As we have said, here too there are different stages of potentiality. When a substance is water it is already in a way potentially light, and when it is air it may still be only potentially in the position proper to it, for its ascent may be hindered, but if the hindrance be removed it actualizes the potentiality and continuously mounts. And likewise the potentially 'such' tends to its actual realization; even as knowledge becomes straightway active if not impeded; and so likewise are the potential dimensions of a thing realized if there be no hindrance. If anyone removes the obstacle he may be said in one sense (but in another not) to cause the movement; for instance if he removes a column from beneath the weight it was supporting, or cuts the string that attached a bladder, under water, to the stone that holds it down, for he incidentally determines the moment at which the potential motion becomes actual, just as the wall from which a ball rebounds determines the direction in which the ball rebounds, though it is the player and not the wall that is the cause of its motion. So it is now clear that in no one of these cases does the thing that is in motion move itself; but it has the passive (though not the active and efficient) principle of movement inherent in itself.

10. From GALILEO's *Dialogue concerning the Two Chief World Systems—Ptolemaic and Copernican* (1632)

[This is an example of the kind of argument by which Galileo endeavoured to refute the doctrine of Aristotle given in the preceding extract. Salviati gives Galileo's views and Simplicio is an Aristotelian.]

Salviati. I consider the upward motion of heavy bodies due to received impetus to be just as natural as their downward motion dependent upon gravity.

Simplicio. This I shall never admit, because the latter has a natural and perpetual internal principle while the former has a finite and constrained external one.

Salviati. If you flinch from conceding to me that the principles of motion of heavy bodies downward and upward, are equally internal and natural, what would you do if I were to tell you that they may be also one and the same?

Simplicio. I leave it to you to judge.

Salviati. Rather, I want you to be the judge. Tell me, do you believe that contradictory internal principles can reside in the same natural body?

Simplicio. Absolutely not.

Salviati. What would you consider to be the natural intrinsic tendencies of earth, lead and gold, and in brief of all very heavy materials? That is, towards what motion do you believe that their internal principle draws them?

Simplicio. Motion towards the centre of heavy things; that is, to the centre of the universe and of the earth, whither they would be conducted if not impeded.

Salviati. So that if the terrestrial globe were pierced by a hole which passed through its centre, a cannon ball dropped through this and moved by its natural and intrinsic principle would be taken to the centre, and all this motion would be spontaneously made and by an intrinsic principle. Is that right?

Simplicio. I take that to be certain.

Salviati. But having arrived at the centre is it your belief that it would pass on beyond, or that it would immediately stop its motion then?

Simplicio. I think it would keep on going a long way.

Salviati. Now wouldn't this motion beyond the centre be upward, and according to what you have said preternatural and constrained? But upon what other principle will you make it depend, other than the very one which has brought the ball to the centre and which you have already called intrinsic and natural?...

Simplicio. I believe there are answers to all these objections, though for the moment I do not remember them.

11. From GALILEO'S *Dialogues concerning Two New Sciences* (1638)

Salviati. Once more, Simplicio is here on time; so let us without delay take up the question of motion. The text of our Author [Galileo] is as follows:

The Motion of Projectiles

In the preceding pages we have discussed the properties of uniform motion and of motion naturally accelerated along planes of all inclinations. I now propose to set forth those properties which belong to a body whose motion is compounded of two other motions, namely, one uniform and one naturally accelerated; these properties, well worth knowing, I propose to demonstrate in a rigid manner. This is the kind of motion seen in a moving projectile; its origin I conceive to be as follows:

Imagine any particle projected along a horizontal plane without friction; then we know, from what has been more fully explained in the preceding pages, that this particle will move along this same plane with a motion which is uniform and perpetual, provided the plane has no limits. But if the plane is limited and elevated, then the moving particle, which we imagine to be a heavy one, will on passing over the edge of the plane acquire, in addition to its previous uniform and perpetual motion, a downward propensity due to its own weight; so that the resulting motion which I call projection (*projectio*), is compounded of one which is uniform and horizontal and of another which is vertical and naturally accelerated. We now proceed to demonstrate some of its properties, the first of which is as follows:

Theorem I, Proposition I

A projectile which is carried by a uniform horizontal motion compounded with a naturally accelerated vertical motion describes a path which is a semi-parabola.

Sagredo. Here, Salviati, it will be necessary to stop a little while for my sake and, I believe, also for the benefit of Simplicio; for it so happens that I have not gone very far in my study of Apollonius and am merely aware of the fact that he treats of the parabola and other conic sections, without an understanding of which I hardly think one will be able to follow the proof of other propositions depending upon them. Since even in this first beautiful theorem the author finds it necessary to prove that the path of a projectile is a parabola, and since, as I imagine, we shall have to deal with only this kind of curves, it will be absolutely necessary to have a thorough acquaintance, if not with all the properties which Apollonius has demonstrated for these figures, at least with those which are needed for the present treatment.

12. From ISAAC NEWTON'S *Principia* (1687)

[The famous Laws of Motion in this extract come near the beginning of the *Principia*, immediately after the Definitions, indicating the similarity in plan of the work to that of Euclid. Much has been written about the true meaning and implications of the laws, on which classical physics is built.]

AXIOMS, OR LAWS OF MOTION

Law I

Every body continues in its state of rest, or of uniform motion in a right line, unless it is compelled to change that state by forces impressed upon it.

Projectiles continue in their motions, so far as they are not retarded by the resistance of the air, or impelled downwards by the force of gravity. A top, whose parts by their cohesion are continually drawn aside from rectilinear motions, does not cease its rotation, otherwise than as it is retarded by the air. The greater bodies of the planets and comets, meeting with less resistance in freer spaces, preserve their motions both progressive and circular for a much longer time.

Law II

The change of motion is proportional to the motive force impressed; and is made in the direction of the right line in which that force is impressed.

If any force generates a motion, a double force will generate double the motion, a triple force triple the motion, whether that force be impressed altogether and at once, or gradually and successively. And this motion (being always directed the same way with the generating force), if the body moved before, is added to or subtracted from the former motion, according as they directly conspire with or are directly contrary to each other; or obliquely joined, when they are oblique, so as to produce a new motion compounded from the determination of both.

Law III

To every action there is always opposed an equal reaction; or, the mutual actions of two bodies upon each other are always equal and directed to contrary parts.

Whatever draws or presses another is as much drawn or pressed by that other. If you press a stone with your finger, the finger is also pressed by the stone. If a horse draws a stone tied to a rope, the horse (if I may so say) will be equally drawn back towards the stone; for the distended rope, by the same endeavour to relax or unbend itself, will draw the horse as much towards the stone as it does the stone towards the horse, and will obstruct the progress of the one as much as it advances that of the other. If a body impinge upon another, and by its force change the motion of the other, that body also (because of the equality of the mutual pressure) will undergo an equal change, in its own motion, towards the contrary part. The changes made by these actions are equal, not in the velocities but in the motions of bodies; that is to say, if the bodies are not hindered by any other impediments. For, because the motions are equally changed, the changes of the velocities made towards contrary parts are inversely proportional to the bodies. This law takes place also in attractions, as will be proved in the next Scholium.

[Newton here explains that the cause of the revolution of a planet is the same as that of the fall of a stone. The centripetal force to which he refers is gravitational attraction.]

The principle of circular motion in free spaces

After this time, we do not know in what manner the ancients explained the question, how the planets came to be retained within certain bounds in these free spaces, and to be drawn off from the rectilinear courses, which, left to themselves, they should have pursued, into regular revolutions in curvilinear orbits. Probably it was to give some sort of satisfaction to this difficulty that solid orbs had been introduced.

The later philosophers pretend to account for it either by the action of certain vortices, as *Kepler* and *Descartes*; or by some other principle of impulse or attraction as *Borelli, Hooke,* and others of our nation; for, from the laws of motion, it is most certain that these effects must proceed from the action of some force or other.

But our purpose is only to trace out the quantity and properties of this force from the phaenomena, and to apply what we discover in some simple cases as principles, by which, in a mathematical way, we may estimate the effects thereof in more involved cases; for it would be endless and impossible to bring every particular to direct and immediate observation.

We said, in a mathematical way, to avoid all questions about the nature or quality of this force, which we would not be understood to determine by any hypothesis; and therefore call it by the general name of a centripetal force, as it is a force which is directed towards some centre; and as it regards more particularly a body in that centre, we call it circumsolar, circumterrestrial, circumjovial; and so in respect of other central bodies.

The action of centripetal forces

That by means of centripetal forces the planets may be retained in certain orbits, we may easily understand, if we consider the motions of projectiles; for a stone that is projected is by the pressure of its own weight forced out of the rectilinear path, which by the initial projection alone it should have pursued, and made to describe a curved line in the air; and through that crooked way is at last brought down to the ground; and the greater the velocity is with which it is projected, the farther it goes before it falls to the earth. We may therefore suppose the velocity to be so increased that it would describe an arc of 1, 2, 5, 10, 100, 1000 miles before it arrived at the earth, till at last, exceeding the limits of the earth, it should pass into space without touching it.

Let *AFB* represent the surface of the earth, *C* its centre, *VD, VE, VF* the curved lines which a body would describe, if projected in an horizontal direction from the top of an high

mountain successively with more and more velocity; and, because
the celestial motions are scarcely retarded by the little or no resistance of the spaces in which they are performed, to keep up the
parity of cases, let us suppose either that there is no air about the
earth, or at least that it is endowed with little or no power of resist-
ing; and for the same reason that the body projected with a less
velocity describes the lesser arc *VD*, and with a greater velocity
the greater arc *VE*, and augmenting the velocity, it goes farther

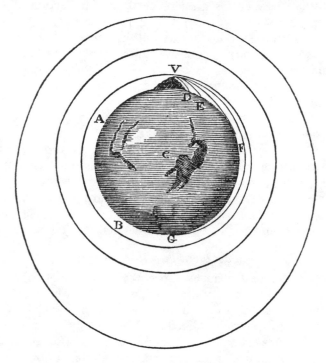

and farther to *F* and *G*, if the velocity was still more and more augmented, it would reach at last quite beyond the circumference of
the earth, and return to the mountain from which it was projected.

And since the areas which by this motion it describes by a radius
drawn to the centre of the earth are (by Prop. 1 Book 1, Princip.
Math.) proportional to the times in which they are described, its
velocity, when it returns to the mountain will be no less than it
was at first; and, retaining the same velocity, it will describe the
same curve over and over, by the same law.

But if we now imagine bodies to be projected in the directions of lines parallel to the horizon from greater heights, as of 5, 10, 100, 1000, or more miles, or rather as many semidiameters of the earth, those bodies, according to their different velocity, and the different force of gravity in different heights, will describe arcs either concentric with the earth, or variously eccentric, and go on revolving through the heavens in those orbits just as the planets do in their orbits.

[This well-known passage comes from the General Scholium at the end of Book III. In it Newton states the objective of science. Following the lead given by Galileo, he maintains that science must be experimental and inductive, in contrast to the deductive method of Aristotle. He says that he frames no hypotheses. The hypotheses he is condemning in particular, are those of Aristotle's occult qualities and of Descartes' mechanical vortices. Nevertheless hypothesis has an important place in scientific method and Newton himself showed an extraordinary imaginative fertility in his Queries at the end of his *Opticks*.]

But hitherto I have not been able to discover the cause of those properties of gravity from phaenomena, and I frame no hypotheses; for whatever is not deduced from the phaenomena is to be called an hypothesis; and hypotheses, whether metaphysical or physical, whether of occult qualities or mechanical, have no place in experimental philosophy. In this philosophy particular propositions are inferred from the phaenomena, and afterwards rendered general by induction. Thus it was that the impenetrability, the mobility, and the impulsive force of bodies, and the laws of motion and of gravitation, were discovered. And to us it is enough that gravity does really exist, and act according to the laws which we have explained, and abundantly serves to account for all the motions of celestial bodies, and of our sca.

CHAPTER 4

THE CIRCULATION OF THE BLOOD

13. From ROBERT BOYLE'S *A Disquisition about the Final Causes of Natural Things* (1688)

And I remember, that when I asked our famous Harvey, in the only discourse I had with him, (which was but a while before he died) what were the things, that induced him to think of a circulation of the blood? he answered me, that when he took notice, that the valves in the veins of so many parts of the body were so placed, that they gave free passage of the blood towards the heart, but opposed the passage of the venal blood the contrary way; he was invited to imagine. that so provident a cause as nature had not so placed so many valves without design; and no design seemed more probable, than that since the blood could not well, because of the interposing valves, be sent by the veins to the limbs, it should be sent through the arteries, and return through the veins, whose valves did not oppose its course that way.

14. From WILLIAM HARVEY'S *An Anatomical Disquisition on the Motion of the Heart and Blood in Animals* (1628)

[This extract is taken from the central chapter in Harvey's book. In it he states the essence of his theory, that the blood, impelled by the heart, moves 'as it were, in a circle'. He finds analogies in the motions of the heavenly bodies and in the cycle of evaporation and condensation of the earth's moisture.]

Thus far I have spoken of the passage of the blood from the veins into the arteries, and of the manner in which it is transmitted and distributed by the action of the heart; points to which some, moved either by the authority of Galen or Columbus, or the reasonings of others, will give in their adhesion. But what remains to be said upon the quantity and source of the blood which thus passes, is of so novel and unheard-of character, that I not only fear injury to myself from the envy of a few, but I tremble lest I have mankind at large for my enemies, so much doth wont and custom, that become as another nature, and doctrine once sown and that hath struck deep root, and respect for antiquity influence all men: Still the die is cast, and my trust is in my love of truth, and the candour that inheres in cultivated minds. And sooth to say, when I surveyed my mass of evidence, whether derived from vivisections, and my various reflections on them, or from the ventricles of the heart and the vessels that enter into and issue from them, the symmetry and size of these conduits,— for nature doing nothing in vain, would never have given them so large a relative size without a purpose,—or from the arrangement and intimate structure of the valves in particular, and of the other parts of the heart in general, with many things besides, I frequently and seriously bethought me, and long revolved in my mind, what might be the quantity of blood which was transmitted, in how short a time its passage might be effected, and the like: and not finding it possible that this could be supplied by the juices of the ingested aliment without the veins on the one hand becoming drained, and the arteries on the other getting ruptured through the excessive charge of blood, unless the blood should somehow find its way from the arteries into the veins, and so return to the right side of the heart; I began to think whether there might not be A MOTION, AS IT WERE, IN A CIRCLE. Now this I afterwards found to be true; and I finally saw that the blood, forced by the action of the left ventricle into the arteries, was distributed to the body at large, and its several parts, in the same manner as it is sent through the lungs, impelled by the right ventricle into the pulmonary artery, and that it then passed through the veins and along the vena cava, and so round to the left ventricle in the manner already indicated. Which motion we may be allowed to call circular, in the same way as Aristotle says that the air and the rain emulate the circular motion of the superior bodies; for the moist earth, warmed by the sun, evaporates; the vapours drawn upwards are condensed, and descending in the form of rain, moisten the earth again; and by this arrangement are generations of living things produced; and in like manner too are tempests and meteors engendered by the circular motion, and by the approach and recession of the sun.

CHAPTER 5

THE PRESSURE OF THE AIR

15. From a letter of EVANGELISTA TORRICELLI to MICHEL-ANGELO RICCI

[Torricelli explains his experiment on the vacuum.]

Florence, 11 June 1644

I have already intimated to you that a certain physical experiment was being performed on the vacuum; not simply to produce a vacuum, but to make an instrument which would show the changes in the air, which is at times heavier and thicker and at times lighter and more rarefied. Many have said that a vacuum cannot be produced, others that it can be produced, but with repugnance on the part of Nature and with difficulty; so far, I know of no one who has said that it can be produced without effort and without resistance on the part of Nature. I reasoned in this way: if I were to find a plainly apparent cause for the resistance which is felt when one needs to produce a vacuum, it seems to me that it would be vain to try to attribute that action, which patently derives from some other cause, to the vacuum; indeed, I find that by making certain very easy calculations, the cause I have proposed (which is the weight of the air) should in itself have a greater effect than it does in the attempt to produce a vacuum. I say this because some Philosopher, seeing that he could not avoid the admission that the weight of the air causes the resistance which is felt in producing a vacuum, did not say that he admitted the effect of the weight of the air, but persisted in asserting that Nature also contributes at least to the abhorrence of a vacuum. We live submerged at the bottom of an ocean of the element air, which by unquestioned experiments is known to have weight, and so much, indeed, that near the surface of the earth where it is most dense, it weighs (volume for volume) about the four-hundredth part of the weight of water. Those who have written about twilight, moreover, have observed that the vaporous and visible air rises above us about fifty or fifty-four miles; I do not, however, believe its height is as great as this, since if it were, I could show that the vacuum would have to offer much greater resistance than it does— even though there is in their favour the argument that the weight referred to by Galileo applies to the air in very low places where men and animals live, whereas that on the tops of high mountains begins to be distinctly rare and of much less weight than the four-hundredth part of the weight of water.

[Torricelli now described his experiments with the apparatus shown in the Figure. A long glass tube, sealed at one end, was filled with mercury, a finger placed over the open end, and then the tube was upturned and the finger removed under mercury in a basin. The mercury stood to a height of 30 inches in the tube, having a vacuum at the top. Torricelli concluded that it was the pressure of the atmosphere that supported the mercury and continued as follows.]

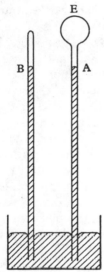

The above conclusion was confirmed by an experiment made at the same time with a vessel *A* and a tube *B*, in which the quicksilver always came to rest at the same level, *AB*. This is an almost certain indication that the force was not within; because if that were so, the vessel *AE* would have had greater force, since within it there was more rarefied material to attract the quicksilver, and a material much more powerful than that in the very small space *B*, on account of its greater rarefaction. I have since tried to consider from this point of view all the kinds of repulsions which are felt in the various effects attributed to vacuum, and thus far I have not encountered anything which does not go (to confirm my opinion).

16. From the account, submitted by MONSIEUR PERIER to MONSIEUR PASCAL, of the experiment performed on the Puy de Dôme, 19 September 1648

[Perier's achievement was to show that the height of the mercury in a barometer is less at the summit of a mountain than at the foot.]

The weather on Saturday last, the nineteenth of this month, was very unsettled. At about five o'clock in the morning, however, it seemed sufficiently clear; and since the summit of the Puy de Dôme was then visible, I decided to go there to make the attempt....

On that day, therefore, at eight o'clock in the morning, we started off all together for the garden of the Minim Fathers, which is almost the lowest spot in the town, and there began the experiment in this manner.

First, I poured into a vessel six pounds of quicksilver which I had rectified during the three days preceding; and having taken glass tubes of the same size, each four feet long and hermetically sealed at one end but open at the other, I placed them in the same vessel and carried out with each of them the usual vacuum experiment. Then, having set them up side by side without lifting them out of the vessel, I found that the quicksilver left in each of them stood at the same level, which was twenty-six inches[1] and three and a half lines above the surface of the quicksilver in the vessel. I repeated this experiment twice at this same spot, in the same tubes, with the same quicksilver, and in the same vessel; and found in each case that the quicksilver in the two tubes stood at the same horizontal level, and at the same height as in the first trial.

That done, I fixed one of the tubes permanently in its vessel for continuous experiment. I marked on the glass the height of the quicksilver, and leaving that tube where it stood, I requested Revd. Father Chastin, one of the brothers of the house, a man as pious as he is capable, and one who reasons very well upon these matters, to be so good as to observe from time to time all day any changes that might occur. With the other tube and a portion of the same quicksilver, I then proceeded with all these gentlemen to the top of the Puy de Dôme, some 500 fathoms above the Convent. There, after I had made the same experiments in the same way that I had made them at the Minims, we found

---

[1] A French inch equals 1.065 U.S./British inches.

that there remained in the tube a height of only twenty-three inches and two lines of quicksilver; whereas in the same tube, at the Minims we had found a height of twenty-six inches and three and a half lines. Thus between the heights of the quicksilver in the two experiments there proved to be a difference of three inches one line and a half. We were so carried away with wonder and delight, and our surprise was so great that we wished, for our own satisfaction, to repeat the experiment. So I carried it out with the greatest care five times more at different points on the summit of the mountain, once in the shelter of the little chapel that stands there, once in the open, once shielded from the wind, once in the wind, once in fine weather, once in the rain and fog which visited us occasionally. Each time I most carefully rid the tube of air; and in all these experiments we invariably found the same height of quicksilver. This was twenty-three inches and two lines, which yields the same discrepancy of three inches, one line and a half in comparison with the twenty-six inches, three lines and a half which had been found at the Minims. This satisfied us fully.

Later, on the way down at a spot called Lafon de L'Arbre, far above the Minims but much farther below the top of the mountain, I repeated the same experiment, still with the same tube, the same quicksilver, and the same vessel, and there found that the height of the quicksilver left in the tube was twenty-five inches. I repeated it a second time at the same spot; and Monsieur Mosnier, one of those previously mentioned, having the curiosity to perform it himself, then did so again, at the same spot. All these experiments yielded the same height of twenty-five inches, which is one inch, three lines and a half less than that which we had found at the Minims, and one inch and ten lines more than we had just found at the top of the Puy de Dôme. It increased our satisfaction not a little to observe in this way that the height of the quicksilver diminished with the altitude of the site.

On my return to the Minims I found that the (quicksilver in the) vessel I had left there in continuous operation was at the same height at which I had left it, that is, at twenty-six inches, three lines and a half; and the Revd. Father Chastin, who had remained there as observer, reported to us that no change had occurred during the whole day, although the weather had been very unsettled, now clear and still, now rainy, now very foggy and now windy.

Here I repeated the experiment with the tube I had carried to the Puy de Dôme, but in the vessel in which the tube used for the continuous experiment was standing. I found that the quicksilver was at the same level in both tubes, and exactly at the height of twenty-six inches three lines and a half, at which it had stood that morning in this same tube, and as it had stood all day in the tube used for the continuous experiment.

I repeated it again a last time, not only in the same tube I had used on the Puy de Dôme, but also with the same quicksilver and in the same vessel that I had carried up the mountain; and again I found the quicksilver at the same height of twenty-six inches, three lines and a half which I had observed in the morning, and thus finally verified the certainty of our results.

17. From OTTO VON GUERICKE'S *New Magdeburg Experiments on the Vacuum* (1672)

[Von Guericke records some of the difficulties he encountered in trying to create a vacuum.]

I filled a cask with water, made it everywhere air-tight, connected it on the lower side with a metal pump wherewith to draw out the water; I reasoned that as I drew out the water the part of the cask above the water would then be 'empty space'. At the first experiment the cask flew to pieces; I then affixed heavier screws, three men succeeded in pumping out the water, but then a sizzling sound was heard, the air filled the space from which water was drawn. Then I tried putting a smaller cask within the larger, so as to avoid the air rushing into the 'empty space', and drawing thence the water. Again the sound, now like the twittering of a bird, was heard and lasted for three days, for the wood was porous and let the air through. Therefore I tried a copper sphere instead; first this burst with a loud report. I attributed this to a probable defect in the spherical shape. With the greatest care I had a perfect sphere constructed. Now finally a vacuum was obtained; opening the cock attached to the sphere, the air rushed in with great violence.

18. From ROBERT BOYLE'S *A Continuation of New Experiments touching the Spring and Weight of Air* (1669)

EXPERIMENT XV

About the greatest height to which water can be raised by Attraction or Sucking Pumps

Having met with an opportunity to borrow a place somewhat convenient to make a Tryal to what height Water may be rais'd by Pumping, I thought fit not to neglect it. . . .

Wherefore, partly because a Tryal of such moment seem'd not to have yet been duely made by any; and partly because the varying weight of the Atmosphere was not (that appears) known, nor (consequently) taken into consideration by the ingenious Monsieur Paschal in his famous Experiment, which yet is but analogous to this; and partly because some very Late as well as Learned Writers have not acquiesc'd in his experiment, but do adhere to the old Doctrine of the Schools, which would have Water raiseable in Pumps to any height, *ob fugam vacui*, (as they speak,) I thought fit to make the best shift I could to make the Tryal, of which I now proceed to give Your Lordship an Account.

The place I borrowed for this purpose was a flat Roof about 30 foot high from the ground, and with Railes along the edges of it. The Tube we made use of should have been of Glass, if we could have procured one long and strong enough. But that being exceeding difficult, especially for me, who was not near a Glass-house, we were fain to cause a Tin-man to make several Pipes of above an inch bore, (for of a great length 'twas alleadg'd they could not be made slenderer,) and as long as he could, of Tin or Laton, as they call thin Plates of Iron Tinn'd over; and these being very carefully soder'd together made up one Pipe, of about one or two and thirty foot long, which being tied to a Pole we tried with Water whether it were stanch, and by the effluxions of that Liquor finding where the Leaks were, we caus'd them to be stopt with Soder, and then for greater security the whole Pipe, especially at the Commissures, was diligently cas'd over with our close black Cement, upon which Plaister of Paris was strewed to keep it from sticking to their hands or cloaths that should manage the Pipe. At the upper part of which was very carefully fastned with the like Cement a strong Pipe of Glass, of between 2 and

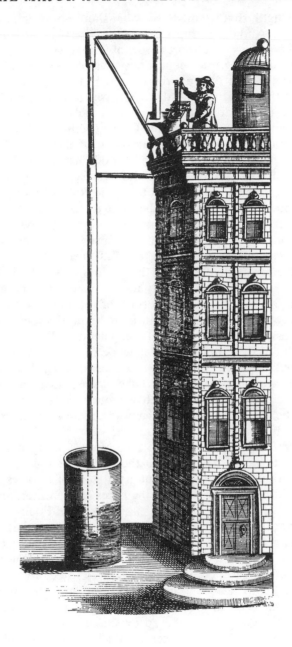

3 foot in length, that we might see what should happen at the top of the water. And to the upper part of this Pipe was (with Cement, and by the means of a short elbow of Tin) very closely fastned another Pipe of the same Metal, consisting of two pieces, making a right Angle with one another, whereof the upper part was parallel to the Horizon, and the other, which was parallel to the Glass-pipe, reacht down to the Engine, which was plac'd on the flat Roof, and was to be with good Cement sollicitously fastned to the lower end of this descending part of the Pipe, whose Horizontal leg was supported by a piece of Wood, nail'd to the above mentioned Railes; as the Tube also was kept from overmuch shaking by a board, (fasten'd to the same Railes,) and having a deep Notch cut in it, for the Tube to be inserted into.

This Apparatus being made, and the whole Tube with its Pole erected along the Wall, and fastned with strings and other helps, and the descending Pipe being carefully cemented on to the Engine, there was plac'd under the bottom of the long Tube a convenient vessel, whereinto so much Water was poured, as reach'd a great way above the orifice of the Pipe, and one was appointed to stand by to pour in more as need should require, that the vessel might be still kept competently full.

After all this the Pump was set on work, but when the water had been raised to a great height, and consequently had a great Pressure against the sides of the Tube, a small Leak or two was either discovered or made, which without moving the Tube we caus'd to be well stoppt, by one that was sent up a Ladder to apply store of Cement where it was requisite.

Wherefore at length we were able after a pretty number of Exuctions, to raise the Water to the middle of the Glass-pipe, above mentioned, but not without great store of bubbles, (made by the Air formerly conceal'd in the pores of the water, and now emerging,) which for a pretty while kept a kind of Foam upon the surface of it, (fresh ones continually succeeding those that broke.) And finding the Engine and Tube as stanch as could be well expected, I thought it a fit season to trie what was the utmost height to which Water could by Suction be elevated, and therefore though the Pump seem'd to have been plyed enough already, yet for further satisfaction, when the Water was within few inches of the top of the Glass, I caused 20 Exuctions more to be nimbly made; to be sure that the water should be raised as high as by our Pump it could be possibly. And having taken notice where the Surface rested, and caus'd a piece of Cement to be stuck near it, (for we could not then come to reach it exactly,) and descending to the Ground where the stagnant water stood, we caus'd a string to be let down, with a weight hanging at the end of it, which we applied to a mark, that had been purposely made at that part of the (Metalline) Tube, which the superficies of the stagnant water had rested at, when the water was elevated to its full height; and the other end of the string being, by him that let it down, applied to that part of the Glass as near as he could guess, where the upper part of the Water reacht, the Weight was pull'd up; and the length of the string, and (consequently) the height of the Cylinder of Water was measur'd, which amounted to 33 foot, and about 6 inches. Which done, I return'd to my lodging, which was not far off, to look upon the Baroscope, to be informed of the present weight of the Atmosphere, which I found to be but moderate, the Quick-silver standing at 29 inches, and between 2 and 3 eights of an inch. This being taken notice of, it was not difficult to compare the success of the Experiment with our Hypothesis. For if we suppose the most received proportion in bulk between Cylinders of Quick-silver and of Water of the same weight, namely that of 1 to 14, the height of the Water ought to have been 34 foot and about two inches, which is about 8 inches greater than we found it.

CHAPTER 6

THE EARLY MICROSCOPISTS

19. From ROBERT HOOKE'S *Micrographia* (1665)

[In this book Hooke records observations with his microscope of a variety of objects, such as the point of a pin, the edge of a razor, the pores of cork, insects, etc.]

OBSERVATION LIII

Of a Flea

The strength and beauty of this small creature, had it no other relation at all to man, would deserve a description.

For its strength, the *Microscope* is able to make no greater discoveries of it than the naked eye, but onely the curious contrivance of its leggs and joints, for the exerting that strength, is very plainly manifested, such as no other creature, I have yet observ'd, has any thing like it; for the joints of it are so adapted, that he can, as 'twere, fold them short one within another, and suddenly stretch, or spring them out to their whole length, that is, of the fore-leggs, the part *A* [see Figure] lies within *B*, and *B* within *C*, parallel to, or side by side each other; but the parts of the two next, lie quite contrary, that is, *D* without *E*, and *E* without *F*, but parallel also; but the parts of the hinder leggs, *G*, *H*, and *I*, bend one within another, like the parts of a double jointed Ruler, or like the foot, legg and thigh of a man; these six leggs he clitches up altogether, and when he leaps, springs them all out, and thereby exerts his whole strength at once.

But, as for the beauty of it, the *Microscope* manifest it to be all over adorn'd with a curiously polish'd suit of sable Armour, neatly jointed, and beset with multitudes of sharp pinns, shaped almost like a Porcupine's Quills, or bright conical steel-bodkins; the head is on either side beautify'd with a quick and round black eye *K*, behind each of which also appears a small cavity, *L*, in which he seems to move to and fro a certain thin film beset with many small transparent hairs, which probably may be his ears; in the forepart of his head, between the two fore-leggs, he has two small long jointed feelers, or rather smellers, *MM* which have four joints, and are hairy, like those of several other creatures; between these it has a small *proboscis*, or *probe*, *NNO*, that seems to consist of a tube *NN*, and a tongue or sucker *O*, which I have perceiv'd him to slip in and out. Besides these, it has also two chaps or biters *PP*, which are somewhat like those of an Ant, but I could not perceive them tooth'd; these were shaped very like the blades of a pair of round top'd Scizers, and were opened and shut just after the same manner; with these instruments does this little busie Creature bite and pierce the skin, and suck out the blood of an Animal, leaving the skin inflamed with a small round red spot. These parts are very difficult to be discovered, because, for the most part, they lye covered between the fore-legs. There are many other particulars, which, being more obvious, and affording no great matter of information, I shall pass by, and refer the Reader to the Figure.

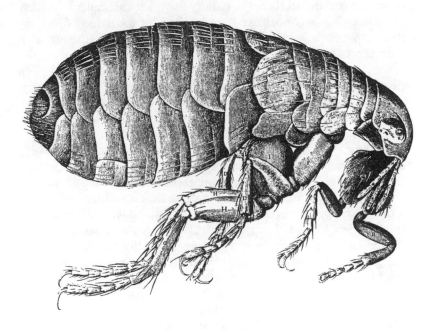

20. From ANTONY VAN LEEUWENHOEK's letters to the Royal Society on his 'Little Animals' (1676–92)

[In these letters Leeuwenhoek describes, with his characteristic, racy ingenuousness, his pioneer explorations of the world of micro-organisms. The first letter recounts the discovery of the protozoa in rain water and the attempts of *Vorticella* to 'get its tail loose' (*Vorticella* anchors itself by its stalk). The second letter is one of several in which Leeuwenhoek stoutly maintains, in face of scepticism, that a drop of water may contain millions of micro-organisms. The third letter deals with the teeming and lively bacteria found in the human mouth.]

From *Letter* 18, 9 October 1676

In the year 1675, about half-way through September (being busy with studying air, when I had much compressed it by means of water), I discovered living creatures in rain, which had stood but a few days in a new tub, that was painted blue within. This observation provoked me to investigate this water more narrowly; and especially because these little animals were, to my eye, more than ten thousand times smaller than the animalcule which Swammerdam has portrayed, and called by the name of Water-flea, or Water-louse, which you can see alive and moving in water with the bare eye.

Of the first sort that I discovered in the said water, I saw, after divers observations, that the bodies consisted of 5, 6, 7, or 8 very clear globules, but without being able to discern any membrane or skin that held these globules together, or in which they were inclosed. When these animalcules bestirred 'emselves, they sometimes stuck out two little horns, which were continually moved, after the fashion of a horse's ears. The part between these little horns was flat, their body else being roundish, save only that it ran somewhat to a point at the hind end; at which pointed end it had a tail, near four times as long as the whole body, and looking as thick, when viewed through my microscope, as a spider's web. At the end of this tail there was a pellet, of the bigness of one of the globules of the body; and this tail I could not perceive to be used by them for their movements in very clear water. These little animals were the most wretched creatures that I have ever seen; for when, with the pellet, they did but hit on any particles or little filaments (of which there are many in water, especially if it hath but stood some days), they stuck intangled in them; and then pulled their body out into an oval, and did struggle, by strongly stretching themselves, to get their tail loose; whereby their whole body then sprang back towards the pellet of the tail, and their tails then coiled up serpentwise, after the fashion of a copper or iron wire that, having been wound close about a round stick, and then taken off, kept all its windings. This motion, of stretching out and pulling together the tail, continued; and I have seen several hundred animalcules, caught fast by one another in a few filaments, lying within the compass of a coarse grain of sand.

I also discovered a second sort of animalcules, whose figure was an oval; and I imagined that their head was placed at the pointed end. These were a little bit bigger than the animalcules first mentioned. Their belly is flat, provided with divers incredibly thin little feet, or little legs, which were moved very nimbly, and which I was able to discover only after sundry great efforts, and wherewith they brought off incredibly quick motions. The upper part of their body was round, and furnished inside with 8, 10, or 12 globules: otherwise these animalcules were very clear. These little animals would change their body into a perfect round, but mostly when they came to lie high and dry. Their body was also very yielding: for if they so much as brushed against a tiny filament, their body bent in, which bend also presently sprang out again; just as if you stuck your finger into a bladder full of water, and then, on removing the finger, the inpitting went away. Yet the greatest marvel was when I brought any of the animalcules on a dry place, for I then saw them change themselves at last into a round, and then the upper part of the body rose up pyramid-like, with a point jutting out in the middle; and after having thus lain moving with their feet for a little while, they burst asunder, and the globules and a watery humour flowed away on all sides, without my being able to discern even the least sign of any skin wherein these globules and the liquid had, to all appearance, been inclosed; and at such times I could discern more globules than when they were alive. This bursting asunder I figure to myself to happen thus: imagine, for example, that you have a sheep's bladder filled with shot, peas, and water; then, if you were to dash it apieces on the ground, the shot, peas, and water would scatter themselves all over the place. . . .

From *Letter* 19, 23 March 1677

Your very welcome letters of the 12th and 22nd *ultimo* have reached me safely. I was glad to see that Mr Boyle and Mr Grew sent me their remembrances: please give these gentlemen, on my behalf, my most respectful greetings. 'Twas also a pleasure to me to see that the other Philosophers liked my observations on water, etc., though they found it hard to conceive of the huge number of little animals present in even a single drop of water. Yet I can't wonder at it, since 'tis difficult to comprehend such things without getting a sight of 'em.

But I have never affirmed, that the animals in water were present in such-and-such a number: I always say, that I imagine I see so many.

My division of the water, and my counting of the animalcules, are done after this fashion. I suppose that a drop of water doth equal a green pea in bigness; and I take a very small quantity of water, which I cause to take on a round figure, of very near the same size as a millet-seed. This latter quantity of water I figure to myself to be the one-hundredth part of the foresaid drop: for I reckon that if the diameter of a millet-seed be taken as 1, then the diameter of a green pea must be quite $4\frac{1}{2}$. This being so, then a quantity of water of the bigness of a millet-seed maketh very nearly the $\frac{1}{91}$ part of a drop, according to the received rules of mathematicks....

This amount of water, as big as a millet-seed, I introduce into a clean little glass tube (whenever I wish to let some curious person or other look at it). This slender little glass tube, containing the water, I divide again into 25 or 30, or more, parts; and I then bring it before my microscope, by means of two silver or copper springs, which I have attached thereto for this purpose, so as to be able to place the little glass tube before my microscope in any desired position, and to be able to push it up or down according as I think fit.

I showed the foresaid animalcules to a certain Gentleman, among others, in the manner just described; and he judged that he saw, in the $\frac{1}{30}$th part of a quantity of water as big as a millet-seed, more than 1000 living creatures. This same Gentleman beheld this sight with great wonder, and all the more because I told him that in this very water there were yet 2 or 3 sorts of even much smaller creatures that were not revealed to his eyes, but which I could see by means of other glasses and a different method (which I keep for myself alone). Now supposing that this Gentleman really saw 1000 animalcules in a particle of water but $\frac{1}{30}$th of the bigness of a millet-seed, that would be 30,000 living creatures in a quantity of water as big as a millet-seed, and consequently 2,730,000 living creatures in one drop of water....

From *Letter 75*, 16 September 1692

In my letter of the 12th of September, 1683, I spake, among other things, of the living creatures that are in the white matter which lieth, or groweth, betwixt or upon one's front teeth or one's grinders. Since that time, and especially in the last two or three years, I have examined this stuff divers times; but to my surprise, I could discern no living creatures in it.

Being unable to satisfy myself about this, I made up my mind to put my back into the job, and to look into the question as carefully as I could. But because I keep my teeth uncommon clean, rubbing them with salt every morning, and after meals generally picking them with a fowl's quill, or pen; I therefore found very little of the said stuff stuck on the outside of my front teeth: and in what I got out from between them, I could find nothing with life in it. Thereupon I took a little of the stuff that was on my frontmost grinders; but though I had two or three shots at these observations, 'twas not till the third attempt that I saw one or two live animalcules. Yet I could well make out some particles lying about that I felt sure must have been animalcules. This put me in a quandary again, seeing that at and about the time when I wrote to you concerning these animalcules, I never failed to see there was life in them: but though now I used just the very same magnifying-glass and apparatus (which I judged to be that best suited to the purpose), yet I couldn't make out any living creatures at all.

Having allowed my speculations to run on this subject for some time, methinks I have now got to the bottom of the dying-off of these animalcules. The reason is, I mostly or pretty near always of a morning drink coffee, as hot as I can, so hot that it puts me into a sweat: beyond this I seldom drink anything save at mealtimes in the middle of the day and in the evening; and by doing so, I find myself in the best of health. Now the animalcules that are in the white matter on the front-teeth, and on the foremost of the back teeth, being unable to bear the hotness of the coffee, are thereby killed: like I've often shown that the animalcules which are in water are made to die by a slight heating.

Accordingly, I took (with the help of a magnifying mirror) the stuff from off and from between the teeth further back in my mouth, where the heat of the coffee couldn't get at it. This stuff I mixt with a little spit out of my mouth (in which there were no air-bubbles), and I did all this in the way I've always done: and then I saw with as great a wonderment as ever before, an unconceivably great number of little animalcules, and in so unbelievably small a quantity of the foresaid stuff, that those who didn't see it with their own eyes could scarce credit it. These animalcules, or most all of them, moved so nimbly among one another, that the whole stuff seemed alive and a-moving.

CHAPTER 7

THE SEVENTEENTH CENTURY

21. From *New Atlantis* by FRANCIS BACON, Lord Verulam (1627)

[The writer of the tale sets sail from Peru for China and Japan by the South Sea, and after many vicissitudes comes upon the island of New Atlantis. Here he is shown an institute of scientific research known as Salomon's House. The first part of this extract, an abridged account of the research being conducted there, is a remarkable prevision of modern applied science.

The second part of the extract reveals Bacon's weakness in his conception of scientific method.]

The End of our Foundation is the Knowledge of Causes, and Secrett Motions of Things; And the Enlarging of the bounds of Humane Empire, to the Effecting of all Things possible.

The Preparations and Instruments are these. We have large and deepe Caves of severall Depths. These Caves we call the Lower Region; And wee use them for all Coagulations, Indurations, Refrigerations, and Conservations of Bodies. We use them likewise for the Imitation of Naturall Mines; And the Producing also of New Artificiall Mettalls, by Compositions and Materialls which we use, and lay ther for many yeares. Wee use them also sometimes, (which may seeme strange,) for Curing of some Diseases, and for Prolongation of Life.

We have High Towers; The Highest about halfe a Mile in Height. Wee use these Towers, according to their severall Heights, and Situations, for Insolation, Refrigeration, Conservation; And for the View of divers Meteors; As Windes, Raine, Snow, Haile; And some of the Fiery Meteors also.

We have great Lakes, both Salt, and Fresh; wherof we have use for the Fish, and Fowle.

We have also a Number of Artificiall Wels, and Fountaines, made in Imitation of the Naturall Sources and Baths.

We have also Great and Spatious Houses, wher wee imitate and demonstrate Meteors.

We have also certaine Chambers, which wee call Chambers of Health, wher wee qualifie the Aire as we thinke good and proper for the Cure of diverse Diseases, and Preservation of Health.

We have also large and various Orchards and Gardens; Wherin we do not so much respect Beauty, as Variety of Ground and Soyle, proper for diverse Trees, and Herbs. And we make (by Art) in the same Orchards and Gardens, Trees and Flowers, to come earlier, or later, then their Seasons; And to come up and beare more speedily then by their Naturall Course they doe. We make them also by Art greater much then their Nature; And their Fruit greater, and sweeter, and of differing Tast, Smell, Colour, and Figure, from their Nature.

We have also Parks, and Enclosures of all Sorts, of Beasts, and Birds; which wee use not onely for View or Rarenesse, but likewise for Dissections and Trialls; That thereby we may take light, what may be wrought upon the Body of Man. We try also all Poysons, and other Medicines upon them, as well of Chyrurgery as Phisicke. By Art likewise, we make them Greater or Taller than their Kinde is; And contrary-wise Dwarfe them and stay their Grouth; Wee make them more Fruitfull and Bearing then their Kind is; and contrary-wise Barren and not Generative. Also we make them differ in Colour, Shape, Activity, many wayes.

Wee have also Particular Pooles, wher we make Trialls upon Fishes, as we have said before of Beasts, and Birds.

Wee have also Places for Breed and Generation of those Kindes of Wormes, and Flies, which are of Speciall Use; Such as are with you your Silkwormes, and Bees.

I will not hold you long with recounting of our Brew-Howses, Bake-Howses, and Kitchins, wher are made diverse Drinks, Breads, and Meats, Rare, and of speciall Effects.

Wee have also diverse Mechanicall Arts, which you have not; And Stuffes made by them; As Papers, Linnen, Silks, Tissues; dainty Works of Feathers of wonderfull Lustre; excellent Dies, and many others.

Wee have also Fournaces of great Diversities, and that keepe great Diversitie of Heates. These diverse Heates wee use, As the Nature of the Operation, which wee intend, requireth.

Wee have also Perspective-Houses, wher wee make Demonstrations of all Lights, and Radiations; And of all Colours. Wee procure meanes of Seeing Objects a-farr off. Wee have also Glasses and Meanes, to see Small and Minute Bodies, perfectly and distinctly; As the Shapes and Colours of Small Flies and Wormes, Graines and Flawes in Gemmes which cannot otherwise be seen, Observations in Urine and Bloud not otherwise to be seen.

Wee have also Sound-Houses, wher wee practise and demonstrate all Sounds and their Generation. Wee have certaine Helps, which sett to the Eare doe further the Hearing greatly. Wee have also meanes to convey Sounds in Trunks and Pipes, in strange Lines, and Distances.

Wee have also Perfume-Houses; wherwith we joyne also Practises of Tast. Wee Multiply Smells, which may seeme strange. Wee make diverse Imitations of Tast likewise, so that they will deceyve any Mans Tast.

Wee have also Engine-Houses, wher are prepared Engines and Instruments for all Sorts of Motions. Wee imitate also Flights of Birds; Wee have some Degrees of Flying in the Ayre. Wee have Shipps and Boates for Going under Water, and Brooking of Seas.

Wee have also Houses of Deceits of the Senses; wher we represent all manner of Feates of Iugling, False Apparitions, Impostures and Illusions.

These are (my Sonne) the Riches of Salomons House.

* * *

For the severall Employments and Offices of our Fellowes; Wee have Twelve that Sayle into Forraine Countries, under the Names of other Nations, (for our owne wee conceale;) Who bring us the Bookes, and Abstracts, and Patternes of Experiments of all other Parts. These wee call Merchants of Light.

Wee have Three that Collect the Experiments which are in all Bookes. These wee call Depredatours.

Wee have Three that Collect the Experiments of all Mechanicall Arts; And also of Liberall Sciences; And also of Practises which are not Brought into Arts. These we call Mystery-Men.

Wee have Three that try New Experiments, such as themselves thinke good. These wee call Pioners or Miners.

Wee have Three that Drawe the Experiments of the Former Foure into Titles, and Tables, to give the better light, for the drawing of Observations and Axiomes out of them. These wee call compilers.

We have Three that bend themselves, Looking into the Experiments of their Fellowes, and cast about how to draw out of them Things of Use, and Practise for Mans life, and Knowledge, as well for Workes, as for Plaine Demonstration of Causes, Meanes of Naturall Divinations, and the easie and cleare Discovery, of the Vertues and Parts of Bodies. These wee call Dowry-men or Benefactours.

Then after diverse Meetings and Consults of our whole Number, to consider of the former Labours and Collections, wee have Three that take care, out of them, to direct New Experiments, of a Higher Light, more Penetrating into Nature then the Former. These wee call Lamps.

Wee have Three others that doe Execute the Experiments so Directed, and Report them. These wee call Inoculatours.

Lastly, wee have Three that raise the former Discoveries by Experiments, into Greater Observations, Axiomes, and Aphorismes. These wee call Interpreters of Nature.

22. From FRANCIS BACON'S *Novum Organum* (1620)

[Bacon explains his new inductive method of making generalisations or laws from the results of observations and experiments in scientific investigation.]

In establishing axioms, another form of induction must be devised than has hitherto been employed; and it must be used for proving and discovering not first principles (as they are called) only, but also. the lesser axioms, and the middle, and indeed all. For the induction which proceeds by simple enumeration is childish; its conclusions are precarious, and exposed to peril from a contradictory instance; and it generally decides on too small a number of facts, and on those only which are at hand. But the induction which is to be available for the discovery and demonstration of sciences and arts, must analyse nature by proper rejections and exclusions; and then, after a sufficient number of negatives, come to a conclusion on the affirmative instances; which has not yet been done or even attempted, save only by Plato, who does indeed employ this form of induction to a certain extent for the purpose of discussing definitions and ideas. But in order to furnish this induction or demonstration well and duly for its work, very many things are to be provided which no mortal has yet thought of; insomuch that greater labour will have to be spent in it than has hitherto been spent on the syllogism. And this induction must be used not only to discover axioms, but also in the formation of notions. And it is in this induction that our chief hope lies.

23. From RENÉ DESCARTES' *Discourse on Method* (1637)

[Descartes describes the discovery of his method, which was the construction of science and philosophy by steps, proceeding from simple self-evident truths to the more complex, in the manner of geometry.]

I was in Germany at the time; the fortune of war (the war that is still going on) had called me there. While I was returning to the army from the Emperor's coronation, the onset of the winter held me up in quarters in which I found no conversation to interest me; and since, fortunately, I was not troubled by any cares or passions, I spent the whole day shut up alone in a stove-heated room, and was at full liberty to discourse with myself about my own thoughts....

I thought the following four [rules] would be enough, provided that I made a firm and constant resolution not to fail even once in the observance of them.

The first was never to accept anything as true if I had not evident knowledge of its being so; that is, carefully to avoid precipitancy and prejudice, and to embrace in my judgement only what presented itself to my mind so clearly and distinctly that I had no occasion to doubt it.

The second, to divide each problem I examined into as many parts as was feasible, and as was requisite for its better solution.

The third, to direct my thoughts in an orderly way; beginning with the simplest objects, those most apt to be known, and ascending little by little, in steps as it were, to the knowledge of the most complex; and establishing an order in thought even when the objects had no natural priority one to another.

And the last, to make throughout such complete enumerations

and such general surveys that I might be sure of leaving nothing out.

Those long chains of perfectly simple and easy reasonings by means of which geometers are accustomed to carry out their most difficult demonstrations had led me to fancy that everything that can fall under human knowledge forms a similar sequence; and that so long as we avoid accepting as true what is not so, and always preserve the right order to deduction of one thing from another, there can be nothing too remote to be reached in the end, or too well hidden to be discovered.

24. From THOMAS HOBBES' *Leviathan* (1651)

[In these two brief extracts, the first taken from chapter VI and the second from chapter XV, Hobbes denies that Good is an absolute value and puts forward the view that moral philosophy is concerned with the natural laws of man's appetites and aversions.]

But whatsoever is the object of any mans Appetite or Desire; that is it, which he for his part calleth *Good*: And the object of his Hate, and Aversion, *Evill*; And of his Contempt, *Vile* and *Inconsiderable*. For these words of Good Evill, and Contemptible, are ever used with relation to the person that useth them: There being nothing simply and absolutely so; nor any common Rule of Good and Evill, to be taken from the nature of the objects themselves;....

For Morall Philosophy is nothing else but the Science of what is *Good*, and *Evill*, in the conversation, and Society of man-kind. *Good*, and *Evill*, are names that signifie our Appetites, and Aversions; which in different tempers, customes, and doctrines of men, are different: And divers men, differ not onely in their Judgement, on the senses of what is pleasant, and unpleasant to the tast, smell, hearing, touch, and sight; but also of what is conformable, or disagreeable to Reason, in the actions of common life. Nay, the same man, in divers times, differs from himselfe; and one time praiseth, that is, calleth Good, what another time he dispraiseth, and calleth Evill: From whence arise Disputes, Controversies, and at last War. And therefore so long a man is in the condition of meer Nature, (which is a condition of War,) as private Appetite is the measure of Good, and Evill: And consequently all men agree on this, that Peace is Good, and therefore also the way, or means of Peace, which (as I have shewed before) are *Justice, Gratitude, Modesty, Equity, Mercy*, & the rest of the Laws of Nature, are good; that is to say, *Morall Vertues*; and their contrarie *Vices*, Evill. Now the science of Vertue and Vice, is Morall Philosophie; and therefore the true Doctrine of the Lawes of Nature, is the true Morall Philosophie.

CHAPTER 8

THE CREATION OF MODERN CHEMISTRY

25. ANTOINE LAVOISIER's sealed note of 1 November 1772

[Lavoisier tries to secure priority for his ideas on combustion—ideas which, he found later, were put forward over a century earlier.]

About eight days ago I discovered that sulphur, in burning, far from losing weight, on the contrary gains in weight; that is to say that from a pound of sulphur one can obtain much more than a pound of vitriolic acid, making allowance for humidity of the air; it is the same with phosphorus; this increase in weight arises from an immense quantity. of air that fixes itself during the combustion and combines with the vapours.

This discovery, which I have established by experiments that I look upon as decisive, has led me to think that what is observed in the combustion of sulphur and phosphorus may well take place with regard to all substances that gain in weight by combustion and calcination: and I am persuaded that the increase in weight of metallic calces is due to the same cause. Experiment has completely confirmed my conjectures: I have carried out the reduction of litharge in closed vessels, using the apparatus of Hales, and observed the liberation, at the moment the calx changed into metal, of a large quantity of air, this air having a volume a thousand times greater than the amount of litharge employed. This discovery appearing to me one of the most interesting of those that have been made since the time of Stahl, I felt compelled to secure my right in it, by depositing it in the hands of the Secretary of the *Académie*, to remain sealed until the time when I shall make my experiments known.

Paris, 1st November, 1772. Lavoisier.

26. From JOSEPH PRIESTLEY'S *Experiments and Observations on Different Kinds of Air* (1775)

[Priestley describes his experiments with 'air purer than the best common air', now known as oxygen.]

There are, I believe, very few maxims in philosophy that have laid firmer hold upon the mind, than that air, meaning atmospherical air (free from various foreign matters, which were always supposed to be dissolved, and intermixed with it) is *a simple elementary substance*, indestructible, and unalterable, at least as much so as water is supposed to be. In the course of my inquiries, I was, however, soon satisfied that atmospherical air is not an unalterable thing; for that the phlogiston with which it becomes loaded from bodies burning in it, and animals breathing it, and various other chemical processes, so far alters and depraves it, as to render it altogether unfit for inflammation, respiration, and other purposes to which it is subservient; and I had discovered that agitation in water, the process of vegetation, and probably other natural processes, by taking out the superflous phlogiston, restore it to its original purity. But I own I had no idea of the possibility of going any farther in this way, and thereby procuring air purer than the best common air.

[Priestley then describes how, by heating *mercurius calcinatus per se* (mercuric oxide) with his burning glass, he obtained a gas which he called *dephlogisticated air* (oxygen).]

On the 8th of this month I procured a mouse, and put it into a glass vessel, containing two ounce-measures of the air from mercurius calcinatus. Had it been common air, a full-grown mouse, as this was, would have lived in it about a quarter of an hour. In this air, however, my mouse lived a full half hour; and though it was taken out seemingly dead, it appeared to have been only exceedingly chilled; for, upon being held to the fire, it presently revived, and appeared not to have received any harm from the experiment.

By this I was confirmed in my conclusion, that the air extracted from mercurius calcinatus, &c. was, *at least*, *as good* as common air; but I did not certainly conclude that it was any *better*; because, though one mouse would live only a quarter of an hour in a given quantity of air, I knew it was not impossible but that another mouse might have lived in it half an hour; so little accuracy is there in this method of ascertaining the goodness of air: and indeed I have never had recourse to it for my own satisfaction, since the discovery of that most ready, accurate, and elegant test that nitrous air furnishes. But in this case I had a view to publishing the most generally-satisfactory account of my experiments that the nature of the thing would admit of.

This experiment with the mouse, when I had reflected upon it some time, gave me so much suspicion that the air into which I had put it was better than common air, that I was induced, the day after, to apply the test of nitrous air to a small part of that very quantity of air which the mouse had breathed so long; so that, had it been common air, I was satisfied it must have been very nearly, if not altogether, as noxious as possible, so as not to be affected by nitrous air; when, to my surprize again, I found that though it had been breathed so long, it was still better than common air. For after mixing it with nitrous air, in the usual proportion of two to one, it was diminished in the proportion of $4\frac{1}{2}$ to $3\frac{1}{2}$; that is, the nitrous air had made it two ninths less than before, and this in a very short space of time; whereas I had never found that, in the longest time, any common air was reduced more than one fifth of its bulk by any proportion of nitrous air, nor more than one fourth by any phlogistic process whatever. Thinking of this extraordinary fact upon my pillow, the next morning I put another measure of nitrous air to the same mixture and, to my utter astonishment, found that it was farther diminished to almost one half of its original quantity. I then put a third measure to it; but this did not diminish it any farther: but, however, left it one measure less than it was even after the mouse had been taken out of it.

Being now fully satisfied that this air, even after the mouse had breathed it half an hour, was much better than common air; and having a quantity of it still left, sufficient for the experiment, viz. an ounce-measure and a half, I put the mouse into it; when I observed that it seemed to feel no shock upon being put into it, evident signs of which would have been visible, if the air had not been very wholesome; but that it remained perfectly at its ease another full half hour, when I took it out quite lively and vigorous. . . .

My reader will not wonder, that, after having ascertained the superior goodness of dephlogisticated air by mice living in it, and the other tests above mentioned, I should have the curiosity to taste it myself. I have gratified that curiosity, by breathing it, drawing it through a glass-syphon, and, by this means, I reduced a large jar full of it to the standard of common air. The feeling of it to my lungs was not sensibly different from that of common air; but I fancied that my breast felt peculiarly light and easy for some time afterwards. Who can tell but that, in time, this pure air may become a fashionable article in luxury. Hitherto only two mice and myself have had the privilege of breathing it.

27. From HENRY CAVENDISH'S *Experiments on Air* (1781)

[Cavendish describes how, by exploding dephlogisticated air (oxygen) and inflammable air (hydrogen), he obtained water. In the following passages he discusses the possible interpretations of the experiments.]

All the foregoing experiments, on the explosion of inflammable air with common and dephlogisticated airs, except those which relate to the cause of the acid found in the water, were made in the summer of the year 1781, and were mentioned by me to Dr PRIESTLEY, who in consequence of it made some experiments of the same kind, as he relates in a paper printed in the preceding volume of the Transactions. During the last summer also, a friend of mine gave some account of them to M. LAVOISIER, as well as of the conclusion drawn from them, that dephlogisticated air is only water deprived of phlogiston; but at that time so far was M. LAVOISIER from thinking any such opinion warranted, that, till he was prevailed upon to repeat the experiment himself, he found some difficulty in believing that nearly the whole of the two airs could be converted into water. It is remarkable, that neither of these gentlemen found any acid in the water produced by the combustion; which might proceed from the latter having burnt the two airs in a different manner from what I did; and from the former having used a different kind of inflammable air, namely, that from charcoal [i.e. carbon monoxide], and perhaps having used a greater proportion of it. . . .

From what has been said there seems the utmost reason to think, that dephlogisticated air is only water deprived of its phlogiston, and that inflammable air, as was before said, is either phlogisticated water, or else pure phlogiston; but in all probability the former....

There are several memoirs of M. LAVOISIER published by the Academy of Sciences, in which he intirely discards phlogiston, and explains those phaenomena which have been usually attributed to the loss or attraction of that substance, by the absorption or expulsion of dephlogisticated air....

It seems therefore, from what has been said, as if the phaenomena of nature might be explained very well on this principle, without the help of phlogiston; and indeed, as adding dephlogisticated air to a body comes to the same thing as depriving it of its phlogiston and adding water to it, and as there are, perhaps, no bodies entirely destitute of water, and as I know no way by which phlogiston can be transferred from one body to another, without leaving it uncertain whether water is not at the same time transferred, it will be very difficult to determine by experiment which of these opinions is the truest; but as the commonly received principle of phlogiston explains all phaenomena, at least as well as M. LAVOISIER'S, I have adhered to that.

28. From ANTOINE LAVOISIER'S *Elementary Treatise on Chemistry* (1789)

[Lavoisier's main purpose in writing his textbook was to disseminate a new chemical nomenclature, drawn up by himself, de Morveau, Berthollet and de Fourcroy.

The following passage illustrates the new nomenclature in relation to his theory of combustion. Lavoisier explains combustion, as we do today, in terms of the chemical combination of the burning substance with oxygen. The heat liberated he calls caloric, and he regards this as released from association with oxygen.]

CHAP. VII

Of the decomposition of oxygen gas by means of metals, and the formation of metallic oxyds

Oxygen has a stronger affinity with metals heated to a certain degree than with caloric; in consequence of which, all metallic bodies, excepting gold, silver, and platina, have the property of decomposing oxygen gas, by attracting its base from the caloric with which it was combined. We have already shown in what manner this decomposition takes place, by means of mercury and iron; having observed, that, in the case of the first, it must be considered as a kind of gradual combustion, whilst, in the latter, the combustion is extremely rapid, and attended with a brilliant flame. The use of the heat employed in these operations is to separate the particles of the metal from each other, and to diminish their attraction of cohesion or aggregation, or, what is the same thing, their mutual attraction for each other.

The absolute weight of metallic substances is augmented in proportion to the quantity of oxygen they absorb; they, at the same time, lose their metallic splendour, and are reduced into an earthy pulverulent matter. In this state metals must not be considered as entirely saturated with oxygen, because their action upon this element is counterbalanced by the power of affinity between it and caloric. During the calcination of metals, the oxygen is therefore acted upon by two separate and opposite powers, that of its attraction for caloric, and that exerted by the metal, and only tends to unite with the latter in consequence of the excess of the latter over the former, which is, in general, very inconsiderable. Wherefore, when metallic substances are oxygenated in atmospheric air, or in oxygen gas, they are not converted into acids like sulphur, phosphorus, and charcoal, but are only changed into intermediate substances, which, though approaching to the nature of salts, have not acquired all the saline properties. The old chemists have affixed the name of *calx* not only to metals in this state, but to every body which has been long exposed to the action of fire without being melted. They have converted this word calx into a generical term, under which they confound calcareous earth, which, from a neutral salt, which it really was before calcination, has been changed by fire into an earthy alkali, by *losing* half of its weight, with metals which, by the same means, have joined themselves to a new substance, whose quantity often *exceeds* half their weight, and by which they have been changed almost into the nature of acids. This mode of classifying substances of so very opposite natures, under the same generic name, would have been quite contrary to our principles of nomenclature, especially as, by retaining the above term for this state of metallic substances, we must have conveyed very false ideas of its nature. We have, therefore, laid aside the expression *metallic calx* altogether, and have substituted in its place the term *oxyd*, from the Greek word ὀξύς.

By this may be seen, that the language we have adopted is both copious and expressive. The first or lowest degree of oxygenation in bodies, converts them into oxyds; a second degree of additional oxygenation constitutes the class of acids, of which the specific names, drawn from their particular bases, terminate in *ous*, as the *nitrous* and *sulphurous* acids; the third degree of oxygenation changes these into the species of acids distinguished by the termination in *ic*, as the *nitric* and *sulphuric* acids; and, lastly, we can express a fourth, or highest degree of oxygenation, by adding the word *oxygenated* to the name of the acid, as has been already done with the *oxygenated muriatic* acid.

We have not confined the term *oxyd* to expressing the combinations of metals with oxygen, but have extended it to signify that first degree of oxygenation in all bodies, which, without converting them into acids, causes them to approach to the nature of salts. Thus, we give the name of *oxyd of sulphur* to that soft substance into which sulphur is converted by incipient combustion; and we call the yellow matter left by phosphorus, after combustion, by the name of *oxyd of phosphorus*. In the same manner, nitrous gas, which is azote in its first degree of oxygenation, is the *oxyd of azote*. We have likewise oxyds in great numbers from the vegetable and animal kingdoms; and I shall show, in the sequel, that this new language throws great light upon all the operations of art and nature.[1]

---

[1] For Lavoisier, nitrous and sulphurous acids were N_2O_3 and SO_2, nitric and sulphuric acids were N_2O_5 and SO_3. We now regard the combination of these oxides with water as the acids, namely HNO_2 and H_2SO_3, and HNO_3 and H_2SO_4 respectively. We do not now use the word oxygenated in the sense Lavoisier recommends and we use the word nitrogen instead of azote.

CHAPTER 9

THE HEROIC AGE OF GEOLOGY

29. From JAMES HUTTON'S *Theory of the Earth* (1795)

[The range of mountains running across the south of Scotland between the counties of Ayr and Berwick consists of almost vertical strata of a hard rock known as schist or schistus. On each side of the range the lower country is composed of horizontal strata of softer sandstone and marl. Hutton made a study of the junction between these vertical and horizontal strata, known as an unconformity, to support and illustrate his theory of interchanging sea and land.]

The river Tiviot has made a wide valley as might have been expected, in running over those horizontal strata of marly or decaying substances; and the banks of this river declining gradually are covered with gravel and soil, and show little of the solid strata of the country. This, however, is not the case with the Jed, which is to the southward of the Tiviot; that river, in many places, runs upon the horizontal strata, and undermines steep banks, which falling shows high and beautiful sections of the regular horizontal strata. The little rivulets also which fall into the Jed have hollowed out deep gallies in the land, and show the uniformity of the horizontal strata.

In this manner I was disposed to look for nothing more than what I had seen among those mineral bodies, when one day, walking in the beautiful valley above the town of Jedburgh, I was surprised with the appearance of vertical strata in the bed of the river, where I was certain that the banks were composed of horizontal strata. I was soon satisfied with regard to this phenomenon, and rejoiced at my good fortune in stumbling upon an object so interesting to the natural history of the earth, and which I had been long looking for in vain.

Here the vertical strata, similar to those that are in the bed of the Tweed, appear; and above those vertical strata, are placed the horizontal beds, which extend along the whole country.

413

The question which we would wish to have solved is this; if the vertical strata had been broken and erected under the super-incumbent horizontal strata; or if, after the vertical strata had been broken and erected, the horizontal strata had been deposited upon the vertical strata, then forming the bottom of the sea.

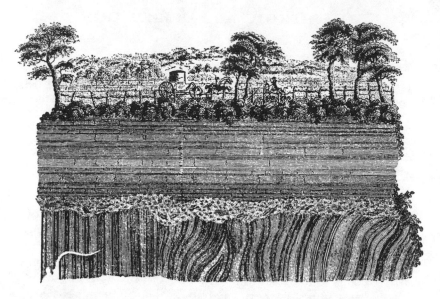

That strata, which are regular and horizontal in one place, should be found bended, broken, or disordered at another, is not uncommon; it is always found more or less in all our horizontal strata. Now, to what length this disordering operation might have been carried, among strata under others, without disturbing the order and continuity of those above, may perhaps be difficult to determine; but here, in this present case, is the greatest disturbance of the under strata, and a very great regularity among those above. Here at least is the most difficult case of this kind to conceive, if we are to suppose that the upper strata had been deposited before those below had been broken and erected.

Let us now suppose that the under strata had been disordered at the bottom of the sea, before the superincumbent bodies were deposited; it is not to be well conceived, that the vertical strata should in that case appear to be cut off abruptly, and present their regular edges immediately under the uniformly deposited substances above. But, in the case now under consideration, there appears the most uniform section of the vertical strata, their ends go up regularly to the horizontal deposited bodies. Now, in whatever state the vertical strata had been in at the time of this event, we can hardly suppose that they could have been so perfectly cut off, without any relict being left to trace that operation. It is much more probable to suppose, that the sea had washed away the relics of the broken and disordered strata, before those that are now superincumbent had been begun to be deposited. But we cannot suppose two such contrary operations in the same place, as that of carrying away the relics of those broken strata, and the depositing of sand and subtile earth in such a regular order. We are therefore led to conclude, that the bottom of the sea, or surface of those erected strata, had been in very different situations at those two periods, when the relics of the disordered strata had been carried away, and when the new materials had been deposited.

If this shall be admitted as a just view of the subject, it will be fair to suppose, that the disordered strata had been raised more or less above the surface of the ocean; that, by the effects of either rivers, winds, or tides, the surface of the vertical strata had been washed bare; and that this surface had been afterwards sunk below the influence of those destructive operations, and thus placed in a situation proper for the opposite effect, the accumulation of matter prepared and put in motion by the destroying causes.

I will not pretend to say that this has all the evidence that should be required, in order to constitute a physical truth, or principle from whence we were to reason farther in our theory; but, as a simple fact, there is more probability for the thing having happened in that manner than in any other; and perhaps this is all that may be attained, though not all that were to be wished on the occasion. Let us now see how far any confirmation may be obtained from the examination of all the attending circumstances in those operations. . . .

In describing the vertical and horizontal strata of the Jed, no mention has been made of a certain puddingstone, which is interposed between the two, lying immediately upon the one and under the other. . . . When we examine the stones and gravel of which it is composed, these appear to have belonged to the vertical strata or schistus mountains. They are in general the hard and solid parts of those indurated strata, worn and rounded by attrition; particularly sand or marl-stone, consolidated and veined with quartz, and many fragments of quartz, all rounded by attrition....

From this it will appear, that the schistus mountains or the vertical strata of indurated bodies had been formed, and had been wasted and worn in the natural operations of the globe, before the horizontal strata were begun to be deposited in those places; the gravel formed of those indurated broken bodies worn round by attrition evince that fact.

[Hutton now describes his further observations of horizontal strata overlying the vertical strata, in the valleys of the Tweed and Teviot.]

It will now be reasonable to suppose that all the schistus which we perceive, whether in the mountains or in the valleys, exposed to our view had been once covered with those horizontal strata which are observed in Berwickshire and Tiviotdale; and that, below all those horizontal strata in the level country, there is at present a body or basis of vertical or inclined schistus, on which the horizontal strata of a secondary order had been deposited. This is the conclusion that I had formed at Jedburgh, before I had seen the confirmation of it in the Tiviot; it is the only one that can be formed according to this view of things; and it must remain in the present state until more evidence be found by which the probability may be either increased or diminished....

We may now come to this general conclusion, that, in this example of horizontal and posterior strata placed upon the vertical schisti, which are prior in relation to the former, we obtain a further view into the natural history of this earth, more than what appears in the simple succession of one stratum above another. We know, in general, that all the solid parts of this earth, which come to our view, have either been formed originally by subsidence at the bottom of the sea, or been transfused in a melted state from the mineral regions among those solid bodies; but here we further learn, that the indurated and erected strata, after being broken and washed by the moving waters, had again been sunk below the sea, and had served as a bottom or basis on which to form a new structure of strata; and also, that those new or posterior strata had been indurated or cemented by the consolidating operations of the mineral region, and elevated from the bottom of the sea into the place of land, or considerably above the general surface of the waters. It is thus that we may investigate particular operations in the general progress of nature, which has for object to renovate the surface of the earth necessarily wasted in the operation of a world sustaining plants and animals.

[Hutton summarises his theory, towards the close of the book, as follows.]

Let us then take a cursory view of this system of things, upon which we have proceeded in our theory, and upon which the constitution of this world seems to depend.

Our solid earth is every where wasted, where exposed to the day. The summits of the mountains are necessarily degraded. The solid and weighty materials of those mountains are every where urged through the valleys, by the force of running water. The soil, which is produced in the destruction of the solid earth, is gradually travelled by the moving water, but is constantly supplying vegetation with its necessary aid. This travelled soil is at last deposited upon the coast, where it forms most fertile countries. But the billows of the ocean agitate the loose materials upon the shore, and wear away the coast, with the endless repetitions of this act of power, or this imparted force. Thus the continent of our earth, sapped in its foundation, is carried away into the deep, and sunk again at the bottom of the sea, from whence it had originated.

We are thus led to see a circulation in the matter of this globe, and a system of beautiful oeconomy in the works of nature. This earth, like the body of an animal, is wasted at the same time that it is repaired. It has a state of growth and augmentation; it has another state, which is that of diminution and decay. This world is thus destroyed in one part, but it is renewed in another; and the operations by which this world is thus constantly renewed, are as evident to the scientific eye, as are those in which it is necessarily destroyed. The marks of the internal fire, by which the rocks beneath the sea are hardened, and by which the land is produced above the surface of the sea, have nothing in them which is doubtful or ambiguous. The destroying operations again, though placed within the reach of our examination, and evident almost to every observer, are no more acknowledged by mankind, than is that system of renovation which philosophy alone discovers.

30. From WILLIAM SMITH'S *Stratigraphical System of Organized Fossils* (1817)

[William Smith describes how strata can be identified and traced by means of the fossils they contain.]

My original method of tracing the Strata by the organized Fossils imbedded therein, is thus reduced to a science not difficult to learn. Ever since the first written account of this discovery was circulated in 1799 it has been closely investigated by my scientific acquaintance in the vicinity of Bath; some of whom search the quarries of different Strata in that district with as much certainty of finding the characteristic Fossils of the respective rocks, as if they were on the shelves of their cabinets. By this new method of searching for organized Fossils with the regularity with which they are imbedded in such a variety of Strata, many new species have been discovered. The Geologist is thus enabled to fix the locality of those previously found; to direct the attentive investigator in his pursuits; and to find in all former cabinets and catalogues numerous proofs of accuracy in this mode of identifying the Strata.

The virtuoso will therefore now enter upon the study and selection of organized Fossils with the twofold advantage of amusement and utility. The various component parts of the soil, and all the subterraneous productions of his estate, become interesting objects of research; the contents of quarries, pits, wells, and other excavations, hitherto thought unworthy of notice, will be scrupulously examined.

The organized Fossils which may be found, will enable him to identify the Strata of his own estate with those of others: thus his lands may be drained with more certainty of success, his buildings substantially improved, and his private and public roads better made, and repaired at less expense....

Many Strata being entirely without organized Fossils, the investigation is much facilitated, by rendering the courses of those Strata which contain them more distinct; and the courses of all the Strata being known, the name of the place where any specimen is found is sufficient to mark its locality in the Strata, and the specimens being filled with the matter in which they are imbedded materially assist in identifying the Stratum to which they belong. In this respect Mineral Conchology has much the advantage of recent; the matter of the Stratum fully compensating in a geological point of view, for any defect in the specimen. Shells are generally without the animals, which are mostly incapable of preservation; fossils frequently represent the animals without the shells (i.e. the interior conformation of the shell). In general, fossil shells are so effectually closed and filled with stony matter, that the hinge, opening, and other characters, cannot be observed.

Numerous Zoophites naturally too tender for preservation, have in their fossil state their shape and most minute organization beautifully retained in limestone, flint, and other solid matter. Thus not only in clays, sands, and rocks, but in the hardest stones, are displayed all the treasures of an ancient deep, which prove the high antiquity and watery origin of the earth; for nothing can more plainly than the Zoophites evince the once fine fluidity of the stoney matter in which they are enveloped, no fluid grosser than water being capable of pervading their pores. The process which converted them and their element into stone seems to have been similar to that of freezing water, which would suddenly fix all the inhabitants of the ocean, each in its place, with all the original form and character. Organized Fossils are to the naturalist as coins to the antiquary; they are the antiquities of the earth; and very distinctly show its gradual regular formation, with the various changes of inhabitants in the watery element.

31. From GEORGES CUVIER's *Essay on the Theory of the Earth* (1821)

[Cuvier mentions his outstanding work of reconstructing extinct animals from their fragmentary remains and then goes on to recount the evidence for his theory of geological catastrophes.]

As an antiquary of a new order, I have been obliged to learn the art of decyphering and restoring these remains, of discovering and bringing together, in their primitive arrangement, the scattered and mutilated fragments of which they are composed, of re-producing, in all their original proportions and characters, the animals to which these fragments formerly belonged, and then of comparing them with those animals which still live on the surface of the earth; an art which is almost unknown, and which pre-supposes, what had scarcely been obtained before, an acquaintance with those laws which regulate the coexistence of the forms by which the different parts of organized beings are distinguished. I had next to prepare myself for these inquiries by others of a far more extensive kind, respecting the animals which still exist. Nothing, except an almost complete review of creation in its present state, could give a character of demonstration to the results of my investigations into its ancient state; but that review has afforded me, at the same time, a great body of rules and affinities which are no less satisfactorily demonstrated; and the whole animal kingdom has been subjected to new laws in consequence of this Essay on a small part of the theory of the earth. . . .

When the traveller passes through those fertile plains where gently-flowing streams nourish in their course an abundant vegetation, and where the soil, inhabited by a numerous population, adorned with flourishing villages, opulent cities, and superb monuments, is never disturbed except by the ravages of war and the oppression of tyrants, he is not led to suspect that nature also has had her intestine wars, and that the surface of the globe has been much convulsed by successive revolutions and various catastrophes. But his ideas change as soon as he digs into that soil which presented such a peaceful aspect, or ascends the hills which border the plain. . . .

The lowest and most level parts of the earth, when penetrated to a very great depth, exhibit nothing but horizontal strata composed of various substances, and containing almost all of them innumerable marine productions. Similar strata, with the same kind of productions, compose the hills even to a great height. Sometimes the shells are so numerous as to constitute the entire body of the stratum. They are almost everywhere in such a perfect state of preservation, that even the smallest of them retain their most delicate parts, their sharpest ridges, and their finest and tenderest processes. They are found in elevations far above the level of every part of the ocean, and in places to which the sea could not be conveyed by any existing cause. They are not only inclosed in loose sand, but are often incrusted and penetrated on all sides by the hardest stones. Every part of the earth, every hemisphere, every continent, every island of any size, exhibits the same phenomenon. We are therefore forcibly led to believe, not only that the sea has at one period or another covered all our plains, but that it must have remained there for a long time, and in a state of tranquillity; which circumstance was necessary for the formation of deposits so extensive, so thick, in part so solid, and containing exuviae so perfectly preserved....

If we examine with greater care these remains of organized bodies, we shall discover, in the midst even of the most ancient secondary strata, other strata that are crowded with animal or vegetable productions, which belong to the land and to fresh water; and amongst the more recent strata, that is, the strata which are nearest the surface, there are some of them in which land animals are buried under heaps of marine productions. Thus the various catastrophes of our planet have not only caused the different parts of our continent to rise by degrees from the basin of the sea, but it has also frequently happened, that lands which had been laid dry have been again covered by the water, in consequence either of these lands sinking down below the level of the sea, or of the sea being raised above the level of the lands. The particular portions of the earth also which the sea has abandoned by its last retreat, had been laid dry once before, and had at that time produced quadrupeds, birds, plants, and all kinds of terrestrial productions; it had then been inundated by the sea, which has since retired from it, and left it to be occupied by its own proper inhabitants.

32. From CHARLES LYELL'S *Principles of Geology* (1830-3)

[Lyell discusses his theory of uniformitarianism, as opposed to theories of catastrophes.]

If we reflect on the history of the progress of geology, as explained in the preceding chapters, we perceive that there have been great fluctuations of opinion respecting the nature of the causes to which all former changes of the earth's surface are referable. The first observers conceived the monuments which the geologist endeavours to decipher to relate to an original state of the earth, or to a period when there were causes in activity, distinct, in kind and degree, from those now constituting the economy of nature. These views were gradually modified, and some of them entirely abandoned in proportion as observations were multiplied, and the signs of former mutations were skilfully interpreted. Many appearances, which had for a long time been regarded as indicating mysterious and extraordinary agency, were finally recognised as the necessary result of the laws now governing the material world; and the discovery of this unlooked-for conformity has at length induced some philosophers to infer, that, during the ages contemplated in geology, there has never been any interruption to the agency of the same uniform laws of change. The same assemblage of general causes, they conceive, may have been sufficient to produce, by their various combinations, the endless diversity of effects, of which the shell of the earth has preserved the memorials; and, consistently with these principles, the recurrence of analogous changes is expected by them in time to come.

Whether we coincide or not in this doctrine, we must admit that the gradual progress of opinion concerning the succession of phenomena in very remote eras, resembles, in a singular manner, that which has accompanied the growing intelligence of every people, in regard to the economy of nature in their own times. In an early state of advancement, when a greater number of natural appearances are unintelligible, an eclipse, an earthquake, a flood, or the approach of a comet, with many other occurrences afterwards found to belong to the regular course of events, are regarded as prodigies. The same delusion prevails as to moral phenomena, and many of these are ascribed to the intervention of demons, ghosts, witches, and other immaterial and supernatural agents. By degrees, many of the enigmas of the moral and physical world are explained, and, instead of being due to extrinsic and irregular causes, they are found to depend on fixed and invariable laws. The philosopher at last becomes convinced of the undeviating uniformity of secondary causes; and, guided by his faith in this principle, he determines the probability of accounts transmitted to him of former occurrences, and often rejects the fabulous tales of former times, on the ground of their being irreconcilable with the experience of more enlightened ages.

Prepossessions in regard to the duration of past time.—As a belief in the want of conformity in the causes by which the earth's crust has been modified in ancient and modern periods was, for a long time, universally prevalent, and that, too, amongst men who were convinced that the order of nature had been uniform for the last several thousand years, every circumstance which could have influenced their minds and given an undue bias to their opinions deserves particular attention. Now the reader may easily satisfy himself, that, however undeviating the course of nature may have been from the earliest epochs, it was impossible for the first cultivators of geology to come to such a conclusion, so long as they were under a delusion as to the age of the world, and the date of the first creation of animate beings. However fantastical some theories of the sixteenth century may now appear to us,—however unworthy of men of great talent and sound judgment,—we may rest assured that, if the same misconception now prevailed in regard to the memorials of human transactions, it would give rise to a similar train of absurdities. Let us imagine, for example, that Champollion, and the French and Tuscan literati when engaged in exploring the antiquities of Egypt, had visited that country with a firm belief that the banks of the Nile were never peopled by the human race before the beginning of the nineteenth century, and that their faith in this dogma was as difficult to shake as the opinion of our ancestors, that the earth was never the abode of living beings until the creation of the present continents, and of the species now existing,—it is easy to perceive what extravagant systems they would frame, while under the influence of this delusion, to account for the monuments discovered in Egypt. The sight of the pyramids, obelisks, colossal statues, and ruined temples, would fill them with such astonishment, that for a time they would be as men spell-bound—wholly incapable of reasoning with sobriety.

CHAPTER 10

THE EIGHTEENTH CENTURY

33. From JOHN LOCKE'S *An Essay concerning Human Understanding* (1690)

[Locke states the basis of his empiricism, that the sole sources of our knowledge are ideas of sensation and ideas of reflection.]

2. Let us then suppose the mind to be, as we say, white paper void of all characters, without any ideas. How comes it to be furnished? Whence comes it by that vast store which the busy and boundless fancy of man has painted on it with an almost endless variety? Whence has it all the *materials* of reason and knowledge? To this I answer, in one word, from EXPERIENCE. In that all our knowledge is founded; and from that it ultimately derives itself. Our observation, employed either about *external sensible objects, or about the internal operations of our minds perceived and reflected on by ourselves, is that which supplies our understandings with all the materials of thinking.* These two are the fountains of knowledge, from whence all the ideas we have, or can naturally have, do spring.

3. First, our Senses, conversant about particular sensible objects, do convey into the mind several distinct perceptions of things, according to those various ways wherein those objects do affect them. And thus we come by those *ideas* we have of *yellow, white, heat, cold, soft, hard, bitter, sweet*, and all those which we call sensible qualities; which when I say the senses convey into the mind, I mean, they from external objects convey into the mind what produces there those perceptions. This great source of most of the ideas we have, depending wholly upon our senses, and derived by them to the understanding, I call SENSATION.

4. Secondly, the other fountain from which experience furnisheth the understanding with ideas is the perception of the operations of our own mind within us, as it is employed about the ideas it has got; which operations, when the soul comes to reflect

on and consider, do furnish the understanding with another set of ideas, which could not be had from things without. And such are *perception, thinking, doubting, believing, reasoning, knowing, willing,* and all the different actings of our own minds; which we being conscious of, and observing in ourselves, do from these receive into our understandings as distinct ideas as we do from bodies affecting our senses. This source of ideas every man has wholly in himself; and though it be not sense, as having nothing to do with external objects, yet it is very like it, and might properly enough be called *internal sense.* But as I call the other Sensation, so I call this REFLECTION, the ideas it affords being such only as the mind gets by reflecting on its own operations within itself. By reflection then, in the following part of this discourse, I would be understood to mean, that notice which the mind takes of its own operations, and the manner of them, by reason whereof there come to be ideas of these operations in the understanding. These two, I say, viz. external material things, as the objects of SENSATION, and the operations of our own minds within, as the objects of REFLECTION, are to me the only originals from whence all our ideas take their beginnings. The term *operations* here I use in a large sense, as comprehending not barely the actions of the mind about its ideas, but some sort of passions arising sometimes from them, such as is the satisfaction or uneasiness arising from any thought.

34. From GEORGE BERKELEY'S *The Principles of Human Knowledge* (1710)

[Berkeley argues that the external world cannot exist without a mind.]

6. Some truths there are so near and obvious to the mind that a man need only open his eyes to see 'em. Such I take this important one to be, viz. that all the choir of heaven and furniture of the earth, in a word all those bodies which compose the mighty frame of the world, have not any subsistence without a mind, that their being is to be perceiv'd or known; that consequently so long as they are not actually perceiv'd by me, or do not exist in my mind or that of any other created spirit, they must either have no existence at all, or else subsist in the mind of some eternal spirit: it being perfectly unintelligible and involving all the absurdity of

abstraction, to attribute to any single part of them an existence independent of a spirit.

To be convinced of which, the reader need only reflect and try to separate in his own thoughts the being of a sensible thing from its being perceived.

7. From what has been said, it follows there is not any other substance than spirit or that which perceives. But for the fuller proof of this point, let it be consider'd, the sensible qualities are colour, figure, motion, smell, taste, and such like, that is, the ideas perceiv'd by sense. Now for an idea to exist in an unperceiving thing is a manifest contradiction, for to have an idea is all one as to perceive, that therefore wherein colour, figure, and the like qualities exist must perceive them; hence 'tis clear there can be no unthinking substance or *substratum* of those ideas.

8. But say you, thô the ideas themselves do not exist without the mind, yet there may be things like them whereof they are copies or resemblances, which things exist without the mind, in an unthinking substance. I answer an idea can be like nothing but an idea, a colour, or figure, can be like nothing but another colour or figure. If we look but ever so little into our thoughts, we shall find it impossible for us to conceive a likeness except only between our ideas. Again, I ask whether those suppos'd originals or external things, of which our ideas are the pictures or representations, be themselves perceivable or no? If they are, then they are ideas and we have gain'd our point; but if you say they are not, I appeal to any one whether it be sense, to assert a colour is like something which is invisible; hard or soft, like something which is intangible, and so of the rest.

35. From DAVID HUME'S *An Enquiry concerning Human Understanding* (1748)

[Hume maintains that the Law of Cause and Effect has no logical basis and that all we can know are our own ideas.]

When any natural object or event is presented, it is impossible for us, by any sagacity or penetration, to discover, or even conjecture, without experience, what event will result from it, or to carry our foresight beyond that object which is immediately present to the memory and senses. Even after one instance or experiment where we have observed a particular event to follow upon another, we are not entitled to form a general rule, or foretell what will happen in like cases; it being justly esteemed an unpardonable temerity to judge of the whole course of nature from one single experiment, however accurate or certain. But when one particular species of event has always, in all instances, been conjoined with another, we make no longer any scruple of foretelling one upon the appearance of the other, and of employing that reasoning, which can alone assure us of any matter of fact or existence. We then call the one object, *Cause*; the other, *Effect*. We suppose that there is some connexion between them; some power in the one, by which it infallibly produces the other, and operates with the greatest certainty and strongest necessity.

It appears, then, that this idea of a necessary connexion among events arises from a number of similar instances which occur of the constant conjunction of these events; nor can that idea ever be suggested by any one of these instances, surveyed in all possible lights and positions. But there is nothing in a number of instances different from every single instance, which is supposed to be exactly similar; except only, that after a repetition of similar instances, the mind is carried by habit, upon the appearance of one event, to expect its usual attendant, and to believe that it will exist. This connexion, therefore, which we *feel* in the mind, this customary transition of the imagination from one object to its usual attendant, is the sentiment or impression from which we form the idea of power or necessary connexion. Nothing farther is in the case. Contemplate the subject on all sides; you will never find any other origin of that idea. This is the sole difference between one instance, from which we can never receive the idea of connexion, and a number of similar instances, by which it is suggested. The first time a man saw the communication of motion by impulse, as by the shock of two billiard balls, he could not pronounce that the one event was *connected*: but only that it was *conjoined* with the other. After he has observed several instances of this nature, he then pronounces them to be *connected*. What alteration has happened to give rise to this new idea of *connexion*? Nothing but that he now *feels* these events to be *connected* in his imagination, and can readily foretell the existence of one from the appearance of the other. When we say, therefore, that one object is connected with another, we mean only that they have acquired a connexion in our thought, and give rise to this inference, by which they become proofs of each other's existence: A conclusion which is somewhat extraordinary, but which seems founded on sufficient evidence....

By what argument can it be proved that the perceptions of the mind must be caused by external objects entirely different from them, though resembling them (if that be possible) and could not arise either from the energy of the mind itself, or from the suggestion of some invisible and unknown spirit, or from some other cause still more unknown to us? It is acknowledged that, in fact, many of these perceptions arise not from anything external, as in dreams, madness, and other diseases. And nothing can be more inexplicable than the manner in which body should so operate upon mind as ever to convey an image of itself to a substance supposed of so different and even contrary a nature.

It is a question of fact, whether the perceptions of the senses be produced by external objects resembling them: how shall this question be determined? By experience, surely; as all other questions of a like nature. But here experience is, and must be, entirely silent. The mind has never anything present to it but the perceptions and cannot possibly reach any experience of their connexion with objects. The supposition of such a connexion is, therefore, without any foundation in reasoning.

36. From IMMANUEL KANT's *Critique of Pure Reason* (1787)

[Kant discusses the limitations of empiricism and introduces his views on *a priori* concepts or categories.]

The illustrious Locke,...meeting with pure concepts of the understanding in experience, deduced them also from experience, and yet proceeded so *inconsequently* that he attempted with their aid to obtain knowledge which far transcends all limits of experience. David Hume recognised that, in order to be able to do this, it was necessary that these concepts should have an *a priori* origin. But since he could not explain how it can be possible that the understanding must think concepts, which are not in themselves connected in the understanding, as being necessarily connected in the object, and since it never occurred to him that the understanding might itself, perhaps, through these concepts, be the author of the experience in which its objects are found, he was constrained to derive them from experience, namely, from a subjective necessity (that is, from *custom*), which arises from repeated association in experience, and which comes mistakenly to be regarded as objective. But from these premises he argued quite consistently. It is impossible, he declared, with these concepts and the principles to which they give rise, to pass beyond the limits of experience. Now this *empirical* derivation, in which both philosophers agree, cannot be reconciled with the scientific *a priori* knowledge which we do actually possess, namely, *pure mathematics* and *general science of nature*; and this fact therefore suffices to disprove such derivation.

While the former of these two illustrious men opened a wide door to *enthusiasm*—for if reason once be allowed such rights, it will no longer allow itself to be kept within bounds by vaguely defined recommendations of moderation—the other gave himself over entirely to *scepticism*, having, as he believed, discovered that what had hitherto been regarded as reason was but an all-prevalent illusion infecting our faculty of knowledge. We now propose to make trial whether it be not possible to find for human reason safe conduct between these two rocks, assigning to her determinate limits, and yet keeping open for her the whole field of her appropriate activities....

If the objects with which our knowledge has to deal were things in themselves, we could have no *a priori* concepts of them. For from what source could we obtain the concepts? If we derived them from the object (leaving aside the question how the object could become known to us), our concepts would be merely empirical, not *a priori*. And if we derived them from the self, that which is merely in us could not determine the character of an object distinct from our representations, that is, could not be a ground why a thing should exist characterised by that which we have in our thought, and why such a representation should not, rather, be altogether empty. But if, on the other hand, we have to deal only with appearances, it is not merely possible, but necessary, that certain *a priori* concepts should precede empirical knowledge of objects. For since a mere modification of our sensibility can never be met with outside us, the objects, as appearances, constitute an object which is merely in us. Now to assert in this manner, that all these appearances, and consequently all objects with which we can occupy ourselves, are one and all in me, that is, are determinations of my identical self, is only another way of saying that there must be a complete unity of them in one and the same apperception. But this unity of possible consciousness also constitutes the form of all knowledge of objects; through it the manifold is thought as belonging to a single object. Thus the mode in which the manifold of sensible representation (intuition) belongs to one consciousness precedes all knowledge of the object as the intellectual form of such knowledge, and itself constitutes a formal *a priori* knowledge of all objects, so far as they are thought (categories). The synthesis of the manifold through pure imagination, the unity of all representations in relation to original apperception, precede all empirical knowledge. Pure concepts of understanding are thus *a priori* possible, and, in relation to experience, are indeed necessary; and this for the reason that our knowledge has to deal solely with appearances, the possibility of which lies in ourselves, and the connection and unity of which (in the representation of an object) are to be met with only in ourselves. Such connection and unity must therefore precede all experience, and are required for the very possibility of it in its formal aspect. From this point of view, the only feasible one, our deduction of the categories has been developed.

37. From DENIS DIDEROT's *Conversation of a Philosopher with the Maréchale de X* (1776)

[In this amusing dialogue Crudeli, who represents Diderot's views, suggests that mind is a product of matter, and that man is a machine.]

I had some business or other with the Maréchal de X. I went to his mansion one morning; he was absent; I had myself announced to Madame la Maréchale. She is a charming woman; she is as beautiful and as devout as an angel; sweetness is clearly expressed on her countenance; and she has, moreover, a tone of voice and a candour in discussion quite in keeping with her expression. She was at her toilette. A chair is drawn up for me; I seat myself and we chat....

The Maréchale: Aren't you Monsieur Crudeli?

Crudeli. Yes, Madame.

The M. Then it's you who believes in nothing?

Cr. The same....

The M. But this world of ours, who made it?

Cr. I ask you that.

The M. God made it.

Cr. And what is God?

The M. A spirit.

Cr. If a spirit can make matter, why could not matter make a spirit?

The M. But why should it make it?

Cr. I see it do it every day. Do you believe that animals have souls?

The M. Certainly I believe it.

Cr. And would you tell me what becomes of the soul of a Peruvian serpent, for example, while it becomes dried, hung in a chimney, exposed to the smoke for a year or two continuously?

The M. I don't care what happens to it; what has it to do with me?

Cr. Madame la Maréchale doesn't know that this dried and smoked serpent revives and is alive again.

The M. I believe none of that.

Cr. It is a clever man, however, Bouguer, who asserts it.

The M. Your clever man has lied about it.

Cr. And if he had spoken truly?

The M. I should have to believe that animals are machines.

Cr. And man, who is only an animal a little more perfect than others....But M. le Maréchal...

The M. Just one more question and it's the last. Are you really quite at peace in your unbelief?

Cr. One could not be more so.

The M. But what if you've deceived yourself?

Cr. If I should have deceived myself?

The M. All that you believed false would be true, and you would be damned. Monsieur Crudeli, it is a dreadful thing to be damned; to burn for all eternity, that's terribly long.

38. From VOLTAIRE'S *A Treatise on Toleration* (1763)

[One of the main ideals of the Enlightenment was toleration. The wars of religion were over but men still tortured and executed each other because of religious prejudice and hysteria. Today men are murdered, on a much larger scale, for ideological and racial reasons. In the following characteristic *conte* Voltaire mocks the absurdity of the conflicts between Christian sects.]

In the early years of the reign of the great Emperor Kam-hi a mandarin of the city of Canton heard from his house a great noise, which proceeded from the next house. He inquired if anybody was being killed, and was told that the almoner of the Danish missionary society, a chaplain from Batavia, and a Jesuit were disputing. He had them brought to his house, put tea and sweets before them, and asked why they quarrelled.

The Jesuit replied that it was very painful for him, since he was always right, to have to do with men who were always wrong; that he had at first argued with the greatest restraint, but had at length lost patience.

The mandarin, with the utmost discretion, reminded them that politeness was needed in all discussion, told them that in China men never became angry, and asked the cause of the dispute.

The Jesuit answered: 'My lord, I leave it to you to decide. These two gentlemen refuse to submit to the decrees of the Council of Trent.'

'I am astonished', said the mandarin. Then, turning to the refractory pair, he said: 'Gentlemen, you ought to respect the opinions of a large gathering. I do not know what the Council of Trent is, but a number of men are always better informed than a single one. No one ought to imagine that he is better than others, and has a monopoly of reason. So our great Confucius teaches; and, believe me, you will do well to submit to the Council of Trent.'

The Dane then spoke. 'My lord speaks with the greatest wisdom', he said; 'we respect great councils, as is proper, and therefore we are in entire agreement with several that were held before the Council of Trent.'

'Oh, if that is the case', said the mandarin, 'I beg your pardon. You may be right. So you and this Dutchman are of the same opinion, against this poor Jesuit.'

'Not a bit', said the Dutchman. 'This fellow's opinions are

almost as extravagant as those of the Jesuit yonder, who has been so very amiable to you. I can't bear them.'

'I don't understand', said the mandarin. 'Are you not all three Christians? Have you not all three come to teach Christianity in our empire? Ought you not, therefore, to hold the same dogmas?'

'It is this way, my lord', said the Jesuit; 'these two are mortal enemies, and are both against me. Hence it is clear that they are both wrong, and I am right.'

'That is not quite clear', said the mandarin; 'strictly speaking, all three of you may be wrong. I should like to hear you all, one after the other.'

The Jesuit then made a rather long speech, during which the Dane and the Dutchman shrugged their shoulders. The mandarin did not understand a word of it. Then the Dane spoke; the two opponents regarded him with pity, and the mandarin again failed to understand. The Dutchman had the same effect. In the end they all spoke together and abused each other roundly. The good mandarin secured silence with great difficulty, and said: 'If you want us to tolerate your teaching here, begin by being yourselves neither intolerant nor intolerable.'

When they went out the Jesuit met a Dominican friar, and told him that he had won, adding that truth always triumphed. The Dominican said: 'Had I been there, you would not have won; I should have convicted you of lying and idolatry.' The quarrel became warm, and the Jesuit and Dominican took to pulling each other's hair. The mandarin, on hearing of the scandal, sent them both to prison. A sub-mandarin said to the judge: 'How long does your excellency wish them to be kept in prison?' 'Until they agree', said the judge. 'Then', said the sub-mandarin, 'they are in prison for life.' 'In that case,' said the judge, 'until they forgive each other.' 'They will never forgive each other', said the other; 'I know them.' 'Then', said the mandarin, 'let them stop there until they pretend to forgive each other.'

39. From *Conversations of Goethe with Eckermann and Soret*
(1823)

[Goethe talks about his theory of colours—now known to be scientifically
worthless.]

Tuesday, 30 December 1823

We then talked about the natural sciences, especially about the
narrow-mindedness with which learned men contend amongst
themselves for priority. 'There is nothing,' said Goethe,'through
which I have learned to know mankind better, than through my
philosophical exertions. It has cost me a great deal, and has been
attended with great annoyance, but I nevertheless rejoice that
I have gained the experience....'

'A Frenchman said to a friend of mine, concerning my theory
of colours,—"We have worked for fifty years to establish and
strengthen the kingdom of Newton, and it will require fifty years
more to overthrow it." The body of mathematicians has endeavoured to make my name so suspected in science that people
are afraid of even mentioning it. Some time ago, a pamphlet fell
into my hands, in which subjects connected with the theory of
colours were treated: the author appeared quite imbued with my
theory, and had deduced everything from the same fundamental
principles. I read the publication with great delight, but, to my
no small surprise, found that the author did not once mention my
name. The enigma was afterwards solved. A mutual friend called
on me, and confessed to me that the clever young author had
wished to establish his reputation by the pamphlet, and had justly
feared to compromise himself with the learned world, if he ventured
to support by my name the views he was expounding. The little
pamphlet was successful, and the ingenious young author has
since introduced himself to me personally, and made his excuses.'

'This circumstance appears to me the more remarkable,' said
I, 'because in everything else people have reason to be proud of
you as an authority, and every one esteems himself fortunate who
has the powerful protection of your public countenance. With
respect to your theory of colours, the misfortune appears to be,
that you have to deal not only with the renowned and universally
acknowledged Newton, but also with his disciples, who are spread
all over the world, who adhere to their master, and whose name is

434

legion. Even supposing that you carry your point at last, you will certainly for a long space of time stand alone with your new theory.'

'I am accustomed to it, and prepared for it', returned Goethe. 'But say yourself,' continued he, 'have I not had sufficient reason to feel proud, when for twenty years I have been forced to own to myself that the great Newton, and all mathematicians and august calculators with him, have fallen into a decided error respecting the theory of colours; and that I, amongst millions, am the only one who knows the truth on this important subject? With this feeling of superiority, it was possible for me to bear with the stupid pretensions of my opponents. People endeavoured to attack me and my theory in every way, and to render my ideas ridiculous; but, nevertheless, I rejoiced exceedingly over my completed work. All the attacks of my adversaries only serve to expose to me the weakness of mankind.'

While Goethe spoke thus, with such a force and a fluency of expression as I have not the power to reproduce with perfect truth, his eyes sparkled with unusual fire; an expression of triumph was observable in them; whilst an ironical smile played upon his lips. The features of his fine countenance were more imposing than ever.

CHAPTER II

THE ATOMIC THEORY

40. From LUCRETIUS' *De Rerum Natura* (On the Nature of Things)

[Lucretius expounded an atomic theory. In the first part of this passage the atoms are referred to as germs.]

> ...no rest is ever found
> For germs throughout the void, but driven on
> In ceaseless varied motion some rebound,
> Leaving large gaps, while some are knit together
> With hardly any interspace at all:
> And these more closely bound with little space
> Locked close by their own intertangled forms,
> These form the rocks, the unyielding iron mass,
> And things like these: but those which spring apart
> Rebounding with great intervals between,
> These give us the thin air...
>   for from our senses far
> The nature of these primal atoms lies.
> Since they're beyond our sight, their motions too
> Must be beyond our ken, and all the more
> Since what you see its movement oft conceals,
> By the very distance from us that it lies.
> Thus oft in the hillside the woolly flocks,
> Cropping the gladsome mead, creep slowly in,
> Where'er the grass with pearly dew invites,
> And lambs full-fed sport round, and butt each other
> In sparkling play: you only see the mass,
> It rests on the green hill a spot of white.

41. From JOHN DALTON'S *A New System of Chemical Philosophy* (1808)

[Dalton introduces the first table of atomic and molecular weights.]

In all chemical investigations, it has justly been considered an important object to ascertain the relative *weights* of the simples which constitute a compound. But unfortunately the enquiry has terminated here; whereas, from the relative weights in the mass, the relative weights of the ultimate particles, or atoms of the bodies might have been inferred, from which their number and weight in various other compounds would appear, in order to assist and guide future investigations, and to correct their results. Now it is one great object of this work, to show the importance and advantage of ascertaining *the relative weights of the ultimate particles, both of simple and compound bodies, the number of simple elementary particles which constitute one compound particle, and the number of less compound particles which enter into the formation of one more compound particle.*

[Dalton then proceeds to give the rules by which he decides how many atoms are contained in a compound particle. These may be termed rules of maximum simplicity. In the case of a substance like water, he assumes that one atom of hydrogen combines with one atom of oxygen. In the cases of the two compounds formed by carbon and oxygen, namely carbonic oxide (carbon monoxide) and carbonic acid (carbon dioxide), he assumes that the former is composed of one atom of carbon and one atom of oxygen, while the latter is composed of one atom of carbon and two atoms of oxygen. And so on.]

From the novelty as well as importance of the ideas suggested in this chapter, it is deemed expedient to give plates, exhibiting the mode of combination in some of the more simple cases. A specimen of these accompanies this first part. The elements or atoms of such bodies as are conceived at present to be simple, are denoted by a small circle, and some distinctive mark; and the combinations consist in the juxtaposition of two or more of these; when three or more particles of elastic fluids are combined together in one, it is to be supposed that the particles of the same kind repel each other, and therefore take their stations accordingly.

ELEMENTS

Simple

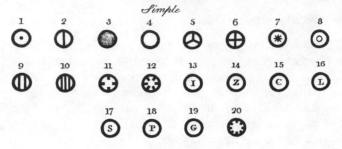

Binary

Ternary

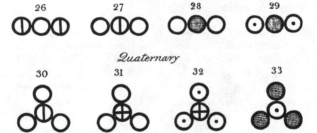

Quaternary

Quinquenary & Sextenary

Septenary

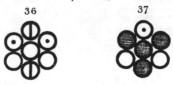

438

Plate IV. This plate contains the arbitrary marks or signs chosen to represent the several chemical elements or ultimate particles.

Fig.

	Fig.			
1 Hydrog. its rel. weight	1	11	Strontites	46
2 Azote	5	12	Barytes	68
3 Carbone or charcoal	5	13	Iron..............	38
4 Oxygen	7	14	Zinc	56
5 Phosphorus	9	15	Copper	56
6 Sulphur	13	16	Lead	95
7 Magnesia	20	17	Silver	100
8 Lime	23	18	Platina	100
9 Soda	28	19	Gold	140
10 Potash	42	20	Mercury	167

21 An atom of water or steam, composed of 1 of oxygen and 1 of hydrogen, retained in physical contact by a strong affinity, and supposed to be surrounded by a common atmosphere of heat; its relative weight 8

22 An atom of ammonia, composed of 1 of azote and 1 of hydrogen 6

23 An atom of nitrous gas, composed of 1 of azote and 1 of oxygen 12

24 An atom of olefiant gas, composed of 1 of carbone and 1 of hydrogen 6

25 An atom of carbonic oxide composed of 1 of carbone and 1 oxygen 12

26 An atom of nitrous oxide, 2 azote + 1 oxygen 17

27 An atom of nitric acid, 1 azote + 2 oxygen 19

28 An atom of carbonic acid, 1 carbone + 2 oxygen 19

29 An atom of carburetted hydrogen, 1 carbone + 2 hydrogen 7

30 An atom of oxynitric acid, 1 azote + 3 oxygen 26

31 An atom of sulphuric acid, 1 sulphur + 3 oxygen 34

32 An atom of sulphuretted hydrogen, 1 sulphur + 3 hydrogen .... 16 / 16

33 An atom of alcohol, 3 carbone + 1 hydrogen 16

34 An atom of nitrous acid, 1 nitric acid + 1 nitrous gas .. 31

35 An atom of acetous acid, 2 carbone + 2 water 26

36 An atom of nitrate of ammonia, 1 nitric acid + 1 ammonia + 1 water 33

37 An atom of sugar, 1 alcohol + 1 carbonic acid 35

42. From AMADEO AVOGADRO'S *Essay on a Manner of Determining the Relative Masses of the Elementary Molecules of Bodies, and the Proportions in which they enter into these Compounds* (1811)

[Avogadro's hypothesis that equal volumes of all gases, at the same temperature and pressure, contain the same number of molecules, is one of the chief foundation stones on which chemistry is built. A molecule of hydrogen is H_2 whereas an atom is H. Avogadro is quite clear about this distinction although his nomenclature is not ideal. He·uses the term *molecule* for either the modern atom or molecule; *simple molecule* or *elementary molecule* for atom; *composite molecule* for molecule; and *integral molecule* for the molecule of a compound.

In the last part of this extract Avogadro points out that Dalton assumed water to be HO instead of H_2O and hence obtained an atomic weight for oxygen only half its true value.]

M. Gay-Lussac has shown in an interesting Memoir (Mémoires de la Société d'Arcueil, Tome II.) that gases always unite in a very simple proportion by volume, and that when the result of the union is a gas, its volume also is very simply related to those of its components. But the quantitative proportions of substances in compounds seem only to depend on the relative number of molecules which combine, and on the number of composite molecules which result. It must then be admitted that very simple relations also exist between the volumes of gaseous substances and the numbers of simple or compound molecules which form them. The first hypothesis to present itself in this connection, and apparently even the only admissible one, is the supposition that the number of integral molecules in any gases is always the same for equal volumes, or always proportional to the volumes. Indeed, if we were to suppose that the number of molecules contained in a given volume were different for different gases, it would scarcely be possible to conceive that the law regulating the distance of molecules could give in all cases relations so simple as those which the facts just detailed compel us to acknowledge between the volume and the number of molecules....

Setting out from this hypothesis, it is apparent that we have the means of determining very easily the relative masses of the molecules of substances obtainable in the gaseous state, and the relative number of these molecules in compounds; for the ratios of the masses of molecules are then the same as those of the densities of the different gases at equal temperature and pressure, and the relative number of molecules in a compound is given at once by the ratio of the volumes of the gases that form it. For example, since the numbers 1·10359 and 0·07321 express the densities of the two gases oxygen and hydrogen compared to that of atmospheric air as unity, and the ratio of the two numbers consequently represents the ratio between the masses of equal volumes of these two gases, it will also represent on our hypothesis the ratio of the masses of their molecules. Thus the mass of the molecule of oxygen will be about 15 times that of the molecule of hydrogen, or more exactly, as 15·074 to 1 . . . .

Dalton, on arbitrary suppositions as to the most likely relative number of molecules in compounds, has endeavoured to fix ratios between the masses of the molecules of simple substances. Our hypothesis, supposing it well-founded, puts us in a position to confirm or rectify his results from precise data, and, above all, to assign the magnitude of compound molecules according to the volumes of the gaseous compounds, which depend partly on the division of molecules entirely unsuspected by this physicist.

Thus Dalton supposes that water is formed by the union of hydrogen and oxygen, molecule to molecule. From this, and from the ratio by weight of the two components, it would follow that the mass of the molecule of oxygen would be to that of hydrogen as $7\frac{1}{2}$ to 1 nearly, or, according to Dalton's evaluation, as 6 to 1. This ratio on our hypothesis is, as we saw, twice as great, namely, as 15 to 1. As for the molecule of water, its mass ought to be roughly expressed by $15 + 2 = 17$ (taking for unity that of hydrogen), if there were no division of the molecule into two; but on account of this division it is reduced to half, $8\frac{1}{2}$, or more exactly 8·537, as may also be found directly by dividing the density of aqueous vapour 0·625 (Gay-Lussac) by the density of hydrogen 0·0732. This mass only differs from 7, that assigned to it by Dalton, by the difference in the values for the composition of water; so that in this respect Dalton's result is approximately correct from the combination of two compensating errors,—the error in the mass of the molecule of oxygen, and his neglect of the division of the molecule.

CHAPTER 12

THE WAVE THEORY OF LIGHT

43. From CHRISTIAAN HUYGENS' *Treatise on Light* (1690)

[Huygens argues that light, like sound, is a form of wave motion.]

It is inconceivable to doubt that light consists in the motion of some sort of matter. For whether one considers its production, one sees that here upon the Earth it is chiefly engendered by fire and flame which contain without doubt bodies that are in rapid motion, since they dissolve and melt many other bodies, even the most solid; or whether one considers its effects, one sees that when light is collected, as by concave mirrors, it has the property of burning as a fire does, that is to say it disunites the particles of bodies. This is assuredly the mark of motion, at least in the true Philosophy, in which one conceives the causes of all natural effects in terms of mechanical motions. This, in my opinion, we must necessarily do, or else renounce all hopes of ever comprehending anything in Physics.

And as, according to this Philosophy, one holds as certain that the sensation of sight is excited only by the impression of some movement of a kind of matter which acts on the nerves at the back of our eyes, there is here yet one reason more for believing that light consists in a movement of the matter which exists between us and the luminous body.

Further, when one considers the extreme speed with which light spreads on every side, and how, when it comes from different regions, even from those directly opposite, the rays traverse one another without hindrance, one may well understand that when we see a luminous object, it cannot be by any transport of matter coming to us from this object, in the way in which a shot or an arrow traverses the air; for assuredly that would too greatly impugn these two properties of light, especially the second of them. It is then in some other way that light spreads; and that which can lead us to comprehend it is the knowledge which we have of the spreading of Sound in the air.

We know that by means of the air, which is an invisible and impalpable body, Sound spreads around the spot where it has been produced, by a movement which is passed on successively from one part of the air to another; and that the spreading of this movement, taking place equally rapidly on all sides, ought to form spherical surfaces ever enlarging and which strike our ears. Now there is no doubt at all that light also comes from the luminous body to our eyes by some movement impressed on the matter which is between the two; since, as we have already seen, it cannot be by the transport of a body which passes from one to the other. If, in addition, light takes time for its passage—which we are now going to examine—it will follow that this movement, impressed on the intervening matter, is successive; and consequently it spreads, as Sound does, by spherical surfaces and waves: for I call them waves from their resemblance to those which are seen to be formed in water when a stone is thrown into it, and which present a successive spreading as circles, though these arise from another cause, and are only in a flat surface.

44. From Isaac Newton's *Opticks* (1704)

[This is part of Query 28, in which Newton argues that light cannot consist of waves because waves spread round corners, whereas light travels in straight lines.]

Query 28.   Are not all Hypotheses erroneous, in which Light is supposed to consist in Pression or Motion, propagated through a fluid Medium?....For Pression or Motion, cannot be propagated in a Fluid in right Lines, beyond an Obstacle which stops part of the Motion, but will bend and spread every way into the quiescent Medium which lies beyond the Obstacle....The Waves on the Surface of stagnating Water, passing by the sides of a broad Obstacle which stops part of them, bend afterwards and dilate themselves gradually into the quiet Water behind the Obstacle. The Waves, Pulses or Vibrations of the Air, wherein Sounds consist, bend manifestly, though not so much as the Waves of Water. For a Bell or a Cannon may be heard beyond a Hill which intercepts the sight of the sounding Body, and Sounds are propagated as readily through crooked Pipes as through streight ones. But Light is never known to follow crooked Passages nor to bend into the Shadow.

[In Query 29 Newton suggests a corpuscular theory of light.]

Query 29.   Are not the Rays of Light very small Bodies emitted from shining Substances? For such Bodies will pass through uniform Mediums in right Lines without bending into the Shadow, which is the Nature of the Rays of Light. They will also be capable of several Properties, and be able to conserve their Properties unchanged in passing through several Mediums, which is another Condition of the Rays of Light.

[In Query 17 Newton supplements the corpuscular theory with a wave theory.]

Query 17.   If a stone be thrown into stagnating Water, the Waves excited thereby continue some time to arise in the place where the Stone fell into the Water, and are propagated from thence in concentrick Circles upon the Surface of the Water to great distances. And the Vibrations or Tremors excited in the Air by percussion, continue a little time to move from the place of percussion in concentrick Spheres to great distances. And in like manner, when a Ray of Light falls upon the Surface of any pellucid Body, and is there refracted or reflected, may not Waves of Vibrations, or Tremors, be thereby excited in the refracting or reflecting Medium at the point of Incidence, and continue to arise there, and to be propagated from thence as long as they continue to arise and be propagated, when they are excited in the bottom of the Eye by the Pressure or Motion of the Finger, or by the Light which comes from the Coal of Fire in the Experiments above mention'd? and are not these Vibrations propagated from the point of Incidence to great distances? And do they not overtake the Rays of Light, and by overtaking them successively, do they not put them into the Fits of easy Reflexion and easy Transmission described above? For if the Rays endeavour to recede from the densest part of the Vibration, they may be alternately accelerated and retarded by the Vibrations overtaking them.

[In Queries 12, 13 and 14 Newton suggests that sensation of colour is determined by wavelength.]

Query 12.   Do not the Rays of Light in falling upon the bottom of the Eye excite Vibrations in the *Tunica Retina*? Which Vibrations, being propagated along the solid Fibres of the optick Nerves into the Brain, cause the Sense of seeing. . . .

Query 13.   Do not several sorts of Rays make Vibrations of several bignesses, which according to their bignesses excite Sensations of several Colours, much after the manner that the Vibrations of the Air, according to their several bignesses excite Sensations of several Sounds? And particularly do not the most refrangible Rays excite the shortest Vibrations for making a Sensation of deep violet, the least refrangible the largest for making a Sensation of deep red, and the several intermediate sorts of Rays, Vibrations of several intermediate bignesses to make Sensations of the several intermediate Colours?

Query 14.   May not the harmony and discord of Colours arise from the proportions of the Vibrations propagated through the Fibres of the optick Nerves into the Brain, as the harmony and discord of Sounds arise from the proportions of the Vibrations of the Air?

45. From Dr THOMAS YOUNG'S *Reply to the animadversions of the Edinburgh Reviewers on some papers published in the Philosophical Transactions* (1804)

[This is the opening of the rather prolix and ponderous reply of Young to the venomous attacks made upon his work by Henry Brougham. It is notable for the statement of the principle of interference at the end of the passsage.]

Welbeck Street, 30 Nov. 1804

A man who has a proper regard for the dignity of his own character, although his sensibility may sometimes be awakened by the unjust attacks of interested malevolence, will esteem it in general more advisable to bear, in silence, the temporary effects of a short-lived injury, than to suffer his own pursuits to be interrupted, in making an effort to repel the invective, and to punish the aggressor. But it is possible that art and malice may be so insidiously combined, as to give to the grossest misrepresentations the semblance of justice and candour; and, especially where the subject of the discussion is of a nature little adapted to the comprehension of the generality of readers, even a man's friends may be so far misled by a garbled extract from his own works, and by the specious mixture of partial truth with essential falsehood, that they may not only be unable to defend him from the unfavourable opinion of others, but may themselves be disposed to suspect, in spite of their partiality, that he has been hasty and inconsiderate at least, if not radically weak and mistaken. In such a case, he owes to his friends such explanations as will enable them to see clearly the injustice of the accusation, and the iniquity of its author: and, if he is in a situation which requires that he should in a certain degree possess the public confidence, he owes to himself and to the public to prove, that the charges of imbecility of mind and perversity of disposition are not more founded with regard to him, than with regard to all who are partakers with him in the unavoidable imperfections of human nature.

Precisely such is my situation. I have at various times communicated to the Royal Society, in a very abridged form, the results of my experiments and investigations, relating to different branches of natural philosophy: and the Council of the Society, with a view perhaps of encouraging patient diligence, has honoured my essays with a place in their Transactions. Several of these essays have been singled out, in an unprecedented manner, from the volumes in which they were printed, and have been made the subjects, in the second and ninth numbers of the Edinburgh Review, not of criticism, but of ridicule and invective; of an attack, not only upon my writings and my literary pursuits, but almost on my moral character. The peculiarity of the style and tendency of this attack led me at once to suspect, that it must have been suggested by some other motive than the love of truth; and I have both internal and external evidence for believing, that the articles in question are, either wholly, or in great measure, the productions of an individual, upon whose mathematical works I had formerly thought it necessary to make some remarks, which, though not favourable, were far from being severe....

I have indeed been accused of insinuating 'that Sir Isaac Newton was but a sorry philosopher'. But it is impossible that an impartial person should read my essays on the subject of light without being sensible that I have as high a respect for his unparalleled talents and acquirements as the blindest of his followers, and the most parasitical of his defenders. I have acknowledged that 'his merits are great beyond all contest or comparison'; that 'his discovery of the composition of white light would alone have immortalised his name'; that the very arguments which tend to overthrow his hypothesis respecting the emanation of light, 'give the strongest proofs of the admirable accuracy of his experiments'; and that a person may, 'with the greatest justice, be attached to every doctrine which is stamped with the Newtonian approbation'. The printer of the Review, feeling perhaps that the last expressions would militate too much in my favour, has thought fit to plunder me of them, by omitting the marks of quotation and to attribute them to my antagonist. But, much as I venerate the name of Newton, I am not therefore obliged to believe that he was infallible. I see, not with exultation, but with regret, that he was liable to err, and that his authority has, perhaps, sometimes even retarded the progress of Science. . . .

It was in May 1801 that I discovered, by reflecting on the beautiful experiments of Newton, a law which appears to me to account for a greater variety of interesting phenomena than any other optical principle that has yet been made known. I shall endeavour to explain this law by a comparison.

Suppose a number of equal waves of water to move upon the surface of a stagnant lake, with a certain constant velocity, and to enter a narrow channel leading out of the lake. Suppose then another similar cause to have excited another equal series of waves, which arrive at the same channel, with the same velocity, and at the same time with the first. Neither series of waves will destroy the other, but their effects will be combined: if they enter the channel in such a manner that the elevations of one series coincide with those of the other, they must together produce a series of greater joint elevations; but if the elevations of one series are so situated as to correspond to the depressions of the other, they must exactly fill up those depressions, and the surface of the water must remain smooth; at least I can discover no alternative, either from theory or from experiment.

Now I maintain that similar effects take place whenever two portions of light are thus mixed; and this I call the general law of the interference of light. I have shown that this law agrees, most accurately, with the measures recorded in Newton's *Optics*, relative to the colours of transparent substances, observed under circumstances which had never before been subjected to calculation, and with a great diversity of other experiments never before explained. This, I assert, is a most powerful argument in favour of the theory which I had before revived.

46. From AUGUSTIN FRESNEL'S *Memoir on the Diffraction of Light* (1819)

[In this opening passage of the memoir Fresnel points out that the wave theory of light is far more economical and convincing in its basic assumptions than the corpuscular or emission theory.]

Before I concern myself particularly with the numerous and varied phenomena included under the common name of diffraction I feel that I must put forward several general considerations on the two systems which, until now, have divided men of science on the nature of light. Newton assumed that particles of light, shot from luminous bodies, come straight to our eyes, where they produce by their onset the sensation of sight. Descartes, Hooke, Huyghens and Euler thought that light resulted from the vibrations of an universal, extremely subtle fluid, agitated by the rapid movements of the particles of the luminous body, in the same way that the air is disturbed by the vibrations of sounding bodies; so that, in this system, it is not the molecules of the fluid in contact with the luminous body which reach the organ of sight, but only the movement which has been impressed upon them.

The first hypothesis has the advantage of leading to more evident consequences because the mathematical analysis is applied more easily; the second, on the contrary, presents in this respect great difficulties. But, in the choice of a system, one must have regard only to the simplicity of the hypotheses; that of the calculus must not have any weight in the balance of probabilities. Nature is not troubled by difficulties of analysis; she shuns only complication of means. She appears to be purposed to make much with little; this is a principle of which new proofs are unceasingly forthcoming from the development of the physical sciences. Astronomy, the glory of the human mind, above all provides a striking confirmation; all Kepler's laws were traced back by the genius of Newton to the single law of gravitation, which then served to explain and even to reveal the most complicated and least apparent perturbations of the planetary movements.

If men have sometimes gone astray in wishing to simplify the elements of a science, it is because they have established systems before collecting a sufficiently large number of facts. An hypothesis of the kind they put forward, very simple when only one class of phenomena is considered, necessitates many other hypotheses when they wish to break out of the tight circle in which they have been confined. If nature has purposed to produce the *maximum* of effects with the *minimum* of causes, it is in the harmony of her laws that she has been able to resolve this great problem.

It is undoubtedly difficult to discover the basis of this admirable economy, namely the simplest causes of the phenomena as seen from an extended viewpoint. But if this general principle of the philosophy of the physical sciences does not lead immediately to the knowledge of truth, it can nevertheless direct the efforts of the human mind, in the rejection of systems which account for the phenomena by too large a number of different causes, and in the adoption by preference of those which, resting on the least number of hypotheses, are consequently the most fertile.

In this respect, the system which makes light consist of the vibrations of an universal fluid has great advantages over that of emission. It enables us to understand how light is susceptible of undergoing so many diverse modifications. I do not mean here those which it experiences momentarily in the bodies which it traverses and which can always be explained by the nature of these media, but I wish to speak of those permanent modifications which it carries with it and which stamp it with new characteristics. One conceives that a fluid, an assemblage of an infinity of mobile molecules subject to a mutual dependence, is capable of a great number of different modifications, by reason of the relative movements which are impressed upon them. The vibrations of the air and the variety of the sensations which they produce on the organ of hearing offer a remarkable parallel.

In the emission system, on the other hand, the motion of each light particle being independent of that of the others, the number of modifications to which it is susceptible appears extremely limited. One can add a movement of rotation to that of transmission; but that is all. As for oscillatory movements, their existence is conceivable only in media which maintain an unequal action on the different sides of the light particles, supposed endowed with different properties. As soon as this action ceases, the oscillations must also cease or become transformed into rotary movements. Thus the rotary movements and the diversity of the sides of a light particle are the only mechanical resources of the emission theory to represent all the permanent modifications of light. They appear very inadequate, if one pays attention to the multitude of phenomena which optics offers. One is more convinced of this on reading the Treatise on experimental and mathematical physics of M. Biot, in which are developed, with great detail and clarity, the principal consequences of Newton's system. One will find there, to account for the phenomena, that it is necessary to accumulate in each light particle a great number of diverse modifications, often very difficult to reconcile with each other.

On the system of waves, the infinite variety of rays of different colours which compose white light proceed quite simply from the difference of wavelength, like the different musical tones from that of sound waves. On the Newtonian theory one cannot attribute this diversity of colours or of sensations produced on the organ of sight to difference of mass or of initial velocity of the light particles, for it would result in dispersion always being proportional to refraction, and experiment proves the contrary. Now it must necessarily be admitted that the particles of the rays of different colours are not of the same nature. There must be then as many different light particles as there are colours, of diverse shades, in the solar spectrum.

After having explained reflection and refraction by the action of repulsive and attractive forces emanating from the surface of a body, Newton, to account for the phenomena of the coloured rings, imagined, in the light particles, fits of easy reflection and of easy transmission, returning periodically at equal intervals. It was natural to suppose that these intervals, like the speed of light, were always the same in the same media, and that, in consequence, under more oblique incidence, the diameter of the rings must diminish, the distance travelled being increased. Experiment shows, on the contrary, that the diameter of the rings increases with obliquity of incidence, and Newton was obliged to conclude that the fits increased in length and to a greater degree than the distance travelled....

Thus the system of emission sufficed so little to explain the phenomena, that each new phenomenon necessitated a new hypothesis....

Not only is the hypothesis of fits improbable by its complication, and difficult to reconcile, in its consequences, with the facts, but it does not even suffice to explain the phenomenon of the coloured rings, for which it was conceived....

In the theory of waves this principle is a consequence of the fundamental hypothesis. One conceives in effect that, when two systems of light waves tend to produce opposed movements at the same point of space, they must mutually enfeeble themselves, and even destroy each other completely if the two impulsions are equal, and that the oscillations must be additive, on the other hand, when they are executed in the same phase. The intensity of the light will depend therefore on the respective positions of the two systems of waves, or, what comes to the same thing, on the difference in the distances travelled, when they emanate from a common source.

CHAPTER 13

THE CONSERVATION AND DISSIPATION OF ENERGY

47. From JAMES JOULE'S lecture, *On Matter, Living Force and Heat* (1847)

[In this popular lecture Joule states his views on the conservation of energy and on the nature of heat. His term 'living force' represents the energy of motion of a body—what is now known as kinetic energy.]

You will at once perceive that the living force of which we have been speaking is one of the most important qualities with which matter can be endowed, and, as such, that it would be absurd to suppose that it can be destroyed, or even lessened, without producing the equivalent of attraction through a given distance of which we have been speaking. You will therefore be surprised to hear that until very recently the universal opinion has been that living force could be absolutely and irrevocably destroyed at any one's option. Thus, when a weight falls to the ground, it has been generally supposed that its living force is absolutely annihilated, and that the labour which may have been expended in raising it to the elevation from which it fell has been entirely thrown away and wasted, without the production of any permanent effect whatever. We might reason, *a priori*, that such absolute destruction of living force cannot possibly take place, because it is manifestly absurd to suppose that the powers with which God has endowed matter can be destroyed any more than that they can be created by man's agency....

How comes it to pass that, though in almost all natural phenomena we witness the arrest of motion and the apparent destruction of living force, we find that no waste or loss of living force has actually occurred? Experiment has enabled us to answer these questions in a satisfactory manner; for it has shown that, wherever living force is apparently destroyed, an equivalent is produced which in process of time may be reconverted into living force. This equivalent is heat. Experiment has shown that wherever living force is apparently destroyed or absorbed, heat is produced. The most frequent way in which living force is thus converted into heat is by means of friction. Wood rubbed against wood or against any hard body, metal rubbed against metal or against any other body—in short, all bodies, solid or even liquid, rubbed against each other are invariably heated, sometimes even so far as to become red-hot. In all these instances the quantity of heat produced is invariably in proportion to the exertion employed in rubbing the bodies together—that is, to the living force absorbed. By fifteen or twenty smart and quick strokes of a hammer on the end of an iron rod of about a quarter of an inch in diameter placed upon an anvil an expert blacksmith will render that end of the iron visibly red-hot. Here heat is produced by the absorption of the living force of the descending hammer in the soft iron; which is proved to be the case from the fact that the iron cannot be heated if it be rendered hard and elastic, so as to transfer the living force of the hammer to the anvil.

The general rule, then, is, that wherever living force is apparently destroyed, whether by percussion, friction, or any similar means, an exact equivalent of heat is restored. The converse of this proposition is also true, namely, that heat cannot be lessened or absorbed without the production of living force, or its equivalent attraction through space. Thus, for instance, in the steam-engine it will be found that the power gained is at the expense of the heat of the fire,—that is, that the heat occasioned by the combustion of the coal would have been greater had a part of it not been absorbed in producing and maintaining the living force of the machinery. It is right, however, to observe that this has not as yet been demonstrated by experiment. But there is no room to doubt that experiment would prove the correctness of what I have said; for I have myself proved that a conversion of heat into living force takes place in the expansion of air, which is analogous to the expansion of steam in the cylinder of the steam-engine. But the most convincing proof of the conversion of heat into living force has been derived from my experiments with the electro-magnetic engine, a machine composed of magnets and bars of iron set in motion by an electrical battery. I have proved by actual experiment that, in exact proportion to the force with which this machine works, heat is abstracted from the electrical battery.

You see, therefore, that living force may be converted into heat, and that heat may be converted into living force, or its equivalent attraction through space. All three, therefore—namely, heat, living force, and attraction through space (to which I might also add light, were it consistent with the scope of the present lecture)—are mutually convertible into one another. In these conversions nothing is lost. The same quantity of heat will always be converted into the same quantity of living force. We can therefore express the equivalency in definite language applicable at all times and under all circumstances. Thus the attraction of 817 lb. through the space of one foot is equivalent to, and convertible into, the living force possessed by a body of the same weight of 817 lb. when moving with the velocity of eight feet per second, and this living force is again convertible into the quantity of heat which can increase the temperature of one pound of water by one degree Fahrenheit. The knowledge of the equivalency of heat to mechanical power is of great value in solving a great number of interesting and important questions. In the case of the steam-engine, by ascertaining the quantity of heat produced by the combustion of coal, we can find out how much of it is converted into mechanical power, and thus come to a conclusion how far the steam-engine is susceptible of further improvements. Calculations made upon this principle have shown that at least ten times as much power might be produced as is now obtained by the combustion of coal. Another interesting conclusion is, that the animal frame, though destined to fulfil so many other ends, is as a machine more perfect than the best contrived steam-engine—that is, is capable of more work with the same expenditure of fuel.

48. From J. R. MAYER's paper, *Remarks on the Forces of Inorganic Nature* (1842)

[On this paper rests Mayer's claim to priority in the enunciation of the principle of the conservation of energy and in the determination of the mechanical equivalent of heat. The following extract is the concluding part of the paper. Mayer's term 'falling force' represents the work a body is capable of performing in falling—what we should now call its potential energy.]

The natural connexion existing between falling force, motion, and heat may be conceived of as follows. We know that heat makes its appearance when the separate particles of a body approach nearer to each other: condensation produces heat. And what applies to the smallest particles of matter, and the smallest intervals between them, must also apply to large masses and to measurable distances. The falling of a weight is a real diminution of the bulk of the earth, and must therefore without doubt be related to the quantity of heat thereby developed; this quantity of heat must be proportional to the greatness of the weight and its distance from the ground. From this point of view we are very easily led to the equations between falling force, motion, and heat, that have already been discussed....

If falling force and motion are equivalent to heat, heat must also naturally be equivalent to motion and falling force. Just as heat appears as an effect of the diminution of bulk and of the cessation of motion, so also does heat disappear as a cause when its effects are produced in the shape of motion, expansion, or raising of weight.

In water-mills, the continual diminution in bulk which the earth undergoes, owing to the fall of the water, gives rise to motion, which afterwards disappears again, calling forth unceasingly a great quantity of heat; and inversely, the steam-engine serves to decompose heat again into motion or the raising of weights. A locomotive engine with its train may be compared to a distilling apparatus; the heat applied under the boiler passes off as motion, and this is deposited again as heat at the axles of the wheels.

We will close our disquisition, the propositions of which have resulted as necessary consequences from the principle 'causa aequat effectum', and which are in accordance with all the phenomena of Nature, with a practical deduction. The solution of the equations subsisting between falling force and motion requires that the space fallen through in a given time, e.g. the first second, should be experimentally determined; in like manner, the solution of the equations subsisting between falling force and motion on the one hand and heat on the other, requires an answer to the question, how great is the quantity of heat which corresponds to a given quantity of motion or falling force? For instance, we must ascertain how high a given weight requires to be raised above the ground in order that its falling force may be equivalent to the raising of the temperature of an equal weight of water from $0°$ to $1°$ C. The attempt to show that such an equation is the expression of a physical truth may be regarded as the substance of the foregoing remarks.

By applying the principles that have been set forth to the relations subsisting between the temperature and the volume of gases, we find that the sinking of a mercury column by which a gas is compressed is equivalent to the quantity of heat set free by the compression; and hence it follows, the ratio between the capacity for heat of air under constant pressure and its capacity under constant volume being taken as $= 1.421$, that the warming of a given weight of water from $0°$ to $1°$ C. corresponds to the fall of an equal weight from the height of about 365 metres.

If we compare with this result the working of our best steam-engines, we see how small a part only of the heat applied under the boiler is really transformed into motion or the raising of weights; and this may serve as justification for the attempts at the profitable production of motion by some other method than the expenditure of the chemical difference between carbon and oxygen—more particularly by the transformation into motion of electricity obtained by chemical means.

49. From SADI CARNOT'S *Reflections on the Motive Power of Heat* (1824)

[Carnot explains that the most convenient form of heat engine is one utilising a gas or vapour such as steam, and that its efficiency depends on the difference in temperature of the steam as it enters and leaves the engine.]

The elastic fluids, gases or vapours, are the means really adapted to the development of the motive power of heat. They combine all the conditions necessary to fulfil this office. They are easy to compress; they can be almost infinitely expanded; variations of volume occasion in them great changes of temperature; and, lastly, they are very mobile, easy to heat and to cool, easy to transport from one place to another, which enables them to produce rapidly the desired effects. We can easily conceive a multitude of machines fitted to develop the motive power of heat through the use of elastic fluids; but in whatever way we look at it, we should not lose sight of the following principles:

  1. The temperature of the fluid should be made as high as

possible, in order to obtain a great fall of caloric, and consequently a large production of motive power.

2. For the same reason the cooling should be carried as far as possible.

3. It should be so arranged that the passage of the elastic fluid from the highest to the lowest temperature should be due to increase of volume; that is, it should be so arranged that the cooling of the gas should occur spontaneously as the effect of rarefaction. The limits of the temperature to which it is possible to bring the fluid primarily, are simply the limits of the temperature obtainable by combustion; they are very high.

The limits of cooling are found in the temperature of the coldest body of which we can easily and freely make use; this body is usually the water of the locality. . . .

It is seldom that in steam-engines the elastic fluid is produced under a higher pressure than six atmospheres—a pressure corresponding to about 160° Centigrade, and it is seldom that condensation takes place at a temperature much under 40°. The fall of caloric from 160° to 40° is 120°. . . .

Coal being capable of producing, by its combustion, a temperature higher than 1000°, and the cold water, which is generally used in our climate, being at about 10°, we can easily procure a fall of caloric of 1000°, and of this only 120° are utilized by steam-engines. Even these 120° are not wholly utilized. There is always considerable loss due to useless re-establishments of equilibrium in the caloric.

It is easy to see the advantages possessed by high-pressure machines over those of lower pressure. This superiority lies essentially in the power of utilizing a greater fall of caloric. The steam produced under a higher pressure is found also at a higher temperature, and as, further, the temperature of condensation remains always about the same, it is evident that the fall of caloric is more considerable.

CHAPTER 14

FIELD PHYSICS

50. MICHAEL FARADAY on Lines of Force and the Field

[In these passages, taken from *Experimental Researches in Electricity*, vol. III, Faraday discusses lines of magnetic force, the idea that a body extends as far as its lines of gravitational force, and that light consists of vibrations in lines of force.]

On a former occasion certain lines about a bar-magnet were described and defined (being those which are depicted to the eye by the use of iron filings sprinkled in the neighbourhood of the magnet), and were recommended as expressing accurately the nature, condition, direction, and amount of the force in any given region either within or outside of the bar. At that time the lines were considered in the abstract. Without departing from or unsettling anything then said, the inquiry is now entered upon of the possible and probable *physical existence* of such lines. . . .

Many powers act manifestly at a distance; their physical nature is incomprehensible to us: still we may learn much that is real and positive about them, and amongst other things something of the condition of the space between the body acting and that acted upon, or between the two mutually acting bodies. Such powers are presented to us by the phaenomena of gravity, light, electricity, magnetism, &c. These when examined will be found to present remarkable differences in relation to their respective lines of forces; and at the same time that they establish the existence of real physical lines in some cases, will facilitate the consideration of the question as applied especially to magnetism. . . .

In this view of a magnet, the medium or space around it is as essential as the magnet itself, being a part of the true and complete magnetic system. There are numerous experimental results which show us that the relation of the lines to the surrounding space can be varied by occupying it with different substances; just as the relation of a ray of light to the space through which it passes can be varied by the presence of different bodies made to occupy that space, or as the lines of electric force are affected by the media through which either induction or conduction takes place. This variation in regard to the magnetic power may be considered as depending upon the aptitude which the surrounding space has to effect the mutual relation of the two external polarities, or to carry onwards the physical line of force....

You are aware of the speculation which I some time since uttered respecting that view of the nature of matter which considers its ultimate atoms as centres of force, and not as so many little bodies surrounded by forces, the bodies being considered in the abstract as independent of the forces and capable of existing without them. In the latter view, these little particles have a definite form and a certain limited size; in the former view such is not the case, for that which represents size may be considered as extending to any distance to which the lines of force of the particle extend: the particle indeed is supposed to exist only by these forces, and where they are it is. The consideration of matter under this view gradually led me to look at the lines of force as being perhaps the seat of the vibrations of radiant phaenomena....

The view which I am so bold as to put forth considers, therefore, radiation as a high species of vibration in the lines of force which are known to connect particles and also masses of matter together. It endeavours to dismiss the æther, but not the vibrations. The kind of vibration which, I believe, can alone account for the wonderful, varied, and beautiful phaenomena of polarization, is not the same as that which occurs on the surface of disturbed water, or the waves of sound in gases or liquids, for the vibrations in these cases are direct, or to and from the centre of action, whereas the former are lateral. It seems to me, that the resultant of two or more lines of force is in an apt condition for that action which may be considered as equivalent to a *lateral* vibration; whereas a uniform medium, like the æther, does not appear apt, or more apt than air or water.

The occurrence of a change at one end of a line of force easily suggests a consequent change at the other. The propagation of light, and therefore probably of all radiant action, occupies *time*; and, that a vibration of the line of force should account for the phaenomena of radiation, it is necessary that such vibration should occupy time also. I am not aware whether there are any data by which it has been, or could be ascertained whether such a power as gravitation acts without occupying time, or whether lines of force being already in existence, such a lateral disturbance of them at one end as I have suggested above, would require time, or must of necessity be felt instantly at the other end.

51. From JAMES CLERK MAXWELL's *A Treatise on Electricity and Magnetism* (1873)

As I proceeded with the study of Faraday, I perceived that his method of conceiving the phenomena was also a mathematical one, though not exhibited in the conventional form of mathematical symbols. I found also that these methods were capable of being expressed in the ordinary mathematical forms, and thus compared with those of the professed mathematicians.

For instance, Faraday, in his mind's eye, saw lines of force traversing all space where the mathematicians saw centres of force attracting at a distance: Faraday saw a medium where they saw nothing but distance: Faraday sought the seat of the phenomena in real actions going on in the medium, they were satisfied that they had found it in a power of action at a distance impressed on the electric fluids....

Great progress has been made in electrical science, chiefly in Germany, by cultivators of the theory of action at a distance.... The great success which these eminent men have attained in the application of mathematics to electrical phenomena, gives, as is natural, additional weight to their theoretical speculations, so that those who, as students of electricity, turn to them as the greatest authorities in mathematical electricity, would probably imbibe, along with their mathematical methods, their physical hypotheses.

These physical hypotheses, however, are entirely alien from the way of looking at things which I adopt, and one object which I have in view is that some of those who wish to study electricity may, by reading this treatise, come to see that there is another way of treating the subject, which is no less fitted to explain the phenomena, and which, though it may appear less definite, corresponds, as I think, more faithfully with our actual knowledge, both in what it affirms and in what it leaves undecided.

52. From HEINRICH HERTZ'S *Electric Waves* (1892)

[Hertz was led to the discovery of electromagnetic waves by Maxwell's theory. His attitude to the theory was that of the present day, that it merely expresses relations between phenomena.]

And now, to be more precise, what is it that we call the Faraday–Maxwell theory? Maxwell has left us as the result of his mature thought a great treatise on Electricity and Magnetism; it might therefore be said that Maxwell's theory is the one which is propounded in that work. But such an answer will scarcely be regarded as satisfactory by all scientific men who have considered the question closely. Many a man has thrown himself with zeal into the study of Maxwell's work, and, even when he has not stumbled upon unwonted mathematical difficulties, has nevertheless been compelled to abandon the hope of forming for himself an altogether consistent conception of Maxwell's ideas. I have fared no better myself. Notwithstanding the greatest admiration for Maxwell's mathematical conceptions, I have not always felt quite certain of having grasped the physical significance of his statements....

To the question, 'What is Maxwell's theory?' I know of no shorter or more definite answer than the following:—Maxwell's theory is Maxwell's system of equations. Every theory which leads to the same system of equations, and therefore comprises the same possible phenomena, I would consider as being a form or special case of Maxwell's theory; every theory which leads to different equations, and therefore to different possible phenomena, is a different theory....

I have endeavoured to avoid from the beginning the introduction of any conceptions which are foreign to this standpoint and which might afterwards have to be removed. I have further endeavoured in the exposition to limit as far as possible the number of those conceptions which are arbitrarily introduced by us, and only to admit such elements as cannot be removed or altered without at the same time altering possible experimental results. It is true, that in consequence of these endeavours, the theory acquires a very abstract and colourless appearance. It is not particularly pleasing to hear general statements made about 'directed changes of state', where we used to have placed before our eyes pictures of electrified atoms. It is not particularly satisfactory to see equations set forth as direct results of observation and experiment, where we used to get long mathematical deductions as apparent proofs of them. Nevertheless, I believe that we cannot, without deceiving ourselves, extract much more from known facts than is asserted in the papers referred to. If we wish to lend more colour to the theory, there is nothing to prevent us from supplementing all this and aiding our powers of imagination by concrete representations of the various conceptions as to the nature of electric polarisation, the electric current, etc. But scientific accuracy requires of us that we should in no wise confuse the simple and homely figure, as it is presented to us by nature, with the gay garment which we use to clothe it. Of our own free will we can make no change whatever in the form of the one, but the cut and colour of the other we can choose as we please.

CHAPTER 15

THE RISE OF ORGANIC CHEMISTRY

53. From a paper by JEAN DUMAS and JUSTUS VON LIEBIG, *Note on the present state of Organic Chemistry* (1837)

[The first glimpse of the structure of organic chemistry, contrasted with the well-organised field of inorganic chemistry, is described in this somewhat sanguine account of the theory of organic radicals, which was read before the French Academy of Sciences.]

Scarcely sixty years have elapsed since the memorable epoch when there appeared, in the midst of this assembly, the first, fertile attempts at chemical theory which we owe to the genius of Lavoisier. This short space of time has sufficed for the most profound problems of inorganic chemistry to be basically considered and we are convinced that this branch of our knowledge possesses nearly all the fundamental ideas required to deal with the means of observation at its command.

Not only is this an incontestable fact but it is one which is easily explained. Inorganic chemistry is concerned with the account of elements, with their binary combinations and their combinations into salts. Now elements can be classified in several, very obvious groups such that if one studies carefully the properties of one member of a group, one can almost always predict the properties of its neighbours. The study of oxygen informs us about sulphur; that of chlorine suffices to initiate us into the smallest details of the properties of iodine, etc.

Thus this task, which appeared at first beyond human power, for it was no less than the study and analysis of thousands of substances of very different appearance and properties, has nevertheless been accomplished in less than half a century and there remain only gaps, here and there, to be filled....

But how can we apply, with a like success, such ideas to organic chemistry? There we meet as many kinds of compounds as in inorganic chemistry, and they are no less diverse. Yet there, in place of fifty-four elements, we find scarcely more than three or four in the great number of known compounds. In short, how, with the aid of the laws of inorganic chemistry, can we explain and classify the very various substances that we obtain from organic bodies, which are nearly all formed of carbon, hydrogen and oxygen only, to which is sometimes added nitrogen?

This is a great and noble problem of natural philosophy, a problem calculated to excite to the highest degree the emulation of chemists; for once resolved, the finest triumphs are promised to science. The mysteries of growth and the mysteries of animal life will unveil themselves before our eyes; we shall seize the clue to all those modifications of matter, so speedy, so sudden and so strange, which take place in animals and plants; nay more, we shall find the means of imitating them in our laboratories.

Well, we do not fear to assert, and it is not a light assertion, that this great and noble problem is today resolved; there remains only the unravelling of all the consequences which its solution entails....

In fact to produce with three or four elements combinations as varied, perhaps more varied, than those which compose the whole inorganic realm, nature has taken a course as simple as unexpected; for with the elements she has made compounds which possess all the properties of the elements themselves.

There lies the whole secret of organic chemistry, we are convinced.

Thus organic chemistry possesses its own elements, which play the role of chlorine or oxygen in inorganic chemistry or, on the other hand, the role of the metals. Cyanogen, amide, benzoyl, the radicals of ammonia, of the fatty substances, of the alcohols and of analogous substances, these are the real elements with which organic chemistry operates and not the actual elements, carbon, hydrogen, oxygen and nitrogen, which appear only when all trace of their organic origin has disappeared.

For us, inorganic chemistry embraces all substances which result from the direct combination of the actual elements.

Organic chemistry, on the other hand, must deal with all substances formed from compounds functioning like the elements.

In inorganic chemistry the radicals are simple; in organic chemistry the radicals are complex. There lies the whole difference....

To discover these radicals, to study them, to characterize them, such has been, for ten years, our daily endeavour. Animated by the same hope, traversing the same route, making use of the same means, it was rarely that we did not study simultaneously the same substances, or substances closely allied, and that we did not regard the facts, which presented themselves to us, from the same point of view. Sometimes, nevertheless, our opinions appeared to diverge and then, both of us carried away by the heat of the combat that we waged with nature, there sprang up between us discussions, the keenness of which we both regret. Who could deny, however, the usefulness and necessity of these discussions? Who could tell how many beautiful researches they have created, and how many they will still create?

54. From EDWARD FRANKLAND'S paper, *On a New Series of Organic Bodies containing Metals* (1852)

[Frankland's paper gave the first clear descriptionof the concept of valency, i.e. the combining-power of an element.]

When the formulae of inorganic chemical compounds are considered, even a superficial observer is impressed with the general symmetry of their construction. The compounds of nitrogen, phosphorus, antimony, and arsenic, especially, exhibit the tendency of these elements to form compounds containing 3 or 5 atoms of other elements; and it is in these proportions that their affinities are best satisfied; thus in the ternal group we have NO_3, NH_3, NI_3, NS_3, PO_3, PH_3, PCl_3, SbO_3, SbH_3, $SbCl_3$, AsO_3, AsH_3, $AsCl_3$, etc.; and in the five-atom group, NO_5, NH_4O, NH_4I, PO_5, PH_4I, etc. Without offering any hypothesis regarding the cause of this symmetrical grouping of atoms, *it is sufficiently evident, from the examples just given, that such a tendency or law prevails, and that, no matter what the character of the uniting atoms may be, the combining-power of the attracting element,* if I may be allowed the term, *is always satisfied by the same number of these atoms.*

55. From AUGUST KEKULÉ'S paper, *The Constitution and Metamorphoses of Chemical Compounds and the Chemical Nature of Carbon* (1858)

[Kekulé states two fundamental facts of organic chemistry, that the atom of carbon has a valency of 4 and that it can link itself to other carbon atoms.]

If we consider only the simplest compounds of carbon (marsh gas, methyl chloride, carbon tetrachloride, chloroform, carbonic acid, phosgene gas, carbon disulphide, prussic acid, etc.), we are struck by the fact that the amount of carbon, which the chemist has recognised as the least possible entering into the composition of a molecule, i.e. as the atom, always combines with four atoms of a monatomic, or two atoms of a diatomic, element; that in general the sum of the chemical units of the elements which combine with one atom of carbon is equal to 4. This leads to the view that carbon is tetratomic...for example,

$$CH_4 \qquad COCl_2 \qquad CO_2 \qquad CNH$$
$$CCl_4 \qquad \qquad \quad CS_2$$
$$CH_3Cl$$
$$CHCl_3$$

For substances which contain more atoms of carbon, we must assume that at least part of the atoms are held by the affinity of carbon and that the carbon atoms themselves are connected, so that naturally a part of the affinity of one for the other will bind an equal part of the affinity of the other....

When we make comparisons between compounds which have an equal number of carbon atoms in the molecule and which can be changed into each other by simple transformations (e.g. alcohol, ethyl chloride, aldehyde, acetic acid, glycolic acid, oxalic acid, etc.) we find that the carbon atoms are arranged in the same way and only the atoms held to the carbon framework are changed.

56. From AUGUST KEKULÉ'S paper, *Studies on Aromatic Compounds* (1865)

[Kekulé announces his conception of the benzene ring, consisting of six carbon atoms linked by alternate single and double bonds or affinity units.]

If we wish to give an account of the atomistic constitution of aromatic compounds, we must take into consideration the following facts:

1. All aromatic compounds, even the simplest, are proportionally richer in carbon than the analogous compounds in the class of the fatty bodies.

2. Among the aromatic compounds, just as in the fatty bodies, there are numerous homologous substances, i.e., those whose differences of composition can be expressed by nCH_2.

3. The simplest aromatic compound contains at least six atoms of carbon.

4. All derivatives of aromatic substances show a certain family similarity; they belong collectively to the group of 'aromatic compounds'. Indeed in more drastic reactions, one part of carbon is often eliminated, but the chief product contains at least six atoms of carbon (benzene, quinone, chloranil, carbolic acid, hydroxyphenic acid, picric acid, etc.). The decomposition stops with the formation of these products if complete destruction of the organic group does not occur.

These facts obviously lead to the conclusion that in all aromatic substances there is contained one and the same atom group, or, if you like, a common nucleus which consists of six carbon atoms. Within this nucleus the carbon atoms are certainly in close combination or in more compact arrangement. To this nucleus, then, more carbon atoms can be added in the same way and according to the same laws as in the case of the fatty bodies.

We must next explain the atomic constitution of this nucleus. Now this can be done very easily by the following hypothesis, which, on the now generally accepted view that carbon is tetratomic, accounts for it in such a simple manner that further development is hardly necessary.

If many carbon atoms combine with one another then it can happen that one affinity unit of one atom binds one affinity unit of the neighbouring atom. As I have shown earlier, this explains homology and, on the whole, the constitution of the fatty bodies. We can further assume that many carbon atoms are linked together through two affinity units; we can also assume that the union occurs alternately through first one and then two affinity units. The first and the last of these views could be expressed as follows:

$$1/1, \quad 1/1, \quad 1/1, \quad 1/1 \text{ etc.}$$
$$1/1, \quad 2/2, \quad 1/1, \quad 2/2 \text{ etc.}$$

The first law of the symmetry of union of the carbon atoms explains the constitution of the fatty bodies, as already mentioned; the second leads to an explanation of the constitution of aromatic substances, or at least of the nucleus which is common to all these substances.

CHAPTER 16

EVOLUTION

57. From JEAN BAPTISTE LAMARCK'S *Zoological Philosophy* (1809)

[Lamarck states his view of the mechanism of evolution and the two laws which govern it; he proceeds to discuss individual species as examples.]

In the preceding chapter we saw that it is now an unquestionable fact that on passing along the animal scale in the opposite direction from that of nature, we discover the existence, in the groups composing this scale, of a continuous but irregular degradation in the organisation of animals, an increasing simplification in their organisation, and, lastly, a corresponding diminution in the number of their faculties.

This well-ascertained fact may throw the strongest light over the actual order followed by nature in the production of all the animals that she has brought into existence, but it does not show us why the increasing complexity of the organisation of animals from the most imperfect to the most perfect exhibits only an *irregular gradation*, in the course of which there occur numerous anomalies or deviations with a variety in which no order is apparent.

Now on seeking the reason of this strange irregularity in the increasing complexity of animal organisation, if we consider the influence that is exerted by the infinitely varied environments of all parts of the world on the general shape, structure and even organisation of these animals, all will then be clearly explained....

Now the true principle to be noted in all this is as follows:

1. Every fairly considerable and permanent alteration in the environment of any race of animals works a real alteration in the needs of that race.

2. Every change in the needs of animals necessitates new activities on their part for the satisfaction of those needs, and hence new habits.

3. Every new need, necessitating new activities for its satis-

faction, requires the animal, either to make more frequent use of some of its parts which it previously used less, and thus greatly to develop and enlarge them; or else to make use of entirely new parts, to which the needs have imperceptibly given birth by efforts of its inner feeling; this I shall shortly prove by means of known facts.

Thus to obtain a knowledge of the true causes of that great diversity of shapes and habits found in the various known animals, we must reflect that the infinitely diversified but slowly changing environment in which the animals of each race have successively been placed, has involved each of them in new needs and corresponding alterations in their habits. This is a truth which, once recognised, cannot be disputed. Now we shall easily discern how the new needs may have been satisfied, and the new habits acquired, if we pay attention to the two following laws of nature, which are always verified by observation.

FIRST LAW

In every animal which has not passed the limit of its development, a more frequent and continuous use of any organ gradually strengthens, develops and enlarges that organ, and gives it a power proportional to the length of time it has been so used; while the permanent disuse of any organ imperceptibly weakens and deteriorates it, and progressively diminishes its functional capacity, until it finally disappears.

SECOND LAW

All the acquisitions or losses wrought by nature on individuals, through the influence of the environment in which their race has long been placed, and hence through the influence of the predominant use or permanent disuse of any organ; all these are preserved by reproduction to the new individuals which arise, provided that the acquired modifications are common to both sexes, or at least to the individuals which produce the young....

...it was part of the plan of organisation of the reptiles, as of other vertebrates, to have four legs in dependence on their skeleton. Snakes ought consequently to have four legs, especially since they are by no means the last order of the reptiles and are farther from the fishes than are the batrachians (frogs, salamanders, etc.).

Snakes, however, have adopted the habit of crawling on the

ground and hiding in the grass; so that their body, as a result of continually repeated efforts at elongation for the purpose of passing through narrow spaces, has acquired a considerable length, quite out of proportion to its size. Now, legs would have been quite useless to these animals and consequently unused. Long legs would have interfered with their need of crawling, and very short legs would have been incapable of moving their body, since they could only have had four. The disuse of these parts thus became permanent in the various races of these animals, and resulted in the complete disappearance of these same parts, although legs really belong to the plan of organisation of the animals of this class....

The bird which is drawn to the water by its need of finding there the prey on which it lives, separates the digits of its feet in trying to strike the water and move about on the surface. The skin which unites these digits at their base acquires the habit of being stretched by these continually repeated separations of the digits; thus in course of time there are formed large webs which unite the digits of ducks, geese, etc., as we actually find them. In the same way efforts to swim, that is to push against the water so as to move about in it, have stretched the membranes between the digits of frogs, sea-tortoises, the otter, beaver, etc.

On the other hand, a bird which is accustomed to perch on trees and which springs from individuals all of whom had acquired this habit, necessarily has longer digits on its feet and differently shaped from those of the aquatic animals that I have just named. Its claws in time become lengthened, sharpened and curved into hooks, to clasp the branches on which the animal so often rests.

We find in the same way that the bird of the water-side which does not like swimming and yet is in need of going to the water's edge to secure its prey, is continually liable to sink in the mud. Now this bird tries to act in such a way that its body should not be immersed in the liquid, and hence makes its best efforts to stretch and lengthen its legs. The long-established habit acquired by this bird and all its race of continually stretching and lengthening its legs, results in the individuals of this race becoming raised as though on stilts, and gradually obtaining long, bare legs, denuded of feathers up to the thighs and often higher still....

Since ruminants can only use their feet for support, and have little strength in their jaws, which only obtain exercise by cutting and browsing on the grass, they can only fight by blows with their heads, attacking one another with their crowns.

In the frequent fits of anger to which the males especially are subject, the efforts of their inner feeling cause the fluids to flow more strongly towards that part of their head; in some there is hence deposited a secretion of horny matter, and in others of bony matter mixed with horny matter, which gives rise to solid protuberances: thus we have the origin of horns and antlers, with which the head of most of these animals is armed.

It is interesting to observe the result of habit in the peculiar shape and size of the giraffe (*Camelo-pardalis*): this animal, the largest of the mammals, is known to live in the interior of Africa in places where the soil is nearly always arid and barren, so that it is obliged to browse on the leaves of trees and to make constant efforts to reach them. From this habit long maintained in all its race, it has resulted that the animal's fore-legs have become longer than its hind-legs, and that its neck is lengthened to such a degree that the giraffe, without standing up on its hind legs, attains a height of six metres.

Among birds, ostriches, which have no power of flight and are raised on very long legs, probably owe their singular shape to analogous circumstances.

The effect of habit is quite as remarkable in the carnivorous mammals as in the herbivores; but it exhibits results of a different kind.

Those carnivores, for instance, which have become accustomed to climbing, or to scratching the ground for digging holes, or to tearing their prey, have been under the necessity of using the digits of their feet: now this habit has promoted the separation of their digits, and given rise to the formation of the claws with which they are armed.

But some of the carnivores are obliged to have recourse to pursuit in order to catch their prey: now some of these animals were compelled by their needs to contract the habit of tearing with their claws, which they are constantly burying deep in the body of another animal in order to lay hold of it, and then make efforts to tear out the part seized. These repeated efforts must have resulted in its claws reaching a size and curvature which would have greatly impeded them in walking or running on stony ground: in such cases the animal has been compelled to make further efforts to draw back its claws, which are so projecting and hooked as to get in its way. From this there has gradually resulted the formation of those peculiar sheaths, into which cats, tigers, lions, etc. withdraw their claws when they are not using them.

Hence we see that efforts in a given direction, when they are long sustained or habitually made by certain parts of a living body, for the satisfaction of needs established by nature or environment, cause an enlargement of these parts and the acquisition of a size and shape that they would never have obtained, if these efforts had not become the normal activities of the animals exerting them. Instances are everywhere furnished by observations on all known animals.

58. From CHARLES DARWIN'S *The Origin of Species* (1859)

[*The Origin of Species* is a sustained argument supported by a wealth of evidence. The following brief extracts are taken from different parts of the book.]

(1) From the *Introduction*

When on board H.M.S. 'Beagle', as naturalist, I was much struck with certain facts in the distribution of the organic beings inhabiting South America, and in the geological relations of the present to the past inhabitants of that continent. These facts, as will be seen in the latter chapters of this volume, seemed to throw some light on the origin of species—that mystery of mysteries, as it has been called by one of our greatest philosophers. On my return home, it occurred to me, in 1837, that something might perhaps be made out on this question by patiently accumulating and reflecting on all sorts of facts which could possibly have any bearing on it. After five years' work I allowed myself to speculate on the subject, and drew up some short notes; these I enlarged in 1844 into a sketch of the conclusions, which then seemed to me probable: from that period to the present day I have steadily pursued the same object. I hope that I may be excused for entering on these personal details, as I give them to show that I have not been hasty in coming to a decision.

(2) From Chapter I, *Variation under Domestication*

[In this chapter Darwin considers the great differences in the varieties of domestic animals and plants, produced by selection on the part of man.]

Believing that it is always best to study some special group, I have, after deliberation, taken up domestic pigeons....

Altogether at least a score of pigeons might be chosen, which, if shown to an ornithologist, and he were told that they were wild birds, would certainly be ranked by him as well-defined species. Moreover, I do not believe that any ornithologist would in this case place the English carrier, the short-faced tumbler, the runt, the barb, pouter, and fantail in the same genus; more especially as in each of these breeds several truly-inherited sub-breeds, or species, as he would call them, could be shown him.

Great as are the differences between the breeds of the pigeon, I am fully convinced that the common opinion of naturalists is correct, namely, that all are descended from the rock-pigeon....

I have seen it gravely remarked, that it was most fortunate that the strawberry began to vary just when gardeners began to attend to this plant. No doubt the strawberry had always varied since it was cultivated, but the slight varieties had been neglected. As soon, however, as gardeners picked out individual plants with slightly larger, earlier, or better fruit, and raised seedlings from them, and again picked out the best seedlings and bred from them, then (with some aid by crossing distinct species) those many admirable varieties of the strawberry were raised which have appeared during the last half-century.

(3) From Chapter III, *Struggle for Existence*

[Having discussed in Chapter II variations occurring in nature, Darwin now considers how these variations give rise to evolution.]

All these results, as we shall more fully see in the next chapter, follow from the struggle for life. Owing to this struggle, variations, however slight and from whatever cause proceeding, if they be in any degree profitable to the individuals of a species, in their infinitely complex relations to other organic beings and to their physical conditions of life, will tend to the preservation of such individuals, and will generally be inherited by the offspring. The offspring, also, will thus have a better chance of surviving, for, of the many individuals of any species which are periodically born, but a small number can survive. I have called this principle, by which each slight variation, if useful, is preserved, by the term Natural Selection, in order to mark its relation to man's power of selection....

A struggle for existence inevitably follows from the high rate at which all organic beings tend to increase. Every being, which during its natural lifetime produces several eggs or seeds, must suffer destruction during some period of its life, and during some season or occasional year, otherwise, on the principle of geometrical increase, its numbers would quickly become so inordinately great that no country could support the product. Hence, as more individuals are produced than can possibly survive, there must in every case be a struggle for existence, either one individual with another of the same species, or with the individuals of distinct species, or with the physical conditions of life....

In the case of every species, many different checks, acting at different periods of life, and during different seasons or years, probably come into play; some one check or some few being generally the most potent; but all will concur in determining the average number or even the existence of the species. In some cases it can be shown that widely-different checks act on the same species in different districts. When we look at the plants and bushes clothing an entangled bank, we are tempted to attribute their proportional numbers and kinds to what we call chance. But how false a view is this! Every one has heard that when an American forest is cut down, a very different vegetation springs up; but it has been observed that ancient Indian ruins in the Southern United States, which must formerly have been cleared of trees, now display the same beautiful diversity and proportion of kinds as in the surrounding virgin forest. What a struggle must have gone on during long centuries between the several kinds of trees, each annually scattering its seeds by the thousand; what war between insect and insect—between insects, snails, and other animals with birds and beasts of prey—all striving to increase, all feeding on each other, or on the trees, their seeds and seedlings, or on the other plants which first clothed the ground and thus checked the growth of the trees!

(4) From Chapter IV, *Natural Selection*

As man can produce, and certainly has produced, a great result by his methodical and unconscious means of selection, what may not natural selection effect?...

Under nature, the slightest differences of structure or constitution may well turn the nicely-balanced scale in the struggle for life, and so be preserved. How fleeting are the wishes and efforts of man! how short his time! and consequently how poor will be his results, compared with those accumulated by Nature during whole geological periods!...

This leads me to say a few words on what I have called Sexual Selection. This form of selection depends, not on a struggle for existence in relation to other organic beings, or to external conditions, but on a struggle between the individuals of one sex, generally the males, for the possession of the other sex. The result is not death to the unsuccessful competitor, but few or no offspring. Sexual selection is, therefore, less rigorous than natural selection. Generally, the most vigorous males, those which are best fitted for their places in nature, will leave most progeny. But in many cases, victory depends not so much on general vigour, as on having special weapons, confined to the male sex. A hornless stag or spurless cock would have a poor chance of leaving numerous offspring. Sexual selection, by always allowing the victor to breed, might surely give indomitable courage, length to the spur, and strength to the wing to strike in the spurred leg, in nearly the same manner as does the brutal cockfighter by the careful selection of his best cocks. How low in the scale of nature the law of battle descends, I know not; male alligators have been described as fighting, bellowing, and whirling round, like Indians in a war-dance, for the possession of the females; male salmons have been observed fighting all day long; male stag-beetles sometimes bear wounds from the huge mandibles of other males; the males of certain hymenopterous insects have been frequently seen by that inimitable observer M. Fabre, fighting for a particular female who sits by, an apparently unconcerned beholder of the struggle, and then retires with the conqueror....

The affinities of all the beings of the same class have sometimes been represented by a great tree. I believe this simile largely speaks the truth. The green and budding twigs may represent existing species; and those produced during former years may represent the long succession of extinct species. At each period of growth all the growing twigs have tried to branch out on all sides, and to overtop and kill the surrounding twigs and branches, in the same manner as species and groups of species have at all times overmastered other species in the great battle for life. The limbs divided into great branches, and these into lesser and lesser branches, were themselves once, when the tree was young, budding twigs; and this connection of the former and present buds by ramifying branches may well represent the classification of all extinct and living species in groups subordinate to groups. Of the many twigs which flourished when the tree was a mere bush, only two or three, now grown into great branches, yet survive and bear the other branches; so with the species which lived during long-past geological periods, very few have left living and modified descendants. From the first growth of the tree, many a limb and branch has decayed and dropped off; and these fallen branches of various sizes may represent those whole orders, families, and genera which have now no living representatives, and which are known to us only in a fossil state. As we here and there see a thin straggling branch springing from a fork low down in a tree, and which by some chance has been favoured and is still alive on its summit, so we occasionally see an animal like the Ornithorhynchus or Lepidosiren, which in some small degree connects by its affinities two large branches of life, and which has apparently been saved from fatal competition by having inhabited a protected station. As buds give rise by growth to fresh buds, and these, if vigorous, branch out and overtop on all sides many a feebler branch, so by generation I believe it has been with the great Tree of Life, which fills with its dead and broken branches the crust of the earth, and covers the surface with its ever-branching and beautiful ramifications.

(5) From Chapter VI, *Difficulties of the Theory*

Long before the reader has arrived at this part of my work, a crowd of difficulties will have occurred to him. Some of them are so serious that to this day I can hardly reflect on them without being in some degree staggered; but, to the best of my judgment, the greater number are only apparent, and those that are real are not, I think, fatal to the theory....

To suppose that the eye with all its inimitable contrivances for adjusting the focus to different distances, for admitting different amounts of light, and for the correction of spherical and chromatic aberration, could have been formed by natural selection, seems, I freely confess, absurd in the highest degree. When it was first said that the sun stood still and the world turned round, the common sense of mankind declared the doctrine false; but the old saying of *Vox populi, vox Dei*, as every philosopher knows, cannot be trusted in science. Reason tells me, that if numerous gradations from a simple and imperfect eye to one complex and perfect can be shown to exist, each grade being useful to its possessor, as is certainly the case; if further, the eye ever varies and the variations be inherited, as is likewise certainly the case; and if such variations should be useful to any animal under changing conditions of life, then the difficulty of believing that a perfect and complex eye could be formed by natural selection, though insuperable by our imagination, should not be considered as subversive of the theory. . . .

In the great class of the Articulata, we may start from an optic nerve simply coated with pigment, the latter sometimes forming a sort of pupil, but destitute of a lens or other optical contrivance. With insects it is now known that the numerous facets on the cornea of their great compound eyes form true lenses, and that the cones include curiously modified nervous filaments. But these organs in the Articulata are so much diversified that Müller formerly made three main classes with seven subdivisions, besides a fourth main class of aggregated simple eyes.

When we reflect on these facts, here given much too briefly, with respect to the wide, diversified, and graduated range of structure in the eyes of the lower animals; and when we bear in mind how small the number of all living forms must be in comparison with those which have become extinct, the difficulty ceases to be very great in believing that natural selection may have converted the simple apparatus of an optic nerve, coated with pigment and invested by transparent membrane, into an optical instrument as perfect as is possessed by any member of the Articulate Class. . . .

To arrive, however, at a just conclusion regarding the formation of the eye, with all its marvellous yet not absolutely perfect characters, it is indispensable that the reason should conquer the imagination; but I have felt the difficulty far too keenly to be surprised at others hesitating to extend the principle of natural selection to so startling a length. . . .

In living bodies, variation will cause the slight alterations, generation will multiply them almost infinitely, and natural selection will pick out with unerring skill each improvement. Let this process go on for millions of years, and during each year on millions of individuals of many kinds, and may we not believe that a living optical instrument might thus be formed as superior to one of glass, as the works of the Creator are to those of man?

(6) From Chapter XI, *On the Geological Succession of Organic Beings*

The extinction of species has been involved in the most gratuitous mystery. Some authors have even supposed that, as the individual has a definite length of life, so have species a definite duration. No one can have marvelled more than I have done at the extinction of species. When I found in La Plata the tooth of a horse embedded with the remains of Mastodon, Megatherium, Toxodon, and other extinct monsters, which all co-existed with still living shells at a very late geological period, I was filled with astonishment; for, seeing that the horse, since its introduction by the Spaniards into South America, has run wild over the whole country and has increased in numbers at an unparalleled rate, I asked myself what could so recently have exterminated the former horse under conditions of life apparently so favourable. But my astonishment was groundless. Professor Owen soon perceived that the tooth, though so like that of the existing horse, belonged to an extinct species. Had this horse been still living, but in some degree rare, no naturalist would have felt the least surprise at its rarity; for rarity is the attribute of a vast number of species of all classes, in all countries. If we ask ourselves why this or that species is rare, we answer that something is unfavourable in its conditions of life; but what that something is we can hardly ever tell. . . .

I have attempted to show that the geological record is extremely imperfect; that only a small portion of the globe has been geologically explored with care; that only certain classes of organic beings have been largely preserved in a fossil state; that the number both of specimens and of species, preserved in our museums, is absolutely as nothing compared with the number of generations which must have passed away even during a single formation....

He who rejects this view of the imperfection of the geological record, will rightly reject the whole theory. For he may ask in vain where are the numberless transitional links which must formerly have connected the closely allied or representative species, found in the successive stages of the same great formation? He may disbelieve in the immense intervals of time which must have elapsed between our consecutive formations.

(7) From Chapter xiv, *Mutual Affinities of Organic Beings*

We have seen that the members of the same class, independently of their habits of life, resemble each other in the general plan of their organisation. This resemblance is often expressed by the term 'unity of type'; or by saying that the several parts and organs in the different species of the class are homologous. The whole subject is included under the general term of Morphology. This is one of the most interesting departments of natural history, and may almost be said to be its very soul. What can be more curious than that the hand of a man, formed for grasping, that of a mole for digging, the leg of the horse, the paddle of the porpoise, and the wing of the bat, should all be constructed on the same pattern, and should include similar bones, in the same relative positions? How curious it is, to give a subordinate though striking instance, that the hind-feet of the kangaroo, which are so well fitted for bounding over the open plains,—those of the climbing, leaf-eating koala, equally well fitted for grasping the branches of trees,—those of the ground-dwelling, insect or root eating, bandicoots,—and those of some other Australian marsupials,—should all be constructed on the same extraordinary type, namely with the bones of the second and third digits extremely slender and enveloped within the same skin, so that they appear like a single toe furnished with two claws. Notwithstanding this similarity of pattern, it is obvious that the hind-feet of these several animals are used for as widely different purposes as it is possible to conceive....

Why should the brain be enclosed in a box composed of such numerous and such extraordinarily shaped pieces of bone, apparently representing vertebrae? As Owen has remarked, the benefit derived from the yielding of the separate pieces in the act of parturition by mammals, will by no means explain the same construction in the skulls of birds and reptiles. Why should similar bones have been created to form the wing and the leg of a bat, used as they are for such totally different purposes, namely flying and walking? Why should one crustacean, which has an extremely complex mouth formed of many parts, consequently always have fewer legs; or conversely, those with many legs have simpler mouths? Why should the sepals, petals, stamens, and pistils, in each flower, though fitted for such distinct purposes, be all constructed on the same pattern?

On the theory of natural selection, we can, to a certain extent, answer these questions....

Organs or parts in this strange condition, bearing the plain stamp of inutility, are extremely common, or even general, throughout nature. It would be impossible to name one of the higher animals in which some part or other is not in a rudimentary condition. In the mammalia, for instance, the males possess rudimentary mammae; in snakes one lobe of the lungs is rudimentary; in birds the 'bastard-wing' may safely be considered as a rudimentary digit, and in some species the whole wing is so far rudimentary that it cannot be used for flight. What can be more curious than the presence of teeth in foetal whales, which when grown up have not a tooth in their heads; or the teeth, which never cut through the gums, in the upper jaws of unborn calves?

(8) From Chapter xv, *Recapitulation and Conclusion*

If then, animals and plants do vary, let it be ever so slightly or slowly, why should not variations or individual differences, which are in any way beneficial, be preserved and accumulated through natural selection, or the survival of the fittest? If man can by patience select variations useful to him, why, under changing and complex conditions of life, should not variations useful to nature's living products often arise, and be preserved or selected? What limit can be put to this power, acting during long ages and rigidly scrutinising the whole constitution, structure, and habits of each creature,—favouring the good and rejecting the bad? I can see no limit to this power, in slowly and beautifully adapting each form to the most complex relations of life. The theory of natural selection, even if we look no farther than this, seems to be in the highest degree probable.

CHAPTER 17

THE GERM THEORY OF DISEASE

59. From JOSEPH LISTER's article, *On a New Method of Treating Compound Fracture, Abscess etc., with Observations on the Conditions of Suppuration* (1867)

[This article was Lister's first published account of his antiseptic methods in surgery.]

The frequency of disastrous consequences in compound fracture [where the broken bone causes an open wound], contrasted with the complete immunity from danger to life or limb, in a simple fracture, is one of the most striking as well as melancholy facts in surgical practice....

In the course of the year 1864 I was much struck with an account of the remarkable effects produced by carbolic acid upon the sewage of the town of Carlisle, the admixture of a very small proportion not only preventing all odour from the lands irrigated with the refuse material, but, as it was stated, destroying the entozoa which usually infest cattle fed upon such pastures.

My attention having for several years been much directed to the subject of suppuration, more especially in its relation to decomposition, I saw that such a powerful antiseptic was peculiarly adapted for experiments with a view to elucidating that subject, and while I was engaged in the investigation the applicability of carbolic acid for the treatment of compound fracture naturally occurred to me.

My first attempt of this kind was made in the Glasgow Royal Infirmary in March, 1865, in a case of compound fracture of the leg. It proved unsuccessful, in consequence, as I now believe, of improper management; but subsequent trials have more than realised my most sanguine anticipations....

Case I.—James G..., aged eleven years, was admitted into the Glasgow Royal Infirmary on August 12th, 1865, with compound fracture of the left leg, caused by the wheel of an empty cart passing over the limb a little below its middle. The wound, which was about an inch and a half long, was close to, but not exactly over, the line of fracture of the tibia. A probe, however, could be passed beneath the integument over the seat of fracture and for some inches beyond it. Very little blood had been extravasated into the tissues.

My house-surgeon, Dr Macfee, acting under my instructions, laid a piece of lint dipped in liquid carbolic acid upon the wound, and applied lateral pasteboard splints padded with cotton wool, the limb resting on its outer side, with the knee bent. It was left undisturbed for four days, when, the boy complaining of some uneasiness, I removed the inner splint and examined the wound. It showed no signs of suppuration, but the skin in its immediate vicinity had a slight blush of redness. I now dressed the sore with lint soaked with water having a small proportion of carbolic acid diffused through it; and this was continued for five days, during which the uneasiness and the redness of the skin disappeared, the sore meanwhile furnishing no pus, although some superficial sloughs caused by the acid were separating. But the epidermis being excoriated by this dressing, I substituted for it a solution of one part of carbolic acid in ten to twenty parts of olive oil, which was used for four days, during which a small amount of imperfect pus was produced from the surface of the sore, but not a drop appeared from beneath the skin. It was now clear that there was no longer any danger of deep-seated suppuration, and simple water-dressing was employed. Cicatrisation proceeded just as in an ordinary granulating sore. At the expiration of six weeks I examined the condition of the bones, and, finding them firmly united, discarded the splints; and two days later the sore was entirely healed, so that the cure could not be said to have been at all retarded by the circumstance of the fracture being compound.

This, no doubt, was a favourable case, and might have done well under ordinary treatment. But the remarkable retardation of suppuration, and the immediate conversion of the compound fracture into a simple fracture with a superficial sore, were most encouraging facts.

60. From LOUIS PASTEUR's paper, *Method for Preventing Rabies after Bites* (1885)

[Pasteur recounts his investigations with dogs, and first trials with human beings, of his method of inoculation against rabies.]

These facts established, this is the means of making a dog refractory to rabies in a relatively short time.

In a series of flasks, in which the air is maintained in a dry state by fragments of potash placed in the bottom of the vessel, is suspended each day a piece of rabid spinal cord, fresh from a rabbit which has died of rabies, the rabies having developed after seven days of incubation. Each day likewise there is inoculated under the skin of the dog a full Pravaz syringe of sterilized broth, in which has been dispersed a small fragment of one of these dried cords, commencing with a cord of an ordinal number sufficiently far from the operational day, to be quite sure that the cord is not at all virulent, as ascertained by preliminary experiments. On following days, the same operation is performed with more recent cords, separated by an interval of two days, until a very virulent cord is reached, placed only a day or two earlier in a flask.

The dog is then refractory to rabies. He can be inoculated with rabies virus under the skin, or even on the surface of the skull by trephining, without rabies showing itself.

By the application of this method I had fifty dogs, of every age and breed, refractory to rabies, without having encountered a single failure, when, unexpectedly, three people from Alsace presented themselves at my laboratory: Théodore Vone, a grocer from Meissengott, near Schlestadt, bitten in the arm, on the 4th July, by his own dog which had become rabid; Joseph Meister, aged nine years, bitten likewise on the 4th July, at 8 o'clock in the morning, by the same dog. This child, thrown to the ground by the dog, had many bites, in the hand, in the legs and in the thighs; some were deep, which made even walking difficult. The chief of these bites had been cauterized with carbolic acid only twelve hours after the accident, on the 4th July at 8 o'clock in the evening, by Dr Weber of Villé.

The third person, who had not been bitten, was the mother of little Joseph Meister.

At the autopsy on the dog, destroyed by his master, the stomach was found to be full of hay, straw and bits of wood. The dog was very rabid. Joseph Meister had been lifted from under him, covered with slaver and blood.

M. Vone had severe contusions in the arm but he assured me that his shirt had not been pierced by the fangs of the dog. As he had nothing to fear, I told him that he could go back to Alsace the same day, which he did. But I kept near at hand little Meister and his mother. The weekly meeting of the Academy of Sciences took place on that very 6th July; I there saw our colleague Dr Vulpian, to whom I recounted what had happened. Dr Vulpian, as well as Dr Grancher, professor at the Faculty of Medicine, were kind enough to come immediately to see little Joseph Meister to ascertain the state and the number of his wounds. He had no fewer than 14.

The advice of our learned colleague and of Dr Grancher was, by the intensity and number of the bites, Joseph Meister was almost certain to succumb to rabies. I communicated then to M. Vulpian and to M. Grancher, the new results which I had obtained in the study of rabies since the lecture which I had given at Copenhagen, one year earlier.

The death of this child seeming inevitable, I decided, not without lively and sore anxiety, as may well be imagined, to try on Joseph Meister the method which had constantly succeeded on dogs.

My fifty dogs, it is true, had not been bitten before I made them refractory to rabies, but I knew that this circumstance could be ignored because I had already obtained a state of immunity to rabies in a great number of dogs after they had been bitten. I had given testimony, this year, concerning this new and important progress to the members of the Commission for Rabies.

Consequently, on the 6th July, at 8 o'clock in the evening, sixty hours after the bites of the 4th July, and in the presence of Drs Vulpian and Grancher, an inoculation was made in a fold raised in the skin of the right hypochondrium[1] of little Meister, consisting of a half Pravaz syringe of a cord of a rabbit, which died of rabies on the 21st June, and which had been kept since then for fifteen days in a flask of dry air.

On the following days new inoculations were made, always in the hypochondria, under the conditions given in the table [shown on page 130].

[1] The abdomen immediately below the ribs.

I brought thus to 13 the number of inoculations and to 10 the number of days of treatment. I shall say later that a smaller number of inoculations would have been sufficient. But it will be understood that in this first attempt I had to proceed with a very special circumspection.

A half Pravaz syringe

7th July	9 a.m.	Cord of 23rd June		Cord of 14 days	
7th „	6 p.m.	„	25th „	„	12 „
8th „	9 a.m.	„	27th „	„	11 „
8th „	6 p.m.	„	29th „	„	9 „
9th „	11 a.m.	„	1st July	„	8 „
10th „	11 a.m.	„	3rd „	„	7 „
11th „	11 a.m.	„	5th „	„	6 „
12th „	11 a.m.	„	7th „	„	5 „
13th „	11 a.m.	„	9th „	„	4 „
14th „	11 a.m.	„	11th „	„	3 „
15th „	11 a.m.	„	13th „	„	2 „
16th „	11 a.m.	„	15th „	„	1 „

Two fresh rabbits were inoculated by trephining with each different cord employed, so as to follow the states of virulence of the cords.

Observation of the rabbits enabled me to establish that the cords of 6th, 7th, 8th, 9th, 10th July were not virulent, for they did not make their rabbits rabid. The cords of 11th, 12th, 14th, 15th, 16th July were quite virulent, and the virulent matter found in them was stronger and stronger from one to the other. Rabies showed itself after seven days of incubation in rabbits of 15th and 16th July; after eight days in those of 12th to 14th; after fifteen days in those of 11th July.

In the last days, I had therefore inoculated Joseph Meister with the most virulent rabies virus, that from a dog strengthened by a host of transferences from rabbit to rabbit; this was a virus which gave rabies to rabbits after seven days of incubation, and after eight or ten days to dogs. I was justified in this procedure by what had happened with the fifty dogs of which I have spoken.

When the state of immunity is reached one can, without causing any inconvenience, inoculate the most virulent virus in any quantity whatsoever. This, it has always appeared to me, has no other effect than to consolidate the refractory state.

Joseph Meister escaped then, not only the rabies which his bites could have caused, but that with which I had inoculated him in the course of the treatment for immunity, more virulent than canine rabies.

The final, very virulent inoculation had also the advantage of limiting the period of apprehension of the consequences of the bites. If rabies could break out, it should show itself more quickly through a more virulent virus than through that of the bites.

From the middle of the month of August I looked forward with confidence to a healthy future for Joseph Meister. Today, even more, after the elapse of three months and three weeks since the accident, his health leaves nothing to be desired.

What interpretation is to be given to the new method which I have just made known for preventing rabies after bites? I have no intention of treating this question today in a complete way. I wish to confine myself to several preliminary details, calculated to make comprehensible the reason for the experiments which I pursued, with the aim of basing the ideas on the best of the possible interpretations.

Reviewing the methods of progressive attenuation of the deadly viruses and the prophylaxy that one can deduce from it; being given, moreover, the influence of the air on the attenuation, the first thought which offers itself to the mind to account for the effects of the method, is that the period of contact with dry air of the rabid cords diminishes progressively the intensity of the virulence of these cords until it is nullified.

From that one would be led to believe that the prophylactic method in question rests on the employment of a virus at first without appreciable activity, then a weak one and then ones more and more virulent.

I shall show ultimately that the facts are in disagreement with this way of looking at the matter. I shall show that the delays in the incubation period of rabies communicated, day by day, to the rabbits, as I have just described, to test the state of virulence of the cords dried in contact with the air, have an effect of impoverishing the quantity of the rabies viruses contained in these cords and not an effect of impoverishing their virulence.

Could one admit that the inoculation of a virus, of a constant virulence, would be capable of leading to the refractory state, if one employed it in quantities very small but daily increased? This is an interpretation of the method which I am studying from an experimental point of view.

One can give the new method yet another interpretation, an interpretation certainly very strange at first sight but which merits every consideration, because it is in harmony with certain results already known, presented by vital phenomena in some lower organisms, notably various pathogenic microbes.

Many microbes appear to give rise in their cultures to substances which have the property of hindering their proper development.

In 1880 I began researches to establish that the microbe of chicken cholera must produce a kind of self-poison. I have not succeeded in demonstrating the presence of such a substance; but I think now that this study must be taken up again—and I will not fail to do so—operating in the presence of pure carbon dioxide.

The microbe of swine fever can be cultivated in very different broths, but the results are often so feeble and so promptly arrested, relatively speaking, that the culture barely reveals feeble silky streaks in the middle of the nutritive medium. One would say, without hesitation, that a substance is produced which arrests the development of the microbe, whether one cultivates it in contact with the air or in a vacuum.

M. Ranlin, my former assistant, today professor at the Faculty of Lyon, in a very remarkable thesis which he upheld at Paris, on 22nd March 1870, established that the growth of *aspergillus niger* develops a substance which partially arrests the production of this mould when the nutritive medium does not contain iron salts.

Could it be that what constitutes the rabies virus is formed of two distinct substances and that, in addition to the one which is living and capable of multiplying in the nervous system, there is another, not living, having the property, when it is present in a suitable proportion, of arresting the development of the first? I shall examine experimentally, in my next Communication, with all the attention it deserves, this third interpretation of prophylaxy in rabies.

I have no need to remark in closing that the most important question to be resolved at the moment is the interval to be observed between the time of the bites and that of commencing treatment. This interval, in the case of Joseph Meister, was two and a half days. But we must expect cases where it is much longer.

Tuesday last, 20th October, with the obliging assistance of MM. Vulpian and Grancher, I had to begin treating a youth of fifteen, bitten very seriously on both hands, six whole days before.

I shall hasten to make known to the Academy what happens in this new trial.

The Academy will hear, not perhaps without emotion the account of the act of courage and of presence of mind of the youth whose treatment I undertook last Tuesday. He is a shepherd, 15 years old, by name Jean Baptiste Jupille, of Villers-Farlay (Jura) who on seeing a dog of suspicious gait and of large size hurl itself on a group of six of his small comrades, all younger than himself, rushed forward, armed with his whip, to confront the animal. The dog seized Jupille by his left hand. Jupille then threw the dog to the ground, keeping it under him, and opened its jaws with his right hand to release his left hand, not without receiving several new bites; then, with the thong of his whip, he bound its muzzle and, seizing one of his sabots, beat it to death.

61. From ROBERT KOCH's lecture, *On Bacteriology and its Results* (1890)

[Koch discusses the creation of medical bacteriology and states his three rules for identifying a specific micro-organism as the cause of a specific disease.]

Bacteriology is a very young science—at least, so far as concerns us medical men. About fifteen years ago there was little more known on the subject than that, in cases of anthrax and relapsing fever, peculiar, strange objects were found in the blood, and that the so-called vibrios occur in cases of infective diseases of wounds. No proof had then been given that these objects were the cause of the respective diseases, and, with the exception of a few investigators, who were looked upon as dreamers, people regarded them rather as curiosities than as possible causes of disease.

Indeed, any other opinion was scarcely possible, because it was not established that the organisms in question were independent and specifically connected with the diseases. Bacteria had been found in putrid fluids, more particularly in the blood of strangulated animals, which could not be distinguished from the anthrax bacillus. Some investigators even thought they were not living organisms at all, but regarded them as crystalloid bodies. Bacteria, identical with the spirillum of relapsing fever, were alleged to exist in sewage and in the mouths of healthy persons; and micrococci, the same as those which are found in cases of infective diseases of wounds, were said to exist in the healthy blood and tissues.

Indeed, with the means of experimental and optical research which were then at command, it was not possible to advance beyond this point, and matters must have remained long enough in that state had not new methods of investigation been devised, which in a moment entirely altered matters, and opened up new paths into the unexplored regions.

The most minute bacteria were rendered visible by the aid of an improved system of microscopic lenses and proper methods of using them, combined with the assistance of the aniline colours as stains; and by the use of these means the special morphology of each organism could be distinguished.

At the same time, by the employment of nutritive media, liquid or solid as required, it was rendered possible to separate the various organisms from one another, and to obtain pure cultivations, by means of which the specific characteristics of each could be ascertained with certainty.

I was soon able to show what these new methods of investigation could effect. By their aid a number of new, well-characterized pathogenic organisms were discovered, and—a thing of special importance—the causal connections between them and the associated diseases were established....

The idea that micro-organisms must be the cause of infectious diseases had already been expressed long since by a few leading men, but the majority did not accept the suggestion in a very kindly way; on the contrary, the first discoveries in this direction were regarded by them with scepticism. Hence it was all the more essential to offer irrefutable evidence at the outset that the micro-organisms found in a case of a certain disease are really its cause. At that time the objection was still rightly made that it might be merely a case of the accidental coincidence of the micro-organisms and the disease, and that the former did not act the part of dangerous parasites, but only of harmless ones, which happened to find those conditions necessary for existence in the diseased organs which were not offered to them in the healthy body. Many persons admitted, indeed, the pathogenic properties of the bacteria, but regarded it as possible that they had only been transformed into pathogenic from other harmless micro-organisms, accidentally or regularly present in the body, under the influence of the morbid process.

But if it can be proved—

Firstly, that the parasite is found in every single case of the disease in question, and under conditions corresponding to the pathological changes and the clinical course of the disease;

Secondly, that it occurs in no other disease as an accidental and non-pathogenic parasite;

Thirdly, that when isolated from the body and propagated through a sufficient number of pure cultivations, it can produce the disease anew;

the microbe under these circumstances cannot be an accidental accompaniment of the disease, and no other relationship between the parasite and the disease can be conceived, except that the former is the cause of the latter.

The chain of proof has been completely provided for a number of diseases, such as anthrax, tuberculosis, erysipelas, tetanus, and several diseases of animals—in general, for almost all those diseases which are communicable.

CHAPTER 18

THE NINETEENTH CENTURY

62. From AUGUSTE COMTE'S *The Positive Philosophy* (1840–2)

[Comte states the essence of his philosophy, that science is the final, positive form of knowledge.]

From the study of the development of human intelligence, in all directions, and through all times, the discovery arises of a great fundamental law, to which it is necessarily subject, and which has a solid foundation of proof, both in the facts of our organization and in our historical experience. The law is this:—that each of our leading conceptions,—each branch of our knowledge,—passes successively through three different theoretical conditions: the Theological, or fictitious; the Metaphysical, or abstract; and the Scientific, or positive. In other words, the human mind, by its nature, employs in its progress three methods of philosophizing, the character of which is essentially different, and even radically opposed: viz., the theological method, the metaphysical, and the positive. Hence arise three philosophies, or general systems of conceptions on the aggregate of phenomena, each of which excludes the others. The first is the necessary point of departure of the human understanding; and the third is its fixed and definitive state. The second is merely a state of transition.

In the theological state, the human mind, seeking the essential nature of beings, the first and final causes (the origin and purpose) of all effects,—in short, Absolute knowledge,—supposes all phenomena to be produced by the immediate action of supernatural beings.

In the metaphysical state, which is only a modification of the first, the mind supposes, instead of supernatural beings, abstract forces, veritable entities (that is, personified abstractions) inherent in all beings, and capable of producing all phenomena. What is called the explanation of phenomena is, in this stage, a mere reference of each to its proper entity.

In the final, the positive state, the mind has given over the vain search after absolute notions, the origin and destination of the universe, and the causes of phenomena, and applies itself to the study of their laws,—that is, their invariable relations of succession and resemblance. Reasoning and observation, duly combined, are the means of this knowledge. What is now understood when we speak of an explanation of facts is simply the establishment of a connection between single phenomena and some general facts, the number of which continually diminishes with the progress of science.

63. From JOHN TYNDALL'S *Apology for the Belfast Address* (1874)

[Tyndall's outspoken advocacy of materialism, in his address as President of the British Association for the Advancement of Science at Belfast in August 1874, called forth a chorus of denunciation, particularly in the religious Press. The following is part of his reply.]

The expression to which the most violent exception has been taken is this: 'Abandoning all disguise, the confession I feel bound to make before you is, that I prolong the vision backward across the boundary of the experimental evidence, and discern in that Matter which we, in our ignorance, and notwithstanding our professed reverence for its Creator, have hitherto covered with opprobrium, the promise and potency of every form and quality of life.' To call it a 'chorus of dissent', as my Catholic critic does, is a mild way of describing the storm of opprobrium with which this statement has been assailed. But the first blast of passion being past, I hope I may again ask my opponents to consent to reason. First of all, I am blamed for crossing the boundary of the experimental evidence. This, I reply, is the habitual action of the scientific mind—at least of that portion of it which applies itself to physical investigation. . . .

The course of life upon earth, as far as Science can see, has been one of amelioration—a steady advance on the whole from the lower to the higher. The continued effort of animated nature is to improve its condition and raise itself to a loftier level. In man improvement and amelioration depend largely upon the growth of conscious knowledge, by which the errors of ignorance are continually moulted, and truth is organised. It is the advance of knowledge that has given a materialistic colour to the philosophy of this age. Materialism is therefore not a thing to be mourned over, but to be honestly considered—accepted if it be wholly true, rejected if it be wholly false, wisely sifted and turned to account if it embrace a mixture of truth and error. Of late years the study of the nervous system, and its relation to thought and feeling, have profoundly occupied enquiring minds. It is our duty not to shirk—it ought rather to be our privilege to accept—the established results of such enquiries, for here assuredly our ultimate weal depends upon our loyalty to the truth. Instructed as to the control which the nervous system exercises over man's moral and intellectual nature, we shall be better prepared, not only to mend their manifold defects, but also to strenghten and purify both. Is mind degraded by this recognition of its dependence? Assuredly not. Matter, on the contrary, is raised to the level it ought to occupy, and from which timid ignorance would remove it.

64. From T. H. HUXLEY's lecture, *On the Physical Basis of Life* (1868)

[In this passage Huxley attacks materialism, which 'weighs like a nightmare' on his contemporaries.]

I have endeavoured, in the first part of this discourse, to give you a conception of the direction towards which modern physiology is tending; and I ask you, what is the difference between the conception of life as the product of a certain disposition of material molecules, and the old notion of an Archaeus governing and directing blind matter within each living body, except this— that here, as elsewhere, matter and law have devoured spirit and spontaneity? And as surely as every future grows out of past and present, so will the physiology of the future gradually extend the realm of matter and law until it is co-extensive with knowledge, with feeling, and with action.

The consciousness of this great truth weighs like a nightmare, I believe, upon many of the best minds of these days. They watch what they conceive to be the progress of materialism, in such fear and powerless anger as a savage feels, when, during an eclipse, the great shadow creeps over the face of the sun. The advancing tide of matter threatens to drown their souls; the tightening grasp of law impedes their freedom; they are alarmed lest man's moral nature be debased by the increase of his wisdom.

If the 'New Philosophy' be worthy of the reprobation with which it is visited, I confess their fears seem to me to be well founded. While, on the contrary, could David Hume be consulted, I think he would smile at their perplexities, and chide them for doing even as the heathen, and falling down in terror before the hideous idols their own hands have raised.

For, after all, what do we know of this terrible 'matter', except as a name for the unknown and hypothetical cause of states of our own consciousness? And what do we know of that 'spirit' over whose threatened extinction by matter a great lamentation is arising, like that which was heard at the death of Pan, except that it is also a name for an unknown and hypothetical cause, or condition, of states of consciousness? In other words, matter and spirit are but names for the imaginary substrata of groups of natural phaenomena.

And what is the dire necessity and 'iron' law under which men groan? Truly, most gratuitously invented bugbears. I suppose if there be an 'iron' law, it is that of gravitation; and if there be a physical necessity, it is that a stone, unsupported, must fall to the ground. But what is all we really know, and can know, about the latter phaenomenon? Simply, that, in all human experience, stones have fallen to the ground under these conditions; that we have not the smallest reason for believing that any stone so circumstanced will not fall to the ground; and that we have, on the contrary, every reason to believe that it will so fall. It is very convenient to indicate that all the conditions of belief have been fulfilled in this case, by calling the statement that unsupported stones will fall to the ground, 'a law of nature'. But when, as commonly happens, we change *will* into *must*, we introduce an idea of necessity which most assuredly does not lie in the observed facts, and has no warranty that I can discover elsewhere. For my part, I utterly repudiate and anathematize the intruder. Fact I know; and Law I know; but what is this Necessity, save an empty shadow of my own mind's throwing?

But, if it is certain that we can have no knowledge of the nature of either matter or spirit, and that the notion of necessity is something illegitimately thrust into the perfectly legitimate conception of law, the materialistic position that there is nothing in the world but matter, force, and necessity, is as utterly devoid of justification as the most baseless of theological dogmas.

65. From ERNST HAECKEL'S *The Riddle of the Universe* (1899)

[Haeckel's book is not a great work but it typifies a school of thought at the end of the nineteenth century.]

The supreme and all-pervading law of nature, the true and only cosmological law, is, in my opinion, *the law of substance*; its discovery and establishment is the greatest intellectual triumph of the nineteenth century, in the sense that all other known laws of nature are subordinate to it. Under the name of 'law of substance' we embrace two supreme laws of different origin and age—the older is the chemical law of the 'conservation of matter', and the younger is the physical law of the 'conservation of energy'. It will be self-evident to many readers, and it is acknowledged by most of the scientific men of the day, that these two great laws are essentially inseparable....

Once modern physics had established the law of substance as far as the simpler relations of inorganic bodies are concerned, physiology took up the story, and proved its application to the entire province of the organic world. It showed that all the vital activities of the organism—without exception—are based on a constant 'reciprocity of force' and a correlative change of material, or metabolism, just as much as the simplest processes in 'lifeless' bodies. Not only the growth and the nutrition of plants and animals, but even their functions of sensation and movement, their sense-action and psychic life, depend on the conversion of potential into kinetic energy, and *vice versa*. This supreme law dominates also those elaborate performances of the nervous system which we call, in the higher animals and man, 'the action of the mind'.

Our monistic view, that the great cosmic law applies throughout the whole of nature, is of the highest moment. For it not only involves, on its positive side, the essential unity of the cosmos and the causal connection of all phenomena that come within our cognizance, but it also, in a negative way, marks the highest intellectual progress, in that it definitely rules out the three central dogmas of metaphysics—God, freedom, and immortality. In assigning mechanical causes to phenomena everywhere, the law of substance comes into line with the universal law of causality....

One of the most distinctive features of the expiring century is the increasing vehemence of the opposition between science and Christianity. That is both natural and inevitable. In the same proportion in which the victorious progress of modern science has surpassed all the scientific achievements of earlier ages has the untenability been proved of those mystic views which would subdue reason under the yoke of an alleged revelation; and the Christian religion belongs to that group. The more solidly modern astronomy, physics, and chemistry have established the sole dominion of inflexible natural laws in the universe at large, and modern botany, zoology, and anthropology have proved the validity of those laws in the entire kingdom of organic nature, so much the more strenuously has the Christian religion, in association with dualistic metaphysics, striven to deny the application of these natural laws in the province of the so-called 'spiritual life'—that is, in one section of the physiology of the brain.

66. From HERBERT SPENCER'S *First Principles* (1862)

[In this passage Spencer argues that Evolution is a single process of integrated differentiation, which is proceeding in all parts of the universe.]

The law of Evolution has been thus far contemplated as holding true of each order of existences, considered as a separate order. But the induction as so presented falls short of that completeness which it gains when we contemplate these several orders of existences as forming together one natural whole. While we think of Evolution as divided into astronomic, geologic, biologic, psychologic, sociologic, &c., it may seem to some extent a coincidence that the same law of metamorphosis holds throughout all its divisions. But when we recognize these divisions as mere conventional groupings, made to facilitate the arrangement and acquisition of knowledge—when we remember that the different existences with which they severally deal are component parts of one Cosmos; we see at once that there are not several kinds of Evolution having certain traits in common, but one Evolution going on everywhere after the same manner. We have repeatedly observed that while any whole is evolving, there is always going on an evolution of the parts into which it divides itself; but we have not observed that this equally holds of the totality of things, which is made up of parts within parts from the greatest down to the smallest. We know that while a physically-cohering aggregate like the human body is getting larger and taking on its general shape, each of its organs is doing the same; that while each organ is growing and becoming unlike others, there is going on a differentiation and integration of its component tissues and vessels; and that even the components of these components are severally increasing and passing into more definitely heterogeneous structures. But we have not duly remarked that while each individual is developing, the society of which he is an insignificant unit is developing too; that while the aggregate mass forming a society is integrating and becoming more definitely heterogeneous, so, too, that total aggregate, the Earth, is continuing to integrate and differentiate; that while the Earth, which in bulk is not a millionth of the Solar System, progresses towards its more concentrated structure, the Solar System similarly progresses.

So understood, Evolution becomes not one in principle only, but one in fact. There are not many metamorphoses similarly carried on, but there is a single metamorphosis universally progressing, wherever the reverse metamorphosis has not set in. In any locality, great or small, where the occupying matter acquires an appreciable individuality, or distinguishableness from other matter, there Evolution goes on; or rather, the acquirement of this appreciable individuality is the commencement of Evolution. And this holds regardless of the size of the aggregate, and regardless of its inclusion in other aggregates.

CHAPTER 19

GENETICS AND NEO-DARWINISM

67. From AUGUST WEISMANN'S *The Germ-Plasm, A Theory of Heredity* (1892)

[This extract is taken from the Preface. Weismann refers to Darwin's theory of pangenesis, that minute particles, called gemmules, are released by all parts of the body and are absorbed by the germ cells, i.e. the male sperms or female ova, thereby explaining the effect of environment on heredity. Weismann denies that such an effect can occur because the germ cells reproduce themselves from generation to generation independently of the rest of the body.]

What first struck me when I began seriously to consider the problem of heredity, some ten years ago, was the necessity for assuming the existence of a special organised and living *hereditary substance*, which in all multicellular organisms, unlike the substance composing the perishable body of the individual, is transmitted from generation to generation. This is the theory of *the continuity of the germ-plasm*. My conclusions led me to doubt the usually accepted view of the *transmission of variations acquired* by the body (soma); and further research, combined with experiments, tended more and more to strengthen my conviction that in point of fact no such transmission occurs. Meanwhile, the investigations of several distinguished biologists—in which I myself have had some share—on the process of fertilisation and conjugation, brought about a complete revolution in our previous ideas as to the meaning of this process, and further led me to see that the germ-plasm is composed of vital units, each of equal value, but differing in character, containing all the primary constituents of an individual. . . .

All my investigations on the problem of heredity were so far only links, to be some day united into a chain which had as yet no existence. The question of the ultimate elements on which to base the theory was the very point on which I remained longest in doubt. The 'pangenesis' of Darwin, as already mentioned, seemed to me to be far too independent of facts, and even now I am of the opinion that the very hypothesis from which it derives its name is untenable. There is now scarcely any doubt that the entire conception of the production of the 'gemmules' by the body-cells, their separation from the latter, and their 'circulation', is in reality wholly imaginary. In this regard I am still quite as much opposed to Darwin's views as formerly, for I believe that all parts of the body do not contribute to produce a germ from which the new individual arises, but that on the contrary, the offspring owes its origin to a peculiar substance of extremely complicated structure, viz., the 'germ-plasm'. This substance can never be formed anew; it can only grow, multiply, and be transmitted from one generation to another.

68. From GREGOR MENDEL's paper, *Plant-Hybridisation* (1865)
[The essential findings of Mendel's experiments are given in this extract.]

...The various forms of Peas selected for crossing showed differences in the length and colour of the stem; in the size and form of the leaves; in the position, colour, and size of the flowers; in the length of the flower stalk; in the colour, form, and size of the pods; in the form and size of the seeds; and in the colour of the seed-coats and of the albumen (cotyledons)....

The Forms of the Hybrids [F_1]

...In the case of each of the seven crosses the hybrid-character resembles that of one of the parental forms so closely that the other either escapes observation completely or cannot be detected with certainty. This circumstance is of great importance in the determination and classification of the forms under which the offspring of the hybrids appear. Henceforth in this paper those characters which are transmitted quite, or almost unchanged in the hybridisation, and therefore in themselves constitute the characters of the hybrid, are termed the *dominant*, and those which become latent in the process *recessive*. The expression 'recessive' has been chosen because the characters thereby designated withdraw or entirely disappear in the hybrids, but nevertheless reappear unchanged in their progeny, as will be demonstrated later on.

It was furthermore shown by the whole of the experiments that it is perfectly immaterial whether the dominant character belong to the seed-bearer or to the pollen-parent; the form of the hybrid remains identical in both cases. This interesting fact was also emphasised by Gärtner, with the remark that even the most practised expert is not in a position to determine in a hybrid which of the two parental species was the seed or the pollen plant....

The First Generation (bred) from the Hybrids [F_2]

In this generation there reappear, together with the dominant characters, also the recessive ones with their peculiarities fully developed, and this occurs in the definitely expressed average proportion of three to one, so that among each four plants of this generation three display the dominant character and one the recessive. This relates without exception to all the characters which were investigated in the experiments....

Expt. 1 Form of seed—From 253 hybrids 7,324 seeds were obtained in the second trial year. Among them were 5,474 round or roundish ones and 1,850 angular wrinkled ones. Therefrom the ratio 2·96 to 1 is deduced.

Expt. 2 Colour of albumen—258 plants yielded 8,023 seeds, 6,022 yellow, and 2,001 green; their ratio, therefore, is as 3·01 to 1....

The Second Generation (bred) from the Hybrids [F_3]

Those forms which in the first generation (F_2) exhibit the recessive character do not further vary in the second generation (F_3) as regards this character; they remain constant in their offspring.

It is otherwise with those which possess the dominant character in the first generation (bred from the hybrids). Of these two-thirds yield offspring which display the dominant and recessive characters in the proportion of 3 to 1, and thereby show exactly the same ratio as the hybrid forms, while only one-third remains with the dominant character constant.

69. From T. H. MORGAN'S *The Theory of the Gene* (1926)

[Morgan was one of the founders of modern genetics, and in this extract he gives a concise statement of his theory of the gene.]

We are now in a position to formulate the theory of the gene. The theory states that the characters of the individual are referable to paired elements (genes) in the germinal material that are held together in a definite number of linkage groups; it states that the members of each pair of genes separate when the germ-cells mature in accordance with Mendel's first law, and in consequence each germ-cell comes to contain one set only; it states that the members belonging to different linkage groups assort independently in accordance with Mendel's second law; it states that an orderly interchange—crossing-over—also takes place, at times, between the elements in corresponding linkage groups; and it states that the frequency of crossing-over furnishes evidence of the linear order of the elements in each linkage group and of the relative position of the elements with respect to each other.

These principles, which, taken together, I have ventured to call the theory of the gene, enable us to handle problems of genetics on a strictly numerical basis, and allow us to predict, with a great deal of precision, what will occur in any given situation. In these respects the theory fulfils the requirements of a scientific theory in the fullest sense.

70. From a statement by the Praesidium of the U.S.S.R. Academy of Sciences (1948)

[In Russia the teaching of western genetics, called 'Weismannite-Morganist idealist teaching' in the following extract, is prohibited and it has been replaced by a form of neo-Lamarckism, first put forward by Michurin.]

Michurin's materialist direction in biology is the only acceptable form of science, because it is based on dialectical materialism, and on the revolutionary principle of changing Nature for the benefit of the people. Weismannite-Morganist idealist teaching is pseudo-scientific, because it is founded on the notion of the divine origin of the world and assumes eternal and unalterable scientific laws. The struggle between the two ideas has taken the form of the ideological class-struggle between socialism and capitalism on the international scale, and between the majority of Soviet scientists and a few remaining Russian scientists who have retained traces of bourgeois ideology, on a smaller scale. There is no place for compromise. Michurinism and Morgano-Weismannism cannot be reconciled.

71. From T. D. LYSENKO's address, *The Situation in Biological Science* (1948)

[Lysenko explains the Soviet view of heredity and attacks neo-Mendelism.]

Contrary to Mendelism-Morganism, with its assertion that the causes of variation in the nature of organisms are unknowable and its denial that directed changes in the nature of plants and animals are possible, I. V. Michurin's motto was: 'We cannot wait for favours from Nature; we must wrest them from her.'

His studies and investigations led I. V. Michurin to the following important conclusion: 'It is possible, with man's intervention, to *force* any form of animal or plant *to change more quickly and in a direction desirable to man*. There opens before man a broad field of activity of the greatest value to him.'

The Michurin teaching flatly rejects the fundamental principle of Mendelism-Morganism that heredity is completely independent of the plants' or animals' conditions of life. The Michurin teaching does not recognize the existence in the organism of a separate hereditary substance which is independent of the body. Changes in the heredity of an organism or in the heredity of any part of its body are the result of changes in the living body itself. And changes of the living body occur as the result of departure from the normal in the type of assimilation and dissimilation, of departure from the normal in the type of metabolism. Changes in organisms or in their separate organs or characters may not always, or not fully, be transmitted to the offspring, but changed germs of newly generated organisms always occur only as the result of changes in the body of the parent organism, as the result of direct or indirect action of the conditions of life upon the development of the organism or its separate parts, among them the sexual or vegetative germs. Changes in heredity, acquisition of new characters and their augmentation and accumulation in successive generations are always determined by the organism's conditions of life. Heredity changes and its complexity increases as the result of the accumulation of new characters and properties acquired by organisms in successive generations.

The organism and the conditions required for its life constitute a unity. Different living bodies require different environmental conditions for their development. By studying the character of these requirements we come to know the qualitative features of the nature of organisms, the qualitative features of heredity. Heredity is *the property of a living body to require definite conditions for its life and development and to respond in a definite way to various conditions....*

The representatives of Mendel-Morgan genetics are not only unable to obtain alterations of heredity in a definite direction, but categorically deny that it is possible to change heredity so that it will adequately correspond to the action of the environmental conditions. The principles of Michurin's teaching, on the other hand, tell us that it is possible to obtain changes in heredity *fully corresponding to the effect of the action of conditions of life.*

A case in point is the experiments to convert spring forms of cereal grains into winter forms, and winter forms into forms of still greater winter habit in regions of Siberia, for example, where the winters are severe. These experiments are not only of theoretical interest. They are of considerable practical value for the production of winter-hardy varieties. We already have winter forms of wheat obtained from spring forms, which are not inferior, as regards frost resistance, to the most frost-resistant varieties known in practical farming. Some are even superior.

Many experiments show that when an old established property of heredity is being eliminated, we do not at once get a fully established, solidified new heredity. In the vast majority of cases, what we get is an organism with a plastic nature, which I. V. Michurin called 'destabilized'.

Plant organisms with a 'destabilized' nature are those in which their conservatism has been eliminated, and their electivity with regard to external conditions is weakened. Instead of conservative heredity, such plants preserve, or there appears in them only a *tendency* to show some preference for certain conditions.

The nature of a plant organism may be destabilized:

1. By grafting, i.e., by uniting the tissues of plants of different breeds;

2. By bringing external conditions to bear upon it at definite moments when the organism undergoes this or that process of its development;

3. By crossbreeding, particularly of forms sharply differing in habitat or origin.

The best biologists, first and foremost I. V. Michurin, have devoted a great deal of attention to the practical value of plant organisms with destabilized heredity. Plastic plant forms with unestablished heredity, obtained by any of the enumerated methods, should be further bred from generation to generation in those conditions, the requirement of which, or adaptability to which, we want to induce and perpetuate in the given organisms.

CHAPTER 20

THE STRUCTURE OF THE ATOM

72. From J. J. THOMSON's paper, *Cathode Rays* (1897)

[In this paper J. J. Thomson describes his quantitative experiments on cathode rays and puts forward the hypothesis that the atoms of the elements are divisible and contain much smaller particles, which he terms corpuscles and which we now know as electrons.]

The experiments discussed in this paper were undertaken in the hope of gaining some information as to the nature of the Cathode Rays. The most diverse opinions are held as to these rays; according to the almost unanimous opinion of German physicists they are due to some process in the æther to which—inasmuch as in a uniform magnetic field their course is circular and not rectilinear—no phenomenon hitherto observed is analogous; another view of these rays is that, so far from being wholly ætherial, they are in fact wholly material, and that they mark the paths of particles of matter charged with negative electricity. It would seem at first sight that it ought not to be difficult to discriminate between views so different, yet experience shows that this is not the case, as amongst the physicists who have most deeply studied the subject can be found supporters of either theory.

The electrified-particle theory has for purposes of research a great advantage over the ætherial theory, since it is definite and its consequences can be predicted; with the ætherial theory it is impossible to predict what will happen under any given circumstances, as on this theory we are dealing with hitherto unobserved phenomena in the æther, of whose laws we are ignorant. . . .

An objection very generally urged against the view that the cathode rays are negatively electrified particles, is that hitherto no deflexion of the rays has been observed under a small electrostatic force, and though the rays are deflected when they pass near electrodes connected with sources of large differences of potential, such as induction-coils or electrical machines, the deflexion in this case is regarded by the supporters of the ætherial theory as due to the discharge passing between the electrodes, and not primarily to the electrostatic field, Hertz made the rays travel between two parallel plates of metal placed inside the discharge-tube, but found that they were not deflected when the plates were connected with a battery of storage cells; on repeating this experiment I at first got the same result, but subsequent experiments showed that the absence of deflexion is due to the conductivity conferred on the rarefied gas by the cathode rays. On measuring this conductivity it was found that it diminished very rapidly as the exhaustion increased; it seemed then that on trying Hertz's experiment at very high exhaustions there might be a chance of detecting the deflexion of the cathode rays by an electrostatic force.

The apparatus used is represented in the Figure.

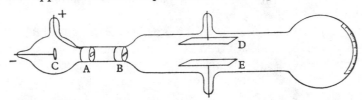

The rays from the cathode C pass through a slit in the anode A, which is a metal plug fitting tightly into the tube and connected with the earth; after passing through a second slit in another earth-connected metal plug B, they travel between two parallel aluminium plates about 5 cm. long by 2 broad and at a distance of 1·5 cm. apart; they then fall on the end of the tube and produce a narrow well-defined phosphorescent patch. A scale pasted on the outside of the tube serves to measure the deflexion of this patch. At high exhaustions the rays were deflected when the two aluminium plates were connected with the terminals of a battery of small storage-cells; the rays were depressed when the upper plate was connected with the positive, the lower with the negative pole. The deflexion was proportional to the difference of potential between the plates, and I could detect the deflexion when the potential-difference was as small as two volts. . . .

[Thomson then goes on to describe how, by measuring also the deflection of the cathode rays in the magnetic field, and by other experiments, he obtained a value of m/e, where m and e are the mass and charge respectively of each cathode ray particle.]

From these determinations we see that the value of m/e is independent of the nature of the gas, and that its value 10^{-7} is very small compared with the value 10^{-4}, which is the smallest value of this quantity previously known, and which is the value for the hydrogen ion in electrolyses.

Thus for the carriers of electricity in the cathode rays m/e is very small compared with its value in electrolyses. The smallness of m/e may be due to the smallness of m or the largeness of e, or to a combination of these two. . . .

The explanation which seems to me to account in the most simple and straightforward manner for the facts is founded on a view of the constitution of the chemical elements which has been favourably entertained by many chemists; this view is that the atoms of the different chemical elements are different aggregations of atoms of the same kind. . . .

If, in the very intense electric field in the neighbourhood of the cathode, the molecules of the gas are dissociated and are split up, not into the ordinary chemical atoms, but into these primordial atoms, which we shall for brevity call corpuscles; and if these corpuscles are charged with electricity and projected from the cathode by the electric field, they would behave exactly like the cathode rays. . . .

Thus on this view we have in the cathode rays matter in a new state, a state in which the subdivision of matter is carried very much further than in the ordinary gaseous state; a state in which all matter—that is, matter derived from different sources such as hydrogen, oxygen, etc.—is of one and the same kind; this matter being the substance from which all the chemical elements are built up.

73. From the paper by E. RUTHERFORD and F. SODDY, *The Cause and Nature of Radioactivity* (1902)

[This paper announced the startling theory that radioactivity consists of the spontaneous breaking up of atoms.]

Turning from the experimental results to their theoretical interpretation, it is necessary to first consider the generally accepted view of the nature of radioactivity. It is well established that this property is the function of the atom and not of the molecule. Uranium and thorium, to take the most definite cases, possess the property in whatever molecular condition they occur, and the former also in the elementary state. So far as the radioactivity of different compounds of different density and states of division can be compared together, the intensity of the radiation appears to depend only on the quantity of active element present. It is not at all dependent on the source from which the element is derived, or the process of purification to which it has been subjected, provided sufficient time is allowed for the equilibrium point to be reached. It is not possible to explain the phenomena by the existence of impurities associated with the radioactive elements, even if any advantage could be derived from the assumption. For these impurities must necessarily be present always to the same extent in different specimens derived from the most widely different sources, and, moreover, they must persist in unaltered amount after the most refined processes of purification. This is contrary to the accepted meaning of the term impurity.

All the most prominent workers in this subject are agreed in considering radioactivity an atomic phenomenon. M. and Mme Curie, the pioneers in the chemistry of the subject, have recently put forward their views (*Comptes Rendus*, CXXXIV, 1902, p. 85). They state that this idea underlies their whole work from the beginning and created their methods of research. M. Becquerel, the original discoverer of the property for uranium, in his announcement of the recovery of the activity of the same element after the active constituent had been removed by chemical treatment, points out the significance of the fact that uranium is giving out cathode-rays. These, according to the hypothesis of Sir William Crookes and Prof. J. J. Thomson, are *material* particles of mass one thousandth of the hydrogen atom.

Since, therefore, radioactivity is at once an atomic phenomenon and accompanied by chemical changes in which new types of matter are produced, these changes must be occurring within the atom, and the radioactive elements must be undergoing spontaneous transformation. The results that have so far been obtained, which indicate that the velocity of this reaction is unaffected by the conditions, makes it clear that the changes in question are different in character from any that have been before dealt with in chemistry. It is apparent that we are dealing with phenomena outside the sphere of known atomic forces. Radio-activity may therefore be considered as a manifestation of subatomic chemical change.

74. From ERNEST RUTHERFORD's Bakerian lecture, *Nuclear Constitution of Atoms* (1920)

[In the first part of this extract Rutherford discusses the establishment of the nuclear theory of the atom; in the second part he describes the artificial disintegration of the nucleus of the nitrogen atom.]

Introduction.—The conception of the nuclear constitution of atoms arose initially from attempts to account for the scattering of α-particles through large angles in traversing thin sheets of matter. Taking into account the large mass and velocity of the α-particles, these large deflexions were very remarkable, and indicated that very intense electric or magnetic fields exist within the atom. To account for these results, it was found necessary to assume that the atom consists of a charged massive nucleus of dimensions very small compared with the ordinarily accepted magnitude of the diameter of the atom. This positively charged nucleus contains most of the mass of the atom, and is surrounded at a distance by a distribution of negative electrons equal in number to the resultant positive charge on the nucleus. Under these conditions, a very intense electric field exists close to the nucleus, and the large deflexion of the α-particle in an encounter with a single atom happens when the particle passes close to the nucleus. Assuming that the electric forces between the α-particle and the nucleus varied according to an inverse square law in the region close to the nucleus, the writer worked out the relations connecting the number of α-particles scattered through any angle with the charge on the nucleus and the energy of the α-particle. Under the central field of force, the α-particle describes a hyperbolic orbit round the nucleus, and the magnitude of the deflexion depends on the closeness of approach to the nucleus. From the data of scattering of α-particles then available, it was deduced that the resultant charge on the nucleus was about $\frac{1}{2}Ae$, where A is the atomic weight and e the fundamental unit of charge. Geiger and Marsden made an elaborate series of experiments to test the correctness of the theory, and confirmed the main conclusions. They found the nucleus charge was about $\frac{1}{2}Ae$, but, from the nature of the experiments, it was difficult to fix the actual value within about 20 per cent. C. G. Darwin worked out completely the deflexion of the α-particle and of the nucleus, taking into account the mass of the latter, and showed that the scattering experiments of Geiger and Marsden could be reconciled with any law of central force, except the inverse square. The nuclear constitution of the atom was thus very strongly supported by the experiments on scattering of α-rays.

Since the atom is electrically neutral, the number of external electrons surrounding the nucleus must be equal to the number of units of resultant charge on the nucleus. It should be noted that, from the consideration of the scattering of X-rays by light elements, Barkla had shown, in 1911, that the number of electrons was equal to about half the atomic weight. This was deduced from the theory of scattering of Sir J. J. Thomson, in which it was assumed that each of the external electrons in an atom acted as an independent scattering unit.

Two entirely different methods had thus given similar results with regard to the number of external electrons in the atom, but the scattering of α-rays had shown in addition that the positive charge must be concentrated on a massive nucleus of small dimensions. It was suggested by Van den Broek that the scattering of α-particles by the atoms was not inconsistent with the possibility that the charge on the nucleus was equal to the atomic number of the atom, i.e., to the number of the atom when arranged in order of increasing atomic weight. The importance of the atomic number in fixing the properties of an atom was shown by the remarkable work of Moseley on the X-ray spectra of the elements. He showed that the frequency of vibration of corresponding lines in the X-ray spectra of the elements depended on the square of a number which varies by unity in successive elements. This relation received an interpretation by supposing that the nuclear charge varied by unity in passing from atom to atom, and was given numerically by the atomic number. I can only emphasise in passing the great importance of Moseley's work, not only in fixing the number of possible elements, and the position of undetermined elements, but in showing that the properties of an atom were defined by a number which varied by unity in successive atoms. This gives a new method of regarding the periodic classification of the elements, for the atomic number, or its equivalent the nuclear charge, is of more fundamental importance than its atomic weight....

Long Range Particles from Nitrogen

In previous papers, I have given an account of the effects produced by close collisions of swift α-particles with light atoms of matter with the view of determining whether the nuclear structure of some of the lighter atoms could be disintegrated by the intense forces brought into play in such close collisions. Evidence was given that the passage of α-particles through dry nitrogen gives rise to swift particles which closely resembled in brilliancy of the scintillations and distance of penetration hydrogen atoms set in motion by close collision with α-particles. It was shown that these swift atoms which appeared only in dry nitrogen and not in oxygen or carbon dioxide could not be ascribed to the presence of water vapour or other hydrogen material but must arise from the collision of α-particles with nitrogen atoms. The number of such scintillations due to nitrogen was small, viz., about 1 in 12 of the corresponding number in hydrogen, but was two to three times the number of natural scintillations from the source....

In a previous paper I have given evidence that the long range particles observed in dry air and pure nitrogen must arise from the nitrogen atoms themselves. It is thus clear that some of the nitrogen atoms are disintegrated by their collision with swift α-particles and that swift atoms of positively charged hydrogen are expelled. It is to be inferred that the charged atom of hydrogen is one of the components of which the nucleus of nitrogen is built up.

CHAPTER 21

THE THEORY OF RELATIVITY

75. From ALBERT EINSTEIN's lecture, *The Theory of Relativity* (1921)

[The following is a brief summary of the theory of relativity in its creator's own words.]

It is a particular pleasure to me to have the privilege of speaking in the capital of the country from which the most important fundamental notions of theoretical physics have issued. I am thinking of the theory of mass motion and gravitation which Newton gave us and the concept of the electromagnetic field, by means of which Faraday and Maxwell put physics on a new basis. The theory of relativity may indeed be said to have put a sort of finishing touch to the mighty intellectual edifice of Maxwell and Lorentz, inasmuch as it seeks to extend field physics to all phenomena, gravitation included.

Turning to the theory of relativity itself, I am anxious to draw attention to the fact that this theory is not speculative in origin; it owes its invention entirely to the desire to make physical theory fit observed fact as well as possible. We have here no revolutionary act but the natural continuation of a line that can be traced through centuries. The abandonment of certain notions connected with space, time, and motion hitherto treated as fundamentals must not be regarded as arbitrary, but only as conditioned by observed facts.

The law of the constant velocity of light in empty space, which has been confirmed by the development of electrodynamics and optics, and the equal legitimacy of all inertial systems[1] (special principle of relativity), which was proved in a particularly incisive manner by Michelson's famous experiment, between them made it necessary, to begin with, that the concept of time should be

---

[1] An inertial system is one in which the laws of mechanics apply, i.e. all systems in uniform relative motion.

made relative, each inertial system being given its own special time. As this notion was developed, it became clear that the connection between immediate experience on one side and co-ordinates and time on the other had hitherto not been thought out with sufficient precision. It is in general one of the essential features of the theory of relativity that it is at pains to work out the relations between general concepts and empirical facts more precisely. The fundamental principle here is that the justification for a physical concept lies exclusively in its clear and unambiguous relation to facts that can be experienced. According to the special theory of relativity, spatial coordinates and time still have an absolute character in so far as they are directly measurable by stationary clocks and bodies. But they are relative in so far as they depend on the state of motion of the selected inertial system. According to the special theory of relativity the four-dimensional continuum formed by the union of space and time (Minkowski) retains the absolute character which, according to the earlier theory, belonged to both space and time separately. The influence of motion (relative to the coordinate system) on the form of bodies and on the motion of clocks, also the equivalence of energy and inert mass, follow from the interpretation of co-ordinates and time as products of measurement.

The general theory of relativity owes its existence in the first place to the empirical fact of the numerical equality of the inertial and gravitational mass of bodies,[1] for which fundamental fact classical mechanics provided no interpretation. Such an interpretation is arrived at by an extension of the principle of relativity to coordinate systems accelerated relatively to one another. The introduction of coordinate systems accelerated relatively to inertial systems involves the appearance of gravitational fields relative to the latter. As a result of this, the general theory of relativity, which is based on the equality of inertia and weight, provides a theory of the gravitational field.

The introduction of coordinate systems accelerated relatively to each other as equally legitimate systems, such as they appear

---

[1] The inertial mass of a body is defined in terms of the acceleration it acquires under a given force, and the gravitational mass by the attraction of other bodies, such as the earth. Their numerical equality is demonstrated by the fact that all bodies fall to the earth with the same acceleration. Einstein expressed this equality in his Principle of Equivalence.

conditioned by the identity of inertia and weight, leads, in conjunction with the results of the special theory of relativity, to the conclusion that the laws governing the arrangement of solid bodies in space, when gravitational fields are present, do not correspond to the laws of Euclidean geometry. An analogous result follows for the motion of clocks. This brings us to the necessity for yet another generalization of the theory of space and time, because the direct interpretation of spatial and temporal coordinates by means of measurements obtainable with measuring rods and clocks now breaks down. That generalization of metric, which had already been accomplished in the sphere of pure mathematics through the researches of Gauss and Riemann, is essentially based on the fact that the metric of the special theory of relativity can still claim validity for small regions in the general case as well.

The process of development here sketched strips the space-time coordinates of all independent reality. The metrically real is now only given through the combination of the space-time coordinates with the mathematical quantities which describe the gravitational field.

There is yet another factor underlying the evolution of the general theory of relativity. As Ernst Mach insistently pointed out, the Newtonian theory is unsatisfactory in the following respect; if one considers motion from the purely descriptive, not from the causal, point of view, it only exists as relative motion of things with respect to one another. But the acceleration which figures in Newton's equations of motion is unintelligible if one starts with the concept of relative motion. It compelled Newton to invent a physical space in relation to which acceleration was supposed to exist. This introduction *ad hoc* of the concept of absolute space, while logically unexceptionable, nevertheless seems unsatisfactory. Hence Mach's attempt to alter the mechanical equations in such a way that the inertia of bodies is traced back to relative motion on their part not as against absolute space but as against the totality of other ponderable bodies. In the state of knowledge then existing, his attempt was bound to fail.

The posing of the problem seems, however, entirely reasonable. This line of argument imposes itself with considerably enhanced force in relation to the general theory of relativity, since, according to that theory, the physical properties of space are affected by ponderable matter. In my opinion the general theory of relativity can solve this problem satisfactorily only if it regards the world as spatially closed. The mathematical results of the theory force one to this view, if one believes that the mean density of ponderable matter in the world possesses some finite value, however small.

CHAPTER 22

THE QUANTUM THEORY

76. From MAX PLANCK's Nobel Prize address, *The Origin and Development of the Quantum Theory* (1920)

When I recall the days of twenty years ago, when the conception of the physical quantum of 'action' was first beginning to disentangle itself from the surrounding mass of available experimental facts, and when I look back upon the long tortuous road which finally led to its disclosure, this development strikes me at times as a new illustration of Goethe's saying, that 'man errs, so long as he is striving'. And all the mental effort of an assiduous investigator must indeed appear vain and hopeless, if he does not occasionally run across striking facts which form incontrovertible proof of the truth he seeks, and show him that after all he has moved at least one step nearer to his objective. The pursuit of a goal, the brightness of which is undimmed by initial failure, is an indispensable condition, though by no means a guarantee, of final success.

In my own case such a goal has been for many years the solution of the question of the distribution of energy in the normal spectrum of radiant heat. The discovery by Gustav Kirchhoff that the quality of the heat radiation produced in an enclosure surrounded by any emitting or absorbing bodies whatsoever, all at the same temperature, is entirely independent of the nature of such bodies, established the existence of a universal function, which depends only upon the temperature and the wave-length, and is entirely independent of the particular properties of the substance. And the discovery of this remarkable function promised a deeper insight into the relation between energy and temperature, which is the principal problem of thermodynamics and therefore also of the entire field of molecular physics.

[Planck describes how he obtained a radiation formula containing two universal constants; the first was 'the value of the electrical elementary charge', i.e. the charge on the electron, and the second he now discusses.]

Much less simple than that of the first was the interpretation of the second universal constant of the radiation law, which, as the product of energy and time (amounting on a first calculation to $6.55.10^{-27}$ erg. sec.) I called the elementary quantum of action. While this constant was absolutely indispensable to the attainment of a correct expression for entropy—for only with its aid could be determined the magnitude of the 'elementary region' or 'range' of probability, necessary for the statistical treatment of the problem—it obstinately withstood all attempts at fitting it, in any suitable form, into the frame of the classical theory. So long as it could be regarded as infinitely small, that is to say for large values of energy or long periods of time, all went well; but in the general case a difficulty arose at some point or other, which became the more pronounced the weaker and the more rapid the oscillations. The failure of all attempts to bridge this gap soon placed one before the dilemma: either the quantum of action was only a fictitious magnitude, and, therefore, the entire deduction from the radiation law was illusory and a mere juggling with formulae, or there is at the bottom of this method of deriving the radiation law some true physical concept. If the latter were the case, the quantum would have to play a fundamental role in physics, heralding the advent of a new state of things, destined, perhaps, to transform completely our physical concepts which since the introduction of the infinitesimal calculus by Leibniz and Newton have been founded upon the assumption of the continuity of all causal chains of events.

Experience has decided for the second alternative. But that the decision should come so soon and so unhesitatingly was due not to the examination of the law of distribution of the energy of heat radiation, still less to my special deduction of this law, but to the steady progress of the work of those investigators who have applied the concept of the quantum of action to their researches.

77. From NIELS BOHR'S article, *Discussion with Einstein on Epistemological Problems in Atomic Physics* (1949)

[At the Solvay conference in Brussels in October 1927, the interpretation of the quantum theory, devised by Bohr and his Copenhagen school, was subjected to the criticism of Einstein. The photon, which is a quantum of light, behaves both as a particle, detectable on a photographic plate, and as waves, detectable by an interference pattern. It therefore differs fundamentally from any object we experience in daily life. Bohr maintained in his theory of complementarity that, although a photon behaves simultaneously both as a particle and waves, we can observe it only in one of its aspects at one time. If we wish to observe it as a particle we must choose one type of apparatus; if as waves, another type.]

The extent to which renunciation of the visualization of atomic phenomena is imposed upon us by the impossibility of their subdivision is strikingly illustrated by the following example to which Einstein very early called attention and often has reverted. If a semi-reflecting mirror is placed in the way of a photon, leaving two possibilities for its direction of propagation, the photon may either be recorded on one, and only one, of two photographic plates situated at great distances in the two directions in question, or else we may, by replacing the plates by mirrors, observe effects exhibiting an interference between the two reflected wave-trains. In any attempt of a pictorial representation of the behaviour of the photon we would, thus, meet with the difficulty: to be obliged to say, on the one hand, that the photon always chooses *one* of the two ways and, on the other hand, that it behaves as if it had passed *both* ways.

It is just arguments of this kind which recall the impossibility of subdividing quantum phenomena and reveal the ambiguity in ascribing customary physical attributes to atomic objects. . . .

These problems were instructively commented upon from different sides at the Solvay meeting, in the same session where Einstein raised his general objections. On that occasion an interesting discussion arose also about how to speak of the appearance of phenomena for which only predictions of statistical character can be made. The question was whether, as to the occurrence of individual effects, we should adopt a terminology proposed by Dirac, that we were concerned with a choice on the part of 'nature' or, as suggested by Heisenberg, we should say that we have to do with a choice on the part of the 'observer' constructing the measuring instruments and reading their recording. Any such terminology would, however, appear dubious since, on the one hand, it is hardly reasonable to endow nature with volition in the ordinary sense, while, on the other hand, it is certainly not possible for the observer to influence the events which may appear under the conditions he has arranged. To my mind, there is no other alternative than to admit that, in this field of experience, we are dealing with individual phenomena and that our possibilities of handling the measuring instruments allow us only to make a choice between the different complementary types of phenomena we want to study.

78. From WERNER HEISENBERG'S lecture, *Recent Changes in the Foundations of Exact Science* (1934)

[Heisenberg maintains that the theory of relativity and the quantum theory have effected permanent changes in the foundations of physical science.]

This immediately raises the more general question of the finality of the changes wrought by modern physics on the foundations of exact science. We have to discuss whether the scientist will once and for all have to renounce all thought of an objective time scale common to all observers, and of objective events in time and space independent of observations on them. Perhaps recent developments represent only a passing crisis. I tend to the opinion, for which there seems to be the strongest evidence, that this renunciation will be final. I would like to begin with an analogy to support this statement. Previous to the beginnings of science in antiquity, the world was conceived as a flat disc, and only the discovery of America and the first circumnavigation of the world destroyed this belief for all time. Of course nobody had ever seen the edge of the world-disc, but just the same this 'end of the world' acquired form and substance through the legends and imaginings of man. We all know the theme of the ever enquiring man who wants to travel to the end of the world. Then, the question of 'the end of the world' had a definite and clear meaning, but the voyages of discovery of Columbus and Magellan made that question meaningless and transformed the ideas linked to it into fairy tales for ever afterwards. For all that, mankind did not renounce the idea of 'the end of the world' as a result of having explored the whole surface of the world—even to-day there are some unexplored parts—but the voyages of Columbus and Magellan gave clear proofs of the necessity to make use of new lines of approach. In accepting the spherical shape of the earth the loss of the old concept was not felt to be a loss. Similarly modern physics has taught us to do without the concepts of an absolute scale of time and of objective events in space and time. The meaning of these two concepts had never been confirmed by direct experience either, at least not as completely as we had believed. They, too, formed a hypothetical 'end of the world'. It must be stressed that the world of ideas which is to be destroyed simultaneously with these concepts of classical physics is much less living than that destroyed by Columbus or Copernicus. Hence the transition to our concept of the universe, wrought by modern physics, is less decisive than that of the fifteenth and sixteenth centuries. The convincing power of the quantum theory is by no means based on the fact that we may have surveyed all methods of measuring the position and velocity of an electron and that we have been unsuccessful in every case in circumventing the uncertainty relations. But the experimental results of say Compton, Geiger, and Bothe are such clear proof of the necessity of making use of the new lines of thought introduced by quantum theory, that the loss of concepts of classical physics no longer appears a loss. The real strength of modern physics, then, rests in its new lines of thought. The hope that new experiments will yet lead us back to objective events in time and space, or to absolute time, are about as well founded as the hope of discovering the end of the world somewhere in the unexplored regions of the Antarctic. This analogy may be further extended; Columbus's discoveries were immaterial to the geography of the Mediterranean countries, and it would be quite wrong to claim that the voyages of discovery of the famous Genoese had made obsolete the positive geographical knowledge of the day. It is equally wrong to speak today of a revolution in physics. Modern physics has changed nothing in the great classical disciplines of, for instance, mechanics, optics, and heat. Only the conception of hitherto unexplored regions, formed prematurely from a knowledge of only certain parts of the world, has undergone a decisive transformation. This conception, however, is always decisive for the future course of research.

CHAPTER 23

COSMOGONY

79. From EDWIN HUBBLE'S *The Observational Approach to Cosmology* (1937)

[Hubble expresses doubts whether the red-shifts in the spectra of the nebulae really indicate that the universe is expanding.]

When Slipher, in his great pioneering work, assembled the first considerable lists of red-shifts, the observations were necessarily restricted to the brighter, nearer nebulae. Consequently, the shifts were moderately small (less than 1 per cent), and they were accepted without question as the familiar velocity-shifts. Attempts were immediately made to study the motions of the nebulae by the same methods used in the study of stellar motions. But later, after the 'velocity-distance relation' had been formulated, and Humason's observations of faint nebulae began to accumulate, the earlier, complete certainty of the interpretation began to fade.

The disturbing features were the facts that the 'velocities' reached enormous values and were precisely correlated with distance. Each million light-years of distance added a hundred miles per second to the 'velocity'. As Humason swept farther and farther out into space he reported 'velocities' of 5,000 miles per second, then 10,000 then 15,000. Finally, near the absolute limit of his spectrograph he recorded red-shifts of 13 and 14 per cent, 'velocities' of about 25,000 miles per second—around the earth in a second, out to the moon in 10 seconds, out to the sun in just over an hour. Red-shifts continue to increase beyond the range of the spectrograph, and, for the faintest nebulae that can be photographed, they are presumably about double the largest recorded shifts—the 'velocities' are about 50,000 miles per second. These quantities we are asked to accept as measuring a general recession of the nebulae, an expansion of the universe itself. The law of red-shifts then reads: the nebulae are receding from the earth, in all directions, with velocities that are proportional to their distances from the earth.

Well, perhaps the nebulae are all receding in this peculiar manner. But the notion is rather startling. The cautious observer naturally examines other possibilities before accepting the proposition even as a working hypothesis. He recalls the alternative formulation of the law of red-shifts—light loses energy in proportion to the distance it travels through space. The law, in this form, sounds quite plausible. Internebular space, we believe, cannot be entirely empty. There must be a gravitational field through which the light-quanta travel for many millions of years before they reach the observer, and there may be some interaction between the quanta and the surrounding medium. The problem invites speculation, and, indeed, has been carefully examined. But no satisfactory, detailed solution has been found. The known reactions have been examined, one after the other—and they have failed to account for the observations. Light *may* lose energy during its journey through space, but if so, we do not yet know how the loss can be explained.

The observer seems to face a dilemma. The familiar interpretation of red-shifts leads to rather startling conclusions. These conclusions can be avoided by an assumption which sounds plausible but which finds no place in our present body of knowledge. The situation can be described as follows. Red-shifts are produced either in the nebulae, where the light originates, or in the intervening space through which the light travels. If the source is in the nebulae, then red-shifts are probably velocity-shifts and the nebulae are receding. If the source lies in the intervening space, the explanation of red-shifts is unknown but the nebulae are sensibly stationary.

80. From ALBERT EINSTEIN's *The Meaning of Relativity* (1950)

[Einstein first refers to the cosmological term, representing a force of repulsion, which he was obliged to insert into the field equations of general relativity to account for a static universe. He then explains why the red-shift must be taken as an indication that the universe is expanding.]

If Hubble's expansion had been discovered at the time of the creation of the general theory of relativity, the cosmologic member would never have been introduced. It seems now so much less justified to introduce such a member into the field equations, since its introduction loses its sole original justification—that of leading to a natural solution of the cosmologic problem....

Some try to explain Hubble's shift of spectral lines by means other than the Doppler effect. There is, however, no support for such a conception in the known physical facts. According to such a hypothesis it would be possible to connect two stars, S_1 and S_2, by a rigid rod. Monochromatic light which is sent from S_1 to S_2 and reflected back to S_1 could arrive with a different frequency (measured by a clock on S_1) if the number of wave lengths of light along the rod should change with time on the way. This would mean that the locally measured velocity of light would depend on time, which would contradict even the special theory of relativity. Further it should be noted that a light signal going to and fro between S_1 and S_2 would constitute a 'clock' which would not be in a constant relation with a clock (e.g. an atomistic clock) in S_1. This would mean that there would exist no metric in the sense of relativity. This not only involves the loss of comprehension of all those relations which relativity has yielded, but it also fails to concur with the fact that certain atomistic forms are not related by 'similarity' but by 'congruence' (the existence of sharp spectral lines, volumes of atoms, etc.)....

Hence one cannot but consider Hubble's discovery as an expansion of the system of stars.

81. From HERMANN BONDI'S *Cosmology* (1952)

[Bondi states the perfect cosmological principle and the theory of continuous creation based upon it.]

For in any theory which contemplates a changing universe, explicit and implicit assumptions must be made about the interactions between distant matter and local physical laws. These assumptions are necessarily of a highly arbitrary nature, and progress on such a basis can only be indefinite and uncertain. It may, however, be questioned whether such speculation is required. If the uniformity of the universe is sufficiently great none of these difficulties arises. The assumption that this is so is known as the perfect cosmological principle. It was introduced by Bondi and Gold (1948) in the form of the statement, that, apart from local irregularities the universe presents the same aspect from any place at any time. It was shown by Bondi and Gold that this single principle forms a sufficient basis for developing without ambiguity a cosmological theory capable of making definite and far-reaching physical statements agreeing with observation....

The fundamental assumption of the theory is that the universe presents on the large scale an unchanging aspect. Since the universe must (on thermodynamic grounds) be expanding, new matter must be continually created in order to keep the density constant. As ageing nebulae drift apart, due to the general motion of expansion, new nebulae are formed in the intergalactic spaces by condensation of newly created matter. Nebulae of all ages hence exist with a certain frequency distribution. Astrophysical estimates of the age of our galaxy do not put it into a very rare class of nebulae.

The theory is deductive in the sense that its conclusions are derived from the cosmological principle, but the very powerful formulation of the principle employed dispenses with the need for additional assumptions.

A different approach has been proposed by Hoyle and will be discussed later in this chapter. In that formulation a suitable modification of the field equations of general relativity is taken as starting-point, so that the conclusions reached are very similar to those of the steady-state theory.

The importance to the theory of the powerful formulation of the cosmological principle makes it highly desirable to examine in detail the arguments for the acceptance of this principle. This examination reveals that the arguments supporting the usual narrow cosmological principle imply the validity of the wider perfect cosmological principle, according to which the large-scale aspect of the universe should not only be independent of the *position* of the observer but also of the *time* of making the observation. . . .

There is little doubt that the continual creation of matter necessary in this theory is the most revolutionary change proposed by it. There is, however, no observational evidence whatever contradicting continual creation at the rate demanded by the perfect cosmological principle. It is easily seen that this is, on the average,

$$3 \times \text{(mean density of matter in the universe)}$$
$$\times \text{Hubble's constant} = 10^{-43} \text{ g./cm.}^3 \text{ sec.}$$

approximately. In other words, on an average the mass of a hydrogen atom is created in each litre of volume every 10^9 years. As will be seen later there are strong arguments showing that the creation rate does not vary widely between different places, so that the average rate given above has universal significance. It is clear that it is utterly impossible to observe directly such a rate of creation. There is therefore no contradiction whatever with the observations, an extreme extrapolation from which forms the principle of conservation of matter. The argument may be stated in the following terms: When observations indicated that matter was at least very nearly conserved it seemed simplest (and therefore most scientific) to assume that the conservation was absolute. But when a wider field is surveyed then it is seen that this apparently simple assumption leads to the great complications discussed in connexion with the formulation of the perfect cosmological principle. The principle resulting in greatest overall simplicity is then seen to be not the principle of conservation of matter but the perfect cosmological principle with its consequence of continual creation. From this point of view continual creation is the simplest and hence the most scientific extrapolation from the observations.

CHAPTER 24

THE TWENTIETH CENTURY

82. From ISAAC NEWTON'S *Opticks* (1704)

[This passage, in which Newton states his belief that the ultimate physical reality consists of massy atoms, is taken from the last Query in his *Opticks*.]

All these things being consider'd, it seems probable to me, that God in the Beginning form'd Matter in solid, massy, hard, impenetrable, movable Particles, of such Sizes and Figures, and with such other Properties, and in such Proportion to Space, as most conduced to the End for which he form'd them; and that these primitive Particles being Solids, are incomparably harder than any porous Bodies compounded of them; even so very hard, as never to wear or break in pieces; no ordinary Power being able to divide what God himself made one in the first Creation. While the Particles continue entire, they may compose Bodies of one and the same Nature and Texture in all Ages: But should they wear away, or break in pieces, the Nature of Things depending on them, would be changed. Water and Earth, composed of old worn Particles and Fragments of Particles, would not be of the same Nature and Texture now, with Water and Earth composed of entire Particles in the Beginning. And therefore, that Nature may be lasting, the Changes of corporeal Things are to be placed only in the various Separations and new Associations and Motions of these permanent Particles; compound Bodies being apt to break, not in the midst of solid Particles, but where those Particles are laid together, and only touch in a few points.

83. From ERNST MACH's lecture, *The Economical Nature of Physical Enquiry* (1882)

[Mach states his view that science is concise description and that atoms are merely imaginary tools for this purpose.]

The communication of scientific knowledge always involves description, that is, a mimetic reproduction of facts in thought, the object of which is to replace and save the trouble of new experience. Again, to save the labour of instruction and of acquisition, concise, abridged description is sought. This is really all that natural laws are. Knowing the value of the acceleration of gravity, and Galileo's laws of descent, we possess simple and compendious directions for reproducing in thought all possible motions of falling bodies. A formula of this kind is a complete substitute for a full table of motions of descent, because by means of the formula the data of such a table can be easily constructed at a moment's notice without the least burdening of the memory....

When we look over a province of facts for the first time, it appears to us diversified, irregular, confused, full of contradictions. We first succeed in grasping only single facts, unrelated with the others. The province, as we are wont to say, is not *clear*. By and by we discover the simple, permanent elements of the mosaic, out of which we can mentally construct the whole province. When we have reached a point where we can discover everywhere the same facts, we no longer feel lost in this province; we comprehend it without effort; it is *explained* for us....

Those elements of an event which we call 'cause and effect' are certain salient features of it, which are important for its mental reproduction. Their importance wanes and the attention is transferred to fresh characters the moment the event or experience in question becomes familiar. If the connexion of such features strikes us as a necessary one, it is simply because the interpolation of certain intermediate links with which we are very familiar, and which possess, therefore, higher authority for us, is often attended with success in our explanations. That *ready* experience fixed in the mosaic of the mind with which we meet new events, Kant calls an innate concept of the understanding (*Verstandesbegriff*).

The grandest principles of physics, resolved into their elements, differ in no wise from the descriptive principles of the natural historian....

When a geometer wishes to understand the form of a curve, he first resolves it into small rectilinear elements. In doing this, however, he is fully aware that these elements are only provisional and arbitrary devices for comprehending in parts what he cannot comprehend as a whole. When the law of the curve is found he no longer thinks of the elements. Similarly, it would not become physical science to see in its self-created, changeable, economical tools, molecules and atoms, realities behind phenomena, forgetful of the lately acquired sapience of her older sister, philosophy, in substituting a mechanical mythology for the old animistic or metaphysical scheme, and thus creating no end of suppositious problems. The atom must remain a tool for representing phenomena, like the functions of mathematics. Gradually, however, as the intellect, by contact with its subject-matter, grows in discipline, physical science will give up its mosaic play with stones and will seek out the boundaries and forms of the bed in which the living stream of phenomena flows. The goal which it has set itself is the *simplest* and *most economical* abstract expression of facts.

84. From HENRI POINCARÉ'S *Science and Hypothesis* (1902)

[Poincaré maintains that scientific theories merely express relations between phenomena and hence that, although superseded by others with different imagery, they may still retain their usefulness.]

The ephemeral nature of scientific theories takes by surprise the man of the world. Their brief period of prosperity ended, he sees them abandoned one after another; he sees ruins piled upon ruins; he predicts that the theories in fashion to-day will in a short time succumb in their turn, and he concludes that they are absolutely in vain. This is what he calls the *bankruptcy of science*.

His scepticism is superficial; he does not take into account the object of scientific theories and the part they play, or he would understand that the ruins may be still good for something. No theory seemed established on firmer ground than Fresnel's, which attributed light to the movements of the ether. Then if Maxwell's theory is to-day preferred, does that mean that Fresnel's work was in vain? No; for Fresnel's object was not to know whether there really is an ether, if it is or is not formed of atoms, if these atoms really move in this way or that; his object was to predict optical phenomena.

This Fresnel's theory enables us to do to-day as well as it did before Maxwell's time. The differential equations are always true, they may be always integrated by the same methods, and the results of this integration still preserve their value. It cannot be said that this is reducing physical theories to simple practical recipes; these equations express relations, and if the equations remain true, it is because the relations preserve their reality. They teach us now, as they did then, that there is such and such a relation between this thing and that; only, the something which we then called *motion*, we now call *electric current*. But these are merely names of the images we substituted for the real objects which Nature will hide for ever from our eyes. The true relations between these real objects are the only reality we can attain, and the sole condition is that the same relations shall exist between these objects as between the images we are forced to put in their place. If the relations are known to us, what does it matter if we think it convenient to replace one image by another?

85. From ALBERT EINSTEIN's lecture, *On the Method of Theoretical Physics* (1933)

[Einstein maintains that physical theories are free inventions of the mind and not unique inductions from the facts.]

Let us now cast an eye over the development of the theoretical system, paying special attention to the relations between the content of the theory and the totality of empirical fact. We are concerned with the eternal antithesis between the two inseparable components of our knowledge, the empirical and the rational, in our department.

We reverence ancient Greece as the cradle of western science. Here for the first time the world witnessed the miracle of a logical system which proceeded from step to step with such precision that every single one of its propositions was absolutely indubitable—I refer to Euclid's geometry. This admirable triumph of reasoning gave the human intellect the necessary confidence in itself for its subsequent achievements. If Euclid failed to kindle your youthful enthusiasm, then you were not born to be a scientific thinker.

But before mankind could be ripe for a science which takes in the whole of reality, a second fundamental truth was needed, which only became common property among philosophers with the advent of Kepler and Galileo. Pure logical thinking cannot yield us any knowledge of the empirical world; all knowledge of reality starts from experience and ends in it. Propositions arrived at by purely logical means are completely empty as regards reality. Because Galileo saw this, and particularly because he drummed it into the scientific world, he is the father of modern physics—indeed, of modern science altogether.

If then, experience is the alpha and the omega of all our knowledge of reality, what is the function of pure reason in science?

A complete system of theoretical physics is made up of concepts, fundamental laws which are supposed to be valid for those concepts and conclusions to be reached by logical deduction. It is these conclusions which must correspond with our separate experiences; in any theoretical treatise their logical deduction occupies almost the whole book.

This is exactly what happens in Euclid's geometry, except that there the fundamental laws are called axioms and there is no question of the conclusions having to correspond to any sort of experience. If, however, one regards Euclidean geometry as the science of the possible mutual relations of practically rigid bodies in space, that is to say, treats it as a physical science, without abstracting from its original empirical content, the logical homogeneity of geometry and theoretical physics becomes complete.

We have thus assigned to pure reason and experience their places in a theoretical system of physics. The structure of the system is the work of reason; the empirical contents and their mutual relations must find their representation in the conclusions of the theory. In the possibility of such a representation lie the sole value and justification of the whole system, and especially of the concepts and fundamental principles which underlie it. Apart from that, these latter are free inventions of the human intellect, which cannot be justified either by the nature of that intellect or in any other fashion *a priori*.

These fundamental concepts and postulates, which cannot be further reduced logically, form the essential part of a theory, which reason cannot touch. It is the grand object of all theory to make these irreducible elements as simple and as few in number as possible, without having to renounce the adequate representation of any empirical content whatever.

The view I have just outlined of the purely fictitious character of the fundamentals of scientfic theory was by no means the prevailing one in the eighteenth and nineteenth centuries. But it is steadily gaining ground from the fact that the distance in thought between the fundamental concepts and laws on one side and, on the other, the conclusions which have to be brought into relation with our experience grows larger and larger, the simpler the logical structure becomes—that is to say, the smaller the number of logically independent conceptual elements which are found necessary to support the structure.

Newton, the first creator of a comprehensive, workable system of theoretical physics, still believed that the basic concepts and laws of his system could be derived from experience. This is no doubt the meaning of his saying, *hypotheses non fingo*.

Actually the concepts of time and space appeared at that time to present no difficulties. The concepts of mass, inertia, and force, and the laws connecting them, seemed to be drawn directly from experience. Once this basis is accepted, the expression for the force of gravitation appears derivable from experience, and it was reasonable to expect the same in regard to other forces.

We can indeed see from Newton's formulation of it that the concept of absolute space, which comprised that of absolute rest, made him feel uncomfortable; he realized that there seemed to be nothing in experience corresponding to this last concept. He was also not quite comfortable about the introduction of forces operating at a distance. But the tremendous practical success of his doctrines may well have prevented him and the physicists of the eighteenth and nineteenth centuries from recognizing the fictitious character of the foundations of his system.

The natural philosophers of those days were, on the contrary, most of them possessed with the idea that the fundamental concepts and postulates of physics were not in the logical sense free inventions of the human mind but could be deduced from experience by 'abstraction'—that is to say, by logical means. A clear recognition of the erroneousness of this notion really only came with the general theory of relativity, which showed that one could take account of a wider range of empirical facts, and that, too, in a more satisfactory and complete manner, on a foundation quite different from the Newtonian. But quite apart from the question of the superiority of one or the other, the fictitious character of fundamental principles is perfectly evident from the fact that we can point to two essentially different principles, both of which correspond with experience to a large extent; this proves at the same time that every attempt at a logical deduction of the basic concepts and postulates of mechanics from elementary experiences is doomed to failure.

If, then, it is true that the axiomatic basis of theoretical physics cannot be extracted from experience but must be freely invented, can we ever hope to find the right way? Nay, more, has this right way any existence outside our illusions? Can we hope to be guided safely by experience at all when there exist theories (such as classical mechanics) which to a large extent do justice to experience, without getting to the root of the matter? I answer without hesitation that there is, in my opinion, a right way, and that we are capable of finding it. Our experience hitherto justifies us in believing that nature is the realization of the simplest conceivable mathematical ideas. I am convinced that we can discover by means of purely mathematical constructions the concepts and the laws connecting them with each other, which furnish the key to the understanding of natural phenomena. Experience may suggest the appropriate mathematical concepts, but they most certainly cannot be deduced from it. Experience remains, of course, the sole criterion of the physical utility of a mathematical construction. But the creative principle resides in mathematics. In a certain sense, therefore, I hold it true that pure thought can grasp reality, as the ancients dreamed.

86. From ARTHUR EDDINGTON'S *The Philosophy of Physical Science* (1938)

[Eddington's main thesis is that 'all the laws of nature that are usually classed as fundamental can be foreseen wholly from epistemological considerations. They correspond to *a priori* knowledge, and are therefore wholly subjective.' In the following passage he banteringly suggests that most physicists pay lip service to this idea but do not really believe it.]

I am about to turn from the scientific to the philosophical setting of scientific epistemology. This is accordingly a suitable place at which to make a comparison with the most commonly accepted view of scientific philosophy. The following statement is fairly typical: That science is concerned with the rational correlation of experience rather than with the discovery of fragments of absolute truth about an external world is a view which is now widely accepted.[1]

I think that the average physicist, in so far as he holds any philosophical view at all about his science, would assent. The phrase 'rational correlation of experience' has a savour of orthodoxy which makes it a safe gambit for applause. The repudiation of more adventurous aims gives a comfortable feeling of modesty—all the more agreeable if we fancy that someone else is being told off. For my own part I accept the statement, provided that 'science' is understood to mean 'physics'. It has taken me nearly twenty years to accept it; but by steady mastication during that period I have managed to swallow it all down bit by bit. Consequently I am rather flabbergasted by the light-hearted way in which this pronouncement, carrying the most profound implications both for philosophy and for physics, is commonly made and accepted.

I have no serious quarrel with the average physicist over his philosophical creed—except that he forgets all about it in practice. My puzzle is why a belief that physics is concerned with the correlation of experience and not with absolute truth about the external world should usually be accompanied by a steady refusal to treat theoretical physics as a description of correlations of experience and an insistence on treating it as a description of the contents of an absolute objective world. If I am in any way heterodox, it is because it seems to me a consequence of accepting the belief, that we shall get nearer to whatever truth is to be found in physics by seeking and employing conceptions suitable for the

[1] Unsigned review, *Phil. Mag.* **25** (1938), 814.

expression of correlations of experience instead of conceptions suitable for the description of an absolute world.

The statement evidently means that the methods of physics are incapable of discovering fragments of absolute truth about an external world; for we should have no right to withold from mankind the absolute truth about the external world if it were within our reach. If the laboratories, built and endowed at great expense, could assist in the discovery of absolute truth about the external world, it would be reprehensible to discourage their use for this purpose. But the assertion that the methods of physics cannot reveal absolute (objective) truth or even fragments of absolute truth, concedes my main point that the knowledge obtained by them is wholly subjective. Indeed it concedes it far too readily; for the assertion is one that ought only to be made after prolonged investigation. As I have pointed out, sciences other than physics and chemistry are not so limited in their scope. The discovery of unmistakeable signs of intelligent life on another planet would be hailed as an epoch-making astronomical achievement; it can scarcely be denied that it would be the discovery of a fragment of absolute truth about the world external to us.

Keeping to physics, the commonly accepted scientific philosophy is that it is not concerned with the discovery of absolute truth about the external world, and its laws are not fragments of absolute truth about the external world, or, as I have put it, they are not laws of the objective world. What then are they, and how is it that we find them in our correlations of experience? Until we can see, by an examination of the procedure of correlation of our observational experience, how these highly complex laws have got into it subjectively, it seems premature to accept a philosophy which cuts us off from all other possible explanations of their origin. This is the examination that we have been conducting.

The end of our journey is rather a bathos after so much toil. Instead of struggling up to a lonely peak, we have reached an encampment of believers, who tell us 'That is what we have been asserting for years'. Presumably they will welcome with open arms the toilworn travellers who have at last found a resting place in the true faith. All the same I am a bit dubious about that welcome. Perhaps the assertion, like many a religious creed, was intended only to be recited and applauded. Anyone who *believes* it is a bit of a heretic.

87. From RAPHAEL DEMOS' article, *Doubts about Empiricism* (1947)

[This is part of a light-hearted, but penetrating, attack on current scientific philosophies which are based on empiricism, i.e. the belief that all knowledge must be derived from sense data.]

My beliefs during the first stage of my philosophical career were a mixed brew of ingredients taken from the Greek and Christian traditions. My tastes were conservative and even reactionary. I believed in the reality of substance, material and mental; I held that there are universal and necessary connections in nature which can be known. In short, I was a naïve objectivist about things and about structures. I was a realist about values too. I believed that there are such traits in nature as good and bad, right and wrong, beauty and ugliness, independently of my preferences. I was convinced further that goodness was a supreme causal agent; that—to paraphrase the familiar quotation—righteousness is power. I believed in God. With Plato (in the *Phaedo*) I maintained that causality is not only efficient but final too; that nature exhibits both a mechanical and a moral order. And these two propositions were, to my view, but twin aspects of the one proposition that nature will not deceive my expectations.

How did I come to believe all this? My convictions came out of a curious amalgam of reason, experience and faith. By experience I meant not only sense-perception but also what has been called religious experience; I also meant feeling as furnishing an acquaintance with values. By reason I meant a power which apprehends principles and laws, natural and moral. By faith I meant a faculty which provides me with my premises. Thus it was by faith primarily that I obtained my belief in God as a moral and personal power within nature and transcending it.

All this and more I cherished naïvely until I was shocked out of it by those of my friends who were philosophers of the scientific camp. They told me that I must believe nothing which did not conform with the canons of scientific method. Above all, I must have no faith in faith. Science is the determined enemy of all authority, tradition and dogmatism. I must submit my beliefs to the test of rational evidence, that is to say, to the test of the senses. I realized that neither religious 'experience' nor feelings are evidence. Thus I gave up faith altogether and modified what

I called experience and reason. Experience now was strictly identified with sensation; and reason meant either the comparison of hypotheses with the data of sense or the analysis of concepts. As my windows to the world became fewer or smaller or otherwise changed in shape, my world too became different. I decided that substance does not exist since it is not a datum of sense. I had secret misgivings of course. The view appeared to me incredible that there were qualia without things which they qualified, that there were events in which nothing materialized. It all seemed like the episode in Peter Pan's Neverland, when, after the Redskins destroyed the nest, the eggs still remained suspended in mid-air. But I swept these misgivings aside as obviously weak-minded and bourgeois. My belief in God was clearly a superstition and had to go the way of beliefs in ghosts and fairies. Values, too, were out; they were nothing more than projections of feelings, just as my religious beliefs were rationalizations of desire. I had been subjectivistic as well as dogmatic and reactionary. Certainly my sins had been many-sided and now I was truly remorseful....

And now I must make a humiliating confession. I could not really believe what I professed to believe. While my good, scientific self envisaged sense-qualia and their configurations, my bad, practical self unconsciously went on behaving as though there were material objects, other minds, necessary connections and the rest. For instance, in the summer I would still make arrangements to prepare against the cold in the winter (by buying coal) even though I knew that I did not know of any necessary connection between past and future. While I believed human beings to be nothing more than collections of material particles, essentially not different from stones, I went on treating my fellow-humans with respect as though they were souls endowed with infinite worth. I realized that I was becoming a divided personality, and a schizophrenic. I became afraid for myself, having recalled the depression to which Hume had been plunged by his empiricism. So I called on a psychoanalyst and laid my troubles before him. When, following the usual technique, he suggested that I start trains of free association with the ideas that bothered me, I babbled: 'matter, madder, hatter' and similar foolishness which I will spare the reader. The doctor's diagnosis of my troubles was all too soothing. He explained that modern empiricism and positivism were but echoes of the Puritan temper in the field of thought. Puritanism, of course, is the great disease of the psyche. Just as religious puritans starved their passions, so had I starved my animal beliefs. Empiricism, he said, is the asceticism of the intellect. He asked me to compare my earlier pagan world, populated with full-blooded substances, to my present austere world of threadbare and anemic sense-data. Empiricism had transformed my flaming garden into a desert. My doctor went so far as to compare me in my present reduced state of belief to one of those victims of bombed-out areas who could carry all their goods in a satchel. The way back to health was obvious: let me overthrow my empirical censor and indulge my natural propensities for belief to the full.

What comforting and yet what poisonous advice! The doctor was actually proposing that I should let my desires guide my beliefs! No, no, I cried, rushing out of his office; better suffer with doubt than prosper with faith. Like John Bunyan I now felt with relief that a great burden had fallen from my shoulders. I was free, free from fear and superstition. To paraphrase one of my masters: there were no longer any spirits, essences, transcendental ideas, immaterial angels and principles to haunt my life with their terrors. The important thing was security. Condemned as I was to inhabit a waste land, at least I knew I was safe; where there is nothing, nothing can hurt me.

Suddenly, I had a bright idea. What had caused me—I asked myself—to reject my earlier common sense and so to be mired in doubt? Why, the conception of scientific method and the generalized doctrine of empiricism. In short, I had been plunged into the darkness of doubt by the glare of certainty. Here, then, is a shaft of light, I said to myself; let me turn to its source for life and heat. Imagine my disappointment when I discovered that the source of the light was not the sun, not even the other side of the moon, but some smoky sickly fire concocted by witches. The doctrines I just cited were nothing more than uncriticized beliefs. My philosophical friends from the naturalistic camp had taken over without questioning the conviction that science is knowledge and that it is the only knowledge; and that sense-perception is our only contact with reality. When I heard them urging everybody to think in philosophy as the scientist does in his field, I smelled an appeal to a new and a more terrible authority....

I had been taught that, unlike religion, science takes nothing on faith. I accepted that statement on faith, but when I examined scientific procedure I was disillusioned. Science establishes its generalizations by an appeal to the inductive principle.[1] Now, what is the evidence for the latter? Not experience surely—Hume's word is final on that point. Scientists check all theories by experience, save the one overarching theory of the uniformity of nature. I was struck with the fact that my naturalist friends proclaimed themselves empiricists when they could offer no empirical evidence for their basic premise. I was even more struck with the fact that while they proclaimed themselves opponents of all uncriticized belief, they had failed to submit their own theory of knowledge to a critical scrutiny. When one of them happened to defend commitment to the inductive principle as a sort of desperate gamble, I was vaguely reminded of Pascal's wager as an argument for the existence of God. Naturalists condescendingly suggest that people resort to the idea of God because it relieves them from anxiety. In the same spirit, it might be suggested that people resort to the principle of the uniformity of nature because it makes them feel at home in the universe, by picturing the latter as tidy and well-behaved. Should I be told that the inductive principle is validated pragmatically because it has worked so far, I would retort that so to argue is to beg the question. What I want to know is whether the principle will work in the future; and, unless I assume the inductive principle, the fact that it has worked in the past goes for nothing. . . .

From Hume down (not to mention Sextus Empiricus) philosophers have made the stupendous assumption that only what is given *to sense* is given. My point at the moment is not that they are wrong, or even that their arguments are unsound; it is that they give no arguments at all. They simply use the words experience and sense-experience as equivalent, without indicating that a definite transition has been made. Surely when the point is so far-reaching in its consequences, it should have been supported by reasons. Of course, all conscious processes are experiences in the initial sense that they are data of conscious awareness. But that is not what is in question. The point rather is that among all the conscious experiences, only sensations are directly cognitive, or at any rate, furnish data for knowledge. Here are sensations with their qualia, feelings with their values,

[1] That it is possible to obtain a valid generalization from particular cases.

visions of God, rational insights: on what basis have these philosophers excluded (as subjective) every experience but that of sense? And yet Plato and others had held sense to be illusory; there was a real issue to be met....

It has been urged in defence of empiricism that sensa satisfy the criterion of intersubjectivity, as other conscious experiences (like feelings and religious insights) do not. But on what grounds is intersubjectivity set up as a criterion? Surely it is reasonable to expect that when somebody like an artist has a unique perspective, his data too would be unique to himself, and still be veridical data. Intersubjectivity (along with other criteria such as simplicity, measurability and predictive potency) is simply an arbitrary rule to control the cognitive behaviour of the scientists; it is part of their *mores*. The tribes of the South Pacific practise polygamy; some nations practise communism; and the scientists have their own peculiar rules to which they conform in their activity of believing....

My investigation of positivism was more brief because the doctrine itself is still in flux. First, I asked myself what proof was offered for the proposition that all meaning is empirical. None that I could find; none had even been attempted so far as I could see. Surely no empirical justification for the positivist doctrine could be found in the traditional-historical use of the term meaning; in fact, many philosophers had used the term to mean non-empirical meaning. Otherwise, what was all the fuss about? Did the positivists maintain that only when meaning is empirical can it be clear and exact? Granted; then the positivists lay down a demand for clarity and exactness. And if so, they are simply pressing upon us their arbitrary preferences and valuations; they are poets, and their place is in the empyrean, not down here in the valleys of prosaic philosophizing. Similarly, my operationalist friends insisted that religion *ought* to follow the example of science and to restrict itself to terms whose meaning is operational. Again, I was unable to fathom the operational meaning of the categorical imperatives proclaimed by my friends....

With this I concluded my review of my new beliefs; the upshot was that I had been made to move out of one church in order to be forced into another. Just as Luther, after denying the authority of the Pope, ended up by affirming the authority of the Bible, my new friends set up their own dogmas in place of those of religion.

Science was the new faith, and its credo could be stated as follows:

1. I believe in the principle of induction, omnipresent and omnipotent.

2. I believe in the uniqueness of sensation as an insight into the real.

3. I believe in Occam's razor as the mediating principle between theories and facts.

Scholium: These three principles are indivisible; three in one, one in three.

4. I believe in the intersubjective communion of the scientists-saints.

5. I believe in the coming of the Kingdom of Knowledge when all events will be systematically explained.

6. I believe in the resurrection of time from the tomb of the past, in a body which is spiritual because it is a construct of the mind.

7. I believe the above articles to be the fixed and final faith forever. Failure to accept them is a failure of nerve.

88. From VLADIMIR ILYICH LENIN'S *Materialism and Empirio-Criticism* (1908)

[Lenin upholds materialism and pours scorn on the idealism of Kant and Mach.]

In his *Ludwig Feuerbach*, Engels declares that the fundamental philosophical trends are materialism and idealism. Materialism regards nature as primary and spirit as secondary; it places being first and thought second. Idealism holds the contrary view. This root distinction between the 'two great camps' into which the philosophers of the 'various schools' of idealism and materialism are divided Engels takes as the cornerstone, and he directly charges with 'confusion' those who use the terms idealism and materialism in any other way....

Engels continues:

'The most telling refutation of this [idealism] as of all other philosophical fancies is practice, viz., experiment and industry. If we are able to prove the correctness of our conception of a natural process by making it ourselves, bringing it into being out of its conditions and using it for our own purposes into the bargain, then there is an end of the Kantian incomprehensible. . . . The chemical substances produced in the bodies of plants and animals remained just such "things-in-themselves" until organic chemistry began to produce them one after another, whereupon the "thing-in-itself" became a thing for us, as, for instance, alizarin, the colouring matter of the madder, which we no longer trouble to grow in the madder roots in the field, but produce more cheaply and simply from coal tar. . . .'

What is the kernel of Engels' objections? Yesterday we did not know that coal tar contained alizarin. Today we learned that it does. The question is, did coal tar contain alizarin yesterday?

Of course it did. To doubt it would be to make a mockery of modern science.

And if that is so, three important epistemological conclusions follow:

(1) Things exist independently of our consciousness, independently of our perceptions, outside of us, for it is beyond doubt that alizarin existed in coal tar yesterday and it is equally beyond doubt that yesterday we knew nothing of the existence of this alizarin and received no sensations from it.

(2) There is definitely no difference in principle between the phenomenon and the thing-in-itself, and there can be no such difference. The only difference is between what is known and what is not yet known. And philosophical inventions of specific boundaries between the one and the other, inventions to the effect that the thing-in-itself is 'beyond' phenomena (Kant), or that we can or must fence ourselves off by some philosophical partition from the problem of a world which in one part or another is still unknown but which exists outside us (Hume)—all this is the sheerest nonsense, *Schrulle*, evasion, invention.

(3) In the theory of knowledge, as in every other branch of science, we must think dialectically, that is, we must not regard our knowledge as ready-made and unalterable, but must determine how *knowledge* emerges from ignorance, how incomplete, inexact knowledge becomes more complete and more exact.

Once we accept the point of view that human knowledge develops from ignorance, we shall find millions of examples of it just as simple as the discovery of alizarin in coal tar, millions of observations not only in the history of science and technology but in the everyday life of each and every one of us that illustrate the transformation of 'things-in themselves' into 'things-for-us', the appearance of 'phenomena' when our sense-organs experience a jolt from external objects, the disappearance of 'phenomena' when some obstacle prevents the action upon our sense-organs of an object which we know to exist. The sole and unavoidable deduction to be made from this—a deduction which all of us make in everyday practice and which materialism deliberately places at the foundation of its epistemology—is that outside us, and independently of us, there exist objects, things and bodies and that our perceptions are images of the external world. Mach's converse theory (that bodies are complexes of sensations) is nothing but pitiful idealist nonsense.

89. From FREDERICK ENGELS' *Dialectics of Nature* (written *c.* 1872–82, first published 1927)

[Engels states the laws of materialist dialectics.]

It is, therefore, from the history of nature and human society that the laws of dialectics are abstracted. For they are nothing but the most general laws of these two aspects of historical development, as well as of thought itself. And indeed they can be reduced in the main to three:

The law of the transformation of quantity into quality and *vice versa;*

The law of the interpenetration of opposites;

The law of the negation of the negation.

All three are developed by Hegel in his idealist fashion as mere laws of *thought*: the first, in the first part of his *Logic*, in the *Doctrine of Being*; the second fills the whole of the second and by far the most important part of his *Logic*, the *Doctrine of Essence*; finally the third figures as the fundamental law for the construction of the whole system. The mistake lies in the fact that these laws are foisted on nature and history as laws of thought, and not deduced from them. This is the source of the whole forced and often outrageous treatment; the universe, willy-nilly, is made out to be arranged in accordance with a system of thought which itself is only the produce of a definite stage of evolution of human thought. If we turn the thing round, then everything becomes simple, and the dialectical laws that look so extremely mysterious in idealist philosophy at once become simple and clear as noonday.

90. Letter by M. POLANYI to *Nature* on *The Cultural Significance of Science* (1940)

[The letter represents one of the many shots fired in the battle between those who advocate the planning of pure science by the State and those who oppose this.]

The admonition in *Nature* of December 28 to abandon 'once and for all the belief that science is set apart from all other social interests as if it possessed a peculiar holiness' expresses very precisely the opposite of what many men of science consider to be their duty. I, for one, can recognize nothing more holy than scientific truth, and consider it a danger to science and to humanity if the pursuit of pure science, regardless of society, is denied by a representative organ of science.

For the last ten years we have been presented by an influential school of thought with phrases about the desirability of a social control of science, accompanied by attacks on the alleged snobbishness and irresponsibility of scientific detachment. The 'social control of science' has proved a meaningless phrase. Science exists only to that extent to which the search for truth is not socially controlled. And therein lies the purpose of scientific detachment. It is of the same character as the independence of the witness, of the jury and of the judge; of the political speaker and the voter; of the writer and teacher and their public; it forms part of the liberties for which every man with an idea of truth and every man with a pride in the dignity of his soul has fought since the beginning of society.

This struggle is today at its height; on its outcome will depend, among other great issues of a kindred nature, whether scientific detachment and the civilization pledged to respect and cultivate pure science shall perish from this earth.

91. From C. D. DARLINGTON'S article, *Freedom and Responsibility in Academic Life* (1957)

[The great majority of scientists in the West are opposed to the planning by the State of research in pure science. It is policy in Great Britain and, perhaps to a lesser degree, in most other non-communist countries, that the universities should have the utmost possible freedom. That this too has its dangers is shown by the following.]

When the chair of physics at Oxford fell vacant in 1865 two candidates offered themselves: Hermann von Helmholtz, a notable German physicist, and Robert Clifton, an agreeable young mathematician with a gift for making instruments and a considerable estate in Lincolnshire. Of the electors two were divines who were not perhaps greatly interested in science and three were scientists who were certainly not interested in experiment. They elected Clifton. The new professor lived to a great age and for just fifty years he was successful in forbidding all new physical experiments in the Clarendon Laboratory. Helmholtz six years later became professor of physics in Berlin. He proved to be one of the great influences in the development of science, an influence stretching beyond his own country and beyond his own time.

The action of a small committee of electors who little knew what they were doing thus had a powerful influence on the development of the physical sciences both in Britain and in Germany and consequently on the course of the great wars of our own century.

BIBLIOGRAPHY
(Part I)

Books marked with an asterisk are recommended for further reading or for the library. I have included also references to which I am particularly indebted; some appear in learned journals and some are not easily purchasable. To save space I have omitted nearly all primary sources, which will be given in the companion volume of extracts.

GENERAL

Science for All, an annotated reading list for the non-specialist. Published for the National Book League by Cambridge, 1958.
Bernal, J. D. *Science in History* (London, 1954).
*Butterfield, H. *The Origins of Modern Science* (London, 1949).
*Conant, J. B. *On Understanding Science* (Mentor Book; orig. publ. 1947).
*Conant, J. B. (ed.) *Harvard Case Histories in Experimental Science* (Harvard, 1957).
Dampier, W. C. *A History of Science and its Relations with Philosophy and Religion*, 4th ed. (Cambridge, 1947).
*Hall, A. R. *The Scientific Revolution 1500–1800* (London, 1954).
Jeans, J. H. *The Growth of Physical Science* (Cambridge, 1947).
Lenard, P. *Great Men of Science* (London, 1933).
*Mason, S. F. *A Short History of the Sciences* (London, 1953).
Pledge, H. T. *Science since 1500* (London, 1940).
*Sherwood Taylor, F. *An Illustrated History of Science* (London, 1955).
*Sherwood Taylor, F. *A Short History of Science* (London, no date).
Singer, C. F. *A Short History of Scientific Ideas to 1900* (Oxford, 1959).
Westaway, F. W. *The Endless Quest* (London, 1934).
Whitehead, A. N. *Science and the Modern World* (Cambridge, 1926).
Wightman, W. P. D. *The Growth of Scientific Ideas* (London, 1950).

CHAPTER I

*Childe, V. Gordon. *Man Makes Himself* (London, 1941).
*Crombie. A. C. *Augustine to Galileo* (London, 1952).
*Farrington, B. *Greek Science* (London, 1944).
*Lilley, S. *Man, Machines and History* (London, 1948).
Samburksy, S. *The Physical World of the Greeks* (London, 1956).
*White, T. H. *The Book of Beasts* (London, 1954).

BIBLIOGRAPHY

CHAPTER 2

*Armitage, A. Sun, Stand Thou Still (London, 1947).
*Crowther, J. G. Six Great Scientists (London, 1955).
Dingle, H. The Scientific Adventure (London, 1952).
Drake, Stillman. Discoveries and Opinions of Galileo (New York, 1957).
Fahie, J. J. Galileo, His Life and Work (London, 1903).
Koestler, A. The Sleepwalkers (London, 1959).

CHAPTER 3

*Andrade, E. N. da C. Isaac Newton (London, 1950).
Cohen, I. B. Isaac Newton's Papers and Letters on Natural Philosophy (Cambridge, 1958).
Koyré, A. 'Galileo and Plato', J. Hist. Ideas (October, 1943).
*More, L. T. Isaac Newton (New York, 1934).
Randall, J. H. 'The Development of Scientific Method in the School of Padua', J. Hist. Ideas (April, 1940).

CHAPTER 4

Bayon, H. P. 'William Harvey', Annals of Science, vols. 3 and 4.
*Chauvois, L. William Harvey (London, 1957).
Fleming, D. 'William Harvey and the Pulmonary Circulation', Isis, vol. 46.
Kilgour, F. C. 'William Harvey', Sci. American (June, 1952).
Pagel, W. 'William Harvey and the Purpose of Circulation', Isis, vol. 42

CHAPTER 5

There is an enormous literature about Pascal; the best life in English is given below.
*Bishop, M. Pascal (London, 1937).
Leavenworth, I. A Methodological Analysis of the Physics of Pascal (New York, 1930).
*More, L. T. The Life and Works of the Honourable Robert Boyle (Oxford, 1944).
*The Physical Treatises of Pascal (Columbia University Press, 1937).

CHAPTER 6

*Dobell, C. Antony van Leeuwenhoek and his 'Little Animals' (London, 1932).
Locy, W. A. The Growth of Biology (London, 1925).
Nordenskiöld, E. The History of Biology (New York, 1928).
Rooseboom, M. 'Leeuwenhoek, the Man', Bull. Brit. Soc. Hist. Sci. (October, 1950).
*Singer, C. J. A Short History of Biology (Oxford, 1931).

BIBLIOGRAPHY

CHAPTER 7

*Bacon, F. *New Atlantis*.

*Clarke, G. N. *Science and Social Welfare in the Age of Newton* (Oxford, 1937).

*Descartes, R. *A Discourse on Method* (Everyman: London, 1949).

Espinasse, M. *Robert Hooke* (London, 1956).

*Farrington, B. *Francis Bacon* (London, 1951).

Gibson, Boyce A. *The Philosophy of Descartes* (London, 1932).

*Hampshire, S. *The Age of Reason* (Mentor Philosophers: New York, 1956).

Lilley, S. 'Cause and Effect in the History of Science', *Centaurus*, vol. 3 (1953).

Nef, J. U. 'The Genesis of Industrialism and of Modern Science', in *Essays in Honor of Conyers Read* (Chicago, 1953).

Ornstein, M. *The Role of Scientific Societies in the Seventeenth Century*, 3rd ed. (Chicago, 1938).

Stace, W. T. *Religion and the Modern Mind* (London, 1953).

*Stimson, D. *Scientists and Amateurs. A History of the Royal Society* (London, 1949).

Wolf, A. *A History of Science, Technology and Philosophy during the Sixteenth and Seventeenth Centuries* (London, 1935).

Zilsel, E. 'The Origin of William Gilbert's Scientific Method', *J. Hist. Ideas*, vol. 2 (1941).

Zilsel, E. 'The Sociological Roots of Science', *Amer. J. Sociology* (January, 1942).

CHAPTER 8

Hartog, P. J. 'The Newer Views of Priestley and Lavoisier', *Annals of Science* (August, 1941).

*Holmyard, E. J. *Makers of Chemistry* (Oxford, 1931).

*McKie, D. *Antoine Lavoisier* (London, 1952).

Meldrum, A. N. *The Eighteenth Century Revolution in Science: The First Phase* (Calcutta, 1929).

*Partington, J. R. *A Short History of Chemistry* (London, 1948).

*Tilden, W. A. *Famous Chemists* (London, 1921).

White, J. H. *The History of the Phlogiston Theory* (London, 1932).

CHAPTER 9

Geikie, A. *The Founders of Geology* (London, 1897).

Hutton, James. '1726–1797'. Part IV of vol. LXIII, section B, of the *Proceedings of the Royal Society of Edinburgh*.

Thomas, H. Hamshaw. 'The Rise of Geology and its Influence on Contemporary Thought', *Annals of Science* (July, 1947).

Zittel, K. A. von. *History of Geology and Palaeontology*, trans. M. M. Ogilvie-Gordon (London, 1905).

CHAPTER 10

*Berlin, I. *The Age of Enlightenment* (Mentor Philosophers: New York, 1956).

Cobban, A. 'The Enlightenment', *The New Cambridge Modern History (1713–63)*.

*Lindsay, A. D. *The Philosophy of Immanuel Kant* (The People's Books: London, 1919).

*Morris, C. R. *Locke, Berkeley, Hume* (Oxford, 1931).

Sherrington, C. *Goethe on Nature and on Science* (Cambridge, 1942).

Wolf, A. *A History of Science, Technology and Philosophy in the Eighteenth Century* (London, 1938).

CHAPTER 11

Dalton, J. *A New System of Chemical Philosophy* (London, 1842).

Gregory, J. C. *A Short History of Atomism* (London, 1931).

Meldrum, A. N. *Avogadro and Dalton* (Aberdeen, 1904).

Meldrum, A. N. *The Development of the Atomic Theory* (Bombay, 1920).

Millington, J. P. *John Dalton* (London, 1906).

Partington, J. R. 'Origins of the Atomic Theory', *Annals of Science*, vol. 4.

CHAPTER 12

Arago, F. *Biographies of Distinguished Scientific Men* (London, 1857).

Bell, A. E. *Christian Huygens* (London, 1947).

Boutry, G. A. 'Augustin Fresnel. His Time, Life and Work, 1788–1827', *Science Progress* (1948).

Crew, H. 'Young's Place in the History of the Wave Theory of Light', *J. Opt. Soc. of America* (January, 1930).

Mach, E. *The Principles of Physical Optics* (London, 1926).

*Newton, I. *Opticks* (Dover Pub. Inc.: New York, 1952).

Whewell, W. *History of the Inductive Sciences* (London, 1837).

*Wood, A. and Oldham, F. *Thomas Young* (Cambridge, 1954).

CHAPTER 13

*Crowther, J. G. *British Scientists of the Nineteenth Century* (London, 1935).

Mach, E. *The History and Root of the Principle of the Conservation of Energy*, trans. P. E. B. Jourdain (Chicago, 1911).

Reynolds, O. 'Memoir of J. P. Joule', *Mem. Proc. Manchester Lit. Phil. Soc.* ser. 4, vol. 6 (1892).

Thompson, S. P. *The Life of Wm. Thomson, Baron Kelvin of Largs* (London, 1910).

*Wood, Alex. *Joule and the Study of Energy* (London, 1925).

BIBLIOGRAPHY

CHAPTER 14

There are several lives of Faraday; the latest and one of the best is given below.

*Bragg, W. *The Story of Electromagnetism* (London, 1941).

Campbell, L. and Garnett, W. *The Life of J. C. Maxwell* (London, 1882).

Commemoration Volume James Clerk Maxwell (Cambridge, 1931).

*Kendall, J. P. *Michael Faraday* (London, 1955).

Martin, T. *Faraday's Discovery of Electromagnetic Induction* (London, 1949).

Turner, J. 'A Note on Maxwell's Interpretation of Some Attempts at Dynamical Explanation', *Annals of Science* (September, 1956).

CHAPTER 15

Findlay, A. *A Hundred Years of Chemistry* (London, 1937).

Lowry, T. M. *Historical Introduction to Chemistry* (London, 1936).

Meyer, E. von. *History of Chemistry* (London, 1891).

*Saunders, B. C. and Clark, R. E. D. *Order and Chaos in the World of Atoms* (London, revised ed., 1948).

Schorlemmer, C. *The Rise and Development of Organic Chemistry* (London, 1894).

*Sherwood Taylor, F. *A History of Industrial Chemistry* (London, 1957).

Walker, O. J. 'August Kekulé and the Benzene Problem', *Annals of Science* (January, 1939).

CHAPTER 16

*Barlow, N. *Charles Darwin and the Voyage of the 'Beagle'* (London, 1945).

*Barlow, N. (ed.) *The Autobiography of Charles Darwin 1809–1882* (London, 1958).

de Beer, Gavin. 'The Darwin–Wallace Centenary', *Endeavour* (April, 1958).

Broom, R. *Finding the Missing Link* (London, 1950).

*Carter, G. S. *A Hundred Years of Evolution* (London, 1957).

*Darwin, C. *The Origin of Species*.

*Darwin, C. *The Voyage of the 'Beagle'*.

Darwin, F. *Charles Darwin* (includes autobiography; London, 1892).

Eiseley, L. *Darwin's Century* (London, 1959).

Fothergill, P. G. *Historical Aspects of Organic Evolution* (London, 1952).

Huxley, J. *Evolution. The Modern Synthesis* (London, 1942).

Huxley, J., Hardy, A. C. and Ford, E. B. (eds.). *Evolution as a Process* (London, 1954).

Irvine, W. *Apes, Angels and Victorians*. A joint biography of Darwin and Huxley (London, 1955).

*Keith, A. *Darwin Revalued* (London, 1955).

*Moore, R. *Charles Darwin* (London, 1957).
Osman Hill, W. C. *Man's Ancestry* (London, 1954).
Simpson, G. G. *The Meaning of Evolution* (London, 1950).
Weiner, J. S. *The Piltdown Forgery* (London, 1955).

CHAPTER 17

Brown, L. 'Robert Koch', *Ann. Med. Hist.* vol. VII (1935, pp. 99, 292, 385).
Bulloch, W. *The History of Bacteriology* (Oxford, 1938).
*Crowther, J. G. *Six Great Doctors* (London, 1957).
*Dubos, R. J. *Louis Pasteur* (London, 1951).
Guthrie, D. *A History of Medicine* (London, 1945).
Walker, K. *The Story of Medicine* (London, 1954).
Webb, G. B. 'Robert Koch', *Ann. Med. Hist.* vol. IV (1932).
Winslow, C. E. A. *The Conquest of Epidemic Disease* (Princeton, 1943).

CHAPTER 18

Barnett, S. A. (ed.). *A Century of Darwin* (London, 1958). (The last chapter contains a good critique of evolutionary ethics.)
Bernal, J. D. *Science and Industry in the Nineteenth Century* (London, 1953).
Brown, A. W. *The Metaphysical Society* (New York, 1947).
Bury, J. B. *The Idea of Progress* (London, 1920).
*Huxley, T. H. and Huxley, J. *Evolution and Ethics 1893–1943* (London, 1947).
Marvin, F. S. *The Century of Hope* (Oxford, 1919).
*Sherwood Taylor, F. *The Century of Science* (London, 1941).
Tyndall, J. *Fragments of Science*, vol. II (London, 1899). (Contains the Belfast address.)
Waddington, C. H. *Science and Ethics* (London, 1942).

CHAPTER 19

*Auerbach, C. *Genetics in the Atomic Age* (London, 1956).
*Darlington, C. D. *The Facts of Life* (London, 1953).
Fisher, R. A. 'Has Mendel's Work been Rediscovered?' *Annals of Science* (April, 1936).
Hardy, A. C. 'Telepathy and Evolutionary Theory', *J. Soc. Psychical Res.* vol. XXXV (1950).
Huxley, J. *Soviet Genetics and World Science* (London, 1949).
Iltis, H. *Life of Mendel*, trans. E. and C. Paul (London, 1932).
Morgan, T. H. *The Theory of the Gene* (Yale, 1926).
Morton, A. G. *Soviet Genetics* (London, 1951).
Zirkle, C. 'Gregor Mendel and his Precursors', *Isis* (June, 1951).

BIBLIOGRAPHY

CHAPTER 20

*Andrade, E. N. da C. *The Atom and its Energy* (London, 1947).

Chalmers, T. W. *Historic Researches* (London, 1949).

*Crowther, J. G. *British Scientists of the Twentieth Century* (London, 1952).

*Curie, Eve. *Madame Curie* (London, 1938).

*Eve, A. S. *Rutherford* (Cambridge, 1939).

*Feather, N. *Lord Rutherford* (London, 1940).

Owen, G. E. 'The Discovery of the Electron', *Annals of Science* (June, 1955).

*Rayleigh, Lord. *The Life of Sir J. J. Thomson* (Cambridge, 1942).

*Soddy, F. *The Story of Atomic Energy* (London, 1949).

*Thomson, G. *The Atom* (Home Univ. Lib. revised ed., 1947).

Thomson, J. J. *Recollections and Reflections* (London, 1936).

Wood, Alex. *The Cavendish Laboratory* (Cambridge, 1946).

CHAPTER 21

*Barnett, L. *The Universe and Dr Einstein* (London, 1950).

*Einstein, A. *Ideas and Opinions* (London, 1954).

*Einstein, A. *Relativity* (London, 15th ed. revised, 1954).

*Frank, P. *Einstein* (London, 1948).

Schilpp, P. A. (ed.). *Albert Einstein: Philosopher-Scientist* (New York, 1951).

CHAPTER 22

*de Broglie, L. *The Revolution in Physics* (London, 1954).

*Heisenberg, W. *Philosophic Problems of Nuclear Science* (London, 1952).

*Planck, M. *Scientific Autobiography and other Papers* (London, 1950).

*Schrödinger, E. *Science and the Human Temperament* (London, 1935).

CHAPTER 23

Gamow, G. *The Creation of the Universe* (New York, 1952).

*Hoyle, Fred. *Frontiers of Astronomy* (London, 1955).

Hubble, E. *The Realm of the Nebulae* (Oxford, 1936).

*Jones, H. Spencer. 'Continuous Creation', *Science News* (Penguin Books), vol. 32 (May, 1954).

*Lovell, A. C. B. *The Individual and the Universe* (Oxford, 1959).

*Lyttleton, R. A. *The Modern Universe* (London, 1956).

*Whitrow, G. J. *The Structure of the Universe* (London, 1949).

CHAPTER 24

*Ayer, A. J. *Language, Truth and Logic* (London, 1936).

*Barber, B. *Science and the Social Order* (London, 1953).

Eddington, A. *The Philosophy of Physical Science* (Cambridge, 1939).

BIBLIOGRAPHY

*Frank, P. *Modern Science and its Philosophy* (Harvard, 1949).

Heath, A. E. (ed.) *Scientific Thought in the Twentieth Century* (London, 1951).

Heisenberg, W. *The Physicist's Conception of Nature* (London, 1958).

Hutten, E. H. *The Language of Modern Physics* (London, 1956).

Jewkes, J., Sawers, D. and Stillerman, R. *The Sources of Invention* (London, 1958).

Needham, J. *Time the Refreshing River* (London, 1943).

Popper, K. R. *The Logic of Scientific Discovery* (London, 1959).

*Russell, E. J. *Science and Modern Life* (London, 1955).

*Toulmin, S. *The Philosophy of Science* (London, 1953).

See also articles in the two journals, *Philosophy of Science* and *British Journal for the Philosophy of Science*.

SOURCES OF THE EXTRACTS

(Part II)

NUMERALS REFER TO EXTRACT NUMBERS

1. ARISTOTLE, *The Physics*, translated by P. H. Wicksteed and F. M. Cornford (Harvard University Press and Heinemann, 1934), Bk. I, ch. I, pp. 11–13.
2. PLUTARCH, *Lives*, Life of Marcellus, in the translation edited by John Dryden and revised by Arthur Hugh Clough, Everyman edition (Dent, London, and Dutton, New York), vol. I, pp. 471–4.
3. T. H. WHITE, *The Book of Beasts* (Jonathan Cape, 1954), pp. 7–9.
4. JONATHAN SWIFT, *Works* (London, 1784), vol. v, pp. 17 and 52.
5. COPERNICUS, 'De Revolutionibus Orbium Coelestium', translated by J. F. Dobson, assisted by S. Brodetsky, from Royal Astronomical Society, *Occasional Notes* (May 1947).
6. GALILEO, letter to Madame Christina of Lorraine, Grand Duchess of Tuscany, 'Concerning the Use of Biblical Quotations in Matters of Science', translated by Stillman Drake; from *Discoveries and Opinions of Galileo* (Doubleday and Company Inc., New York, 1957), pp. 175, 177, 181–2.
7. GALILEO, *Dialogue concerning the Two Chief World Systems— Ptolemaic and Copernican*, translated by Stillman Drake (University of California Press, 1935), pp. 321–6.
8. KEPLER, letters to Herwart; the first letter is from A. Koestler, *The Sleepwalkers* (Hutchinson, 1959), p. 340; the second letter is from Carola Baumgardt, *Johannes Kepler: Life and Letters* (Gollancz, 1952), p. 74.
9. ARISTOTLE, *The Physics*, as extract I, Bk. VIII, ch. IV, pp. 315–17.
10. GALILEO, *Dialogue concerning the Two Chief World Systems— Ptolemaic and Copernican*, as extract 7, pp. 235–7.
11. GALILEO, *Dialogues concerning Two New Sciences*, translated by H. Crew and A. de Salvio (Evanston, Northwestern University Press), pp. 244–5.
12. NEWTON, *Principia*, in the translation from the Latin by Andrew Motte of 1729, revised by Florian Cajori (University of California Press, 1934), pp. 13–14, 550–2, 547.
13. ROBERT BOYLE, *A Disquisition about the Final Causes of Natural Things* (London, 1688), pp. 157–8.
14. WILLIAM HARVEY, *Exercitatio Anatomica de Motu Cordis et Sanguinis in Animalibus*, translated by Robert Willis (London, 1847), pp. 45–6.
15. TORRICELLI, letter to Michelangelo Ricci, from *The Physical Treatises of Pascal*, translated by I. H. B. and A. G. H. Spiers (Columbia University Press, New York, 1937), pp. 163–4.

16. MONSIEUR PERIER'S account to Monsieur Pascal, of the experiment performed on the Puy de Dôme, 19 September 1648, as extract 15, pp. 103–6.

17. OTTO VON GUERICKE, *Experimenta nova (ut vocantur) Magdeburgica de Vacuo Spatio*, translated by Martha Ornstein, from her book, *The Role of Scientific Societies in the Seventeenth Century* (University of Chicago Press, 1938), p. 51.

18. ROBERT BOYLE, *A Continuation of New Experiments touching the Spring and Weight of Air* (Oxford, 1669), pp. 42–5.

19. ROBERT HOOKE, *Micrographia* (London, 1665), Observation LIII.

20. LEEUWENHOEK'S letters, from Clifford Dobell, *Antony van Leeuwenhoek and his Little Animals* (John Bale, Sons and Danielsson Ltd, London, 1932), pp. 117–20, 168–70, 247–9.

21. FRANCIS BACON, *New Atlantis* (Cambridge, 1900), pp. 34–45 (abridged).

22. FRANCIS BACON, *Novum Organum*, translated by R. Ellis and James Spedding (Routledge, n.d.), p. 127.

23 RENÉ DESCARTES, 'Discours de la Méthode', translated by Elizabeth Anscombe and P. T. Geach in *Descartes, Philosophical Writings* (Nelson, 1954), pp. 15, 20–1.

24. THOMAS HOBBES, *Leviathan*, Everyman edition (Dent, London, and Dutton, New York), pp. 24, 82–3.

25. LAVOISIER, sealed note; from D. McKie, *Antoine Lavoisier* (Constable, 1952), pp. 74–5.

26. JOSEPH PRIESTLEY, 'Experiments and Observations on Different Kinds of Air', from Alembic Club Reprint No. 7, *The Discovery of Oxygen*, Part 1 (W. F. Clay, Edinburgh, 1894), pp. 6–7, 16–17, 54.

27. HENRY CAVENDISH, 'Experiments on Air', *Philosophical Transactions*, **74** (1784), pp. 134–5, 140, 150, 151–2.

28. LAVOISIER, *Traité Élémentaire de Chimie*, translated by R. Kerr (Edinburgh, 1790), pp. 78–81.

29. JAMES HUTTON, *Theory of the Earth* (Edinburgh, 1795); the first passage is from vol. 1, pp. 431 f., and the second passage from vol. 2, pp. 561–3.

30. WILLIAM SMITH, *Stratigraphical System of Organized Fossils* (London, 1817), pp. v, ix-x.

31. GEORGES CUVIER, *Discours sur la Théorie de la Terre*, translated by R. Jameson (Edinburgh, 1822), pp. 1–2, 6–7, 7–8, 14.

32. CHARLES LYELL, *Principles of Geology*, 10th ed. (John Murray, London, 1867), pp. 90–2.

33. JOHN LOCKE, *An Essay concerning Human Understanding*, Everyman edition (Dent, London, and Dutton, New York), pp. 26–7.

34. GEORGE BERKELEY, *The Principles of Human Knowledge*, 2nd ed., edited by T. E. Jessop (A. Brown and Sons Ltd, London, 1937), pp. 30–1.

35. DAVID HUME, 'An Enquiry concerning Human Understanding, from *David Hume, Theory of Knowledge*, ed. by D. C. Yalden-Thomson (Nelson, 1951), pp. 76–7, 159.

36. IMMANUEL KANT, *Critique of Pure Reason*, translated by N. K. Smith (Macmillan, 1953), pp. 127–8, 149–50.

37. DENIS DIDEROT, 'Conversation of a Philosopher with the Maréchale de X', from *Diderot, Selected Writings*, translated by Jean Stewart and Jonathan Kemp (Lawrence and Wishart, London, 1937), pp. 218, 228–9.

38. VOLTAIRE, 'A Treatise on Toleration', from *Selected Works of Voltaire*, translated by Joseph McCabe (Watts and Co., 1948), pp. 204–6.

39. *Conversations of Goethe with Eckermann and Soret*, translated by J. Oxenford (London, 1850), vol. I, pp. 106, 107, 108–10.

40. LUCRETIUS, *De Rerum Natura*, translated by Sir Robert Allison (Hatchards, 1925), pp. 46 and 53.

41. JOHN DALTON, *A New System of Chemical Philosophy* (London, 1808); the extract is taken from the edition of 1842, pp. 212–16.

42. AVOGADRO, 'Essay on a Manner of Determining the Relative Masses of the Elementary Molecules of Bodies, and the Proportions in which they enter into these Compounds', from Alembic Club Reprint No. 4, *Foundations of the Molecular Theory* (W. F. Clay, Edinburgh, 1893), pp. 28–9, 30, 33–4.

43. CHRISTIAAN HUYGENS, *Traité de la Lumière*, translated by S. P. Thompson (Macmillan, 1912), pp. 3–4.

44. NEWTON, *Opticks*, Dover Publications Inc. 1952 (based on the 4th ed. of 1730), pp. 362–3, 370, 347–8, 345–6.

45. THOMAS YOUNG, 'Reply to the animadversions of the Edinburgh Reviewers on some papers published in the Philosophical Transactions', from *Miscellaneous Works of the late Thomas Young*, ed. George Peacock (John Murray, 1855), vol. I, pp. 193–5, 201, 202–3.

46. FRESNEL, 'Mémoire sur la Diffraction de la Lumière', couronné par l'Académie des Sciences (1819), translated by A.E.E.M.; from *Œuvres complètes d'Augustin Fresnel*, tome premier (Paris, 1866), pp. 247–59.

47. JOULE, lecture 'On Matter, Living Force and Heat' (1847), from *The Scientific Papers of James Prescott Joule* (London, 1884), pp. 268–71.

48. J. R. MAYER, 'Remarks on the Forces of Inorganic Nature' (1842), translated from the German by G. C. Foster; from *Philosophical Transactions*, November 1862, pp. 375–7.

49. SADI CARNOT, *Réflexions sur la puissance motrice du feu* (1824), translated by R. H. Thurston (Macmillan, 1890), pp. 111–12, 113, 114–15.

50. FARADAY, 'On Lines of Force and the Field', taken from Michael

Faraday, *Experimental Researches in Electricity* (London, vol. III, 1855), pp. 438, 426, 447, 451.

51. J. C. MAXWELL, *A Treatise on Electricity and Magnetism* (Oxford, 1873), from the preface.
52. HEINRICH HERTZ, *Electric Waves*, published in German in 1892 and translated by D. E. Jones (Macmillan, 1893), pp. 20, 21, 27–8.
53. J. DUMAS and J. VON LIEBIG, 'Note sur l'état actuel de la Chimie organique', translated by A.E.E.M., from *Comptes Rendus de l'Académie des Sciences* (1837), tome V, pp. 567–70.
54. EDWARD FRANKLAND, 'On a New Series of Organic Bodies containing Metals', from *Philosophical Transactions*, **142** (1852), p. 440.
55. AUGUST KEKULÉ, *Über die Konstitution und die Metamorphosen der chemischen Verbindungen und über die chemische Natur des Kohlenstoffs*, translated by A.E.E.M., from Ostwald's *Klassiker*, no. 145, pp. 22–5.
56. AUGUST KEKULÉ, 'Untersuchungen über aromatische Verbindungen', translated by A.E.E.M., from Ostwald's *Klassiker*, no. 145, pp. 31–2.
57. J. B. LAMARCK, *Philosophie Zoologique*, translated by H. Elliott (Macmillan, 1914), pp. 107, 112–13, 117–18, 119–20, 122–3.
58. CHARLES DARWIN, *The Origin of Species* (John Murray, 6th ed. 1892), pp. 1, 14, 16, 28, 45, 46–7, 54, 60, 64, 97–8, 124, 134, 135, 136, 137, 276, 293, 294, 358–9, 360–1, 372–3, 387.
59. JOSEPH LISTER, 'On a New Method of Treating Compound Fracture, Abscess etc., with Observations on the Conditions of Suppuration', from *The Lancet*, 16 March 1867, p. 362.
60. LOUIS PASTEUR, 'Méthode pour prévenir la rage après morsure', translated by A.E.E.M., *Comptes Rendus de l'Académie des Sciences*, 26 October 1885, pp. 767–72.
61. ROBERT KOCH, *On Bacteriology and its Results*, translated by T. W. Hime (Baillière, Tindall and Cox, 1890), pp. 3–4, 11–12.
62. AUGUSTE COMTE, *Cours de Philosophie Positive*, translated by Harriet Martineau (London, 1853), pp. 1–2.
63. JOHN TYNDALL, 'Apology for the Belfast Address', from John Tyndall, *Fragments of Science* (Longmans Green and Co., 1899), vol. II, pp. 207–8, 218–19.
64. T. H. HUXLEY, 'On the Physical Basis of Life', from *Collected Essays*, vol. I (Macmillan, 1894), pp. 159–60.
65. ERNST HAECKEL, *Die Welträtsel*, translated by J. McCabe (Watts and Co., 1904), pp. 74, 82–3, 109.
66. HERBERT SPENCER, *First Principles* (Williams and Norgate, 6th ed. 1908), pp. 438–9.
67. AUGUST WEISMANN, *The Germ-Plasm, A Theory of Heredity*, translated from the German by W. N. Parker and H. Ronnfeldt (London, 1893), pp. xi, xii–xiii.

68. GREGOR MENDEL, paper on *Plant-Hybridisation*, translated by C. T. Druery and W. Bateson, *Mendel's Principles of Heredity* (Cambridge, 1909), pp. 309, 342–3, 344, 347.

69. T. H. MORGAN, *The Theory of the Gene* (Yale University Press, revised edition 1928), p. 25.

70. Statement by the Praesidium of the U.S.S.R. Academy of Sciences, quoted by Julian Huxley, *Nature, Lond.*, **163** (18 June 1949), pp. 935–6.

71. T. D. LYSENKO, 'The Situation in Biological Science', from *Agrobiology* (Foreign Languages Publishing House, Moscow, 1954), pp. 532–3, 537.

72. J. J. THOMSON, 'Cathode Rays', *Philosophical Magazine*, October 1897, pp. 293–4, 296–7, 310, 311, 312.

73. E. RUTHERFORD and F. SODDY, 'The Cause and Nature of Radioactivity', *Philosophical Magazine*, September 1902, pp. 394–5.

74. E. RUTHERFORD, 'Nuclear Constitution of Atoms', *Proceedings of the Royal Society*, Ser. A, **97** (1920), pp. 374–5, 379–80, 385.

75. ALBERT EINSTEIN, 'Lecture on the Theory of Relativity', from Albert Einstein, *Ideas and Opinions* (Alvin Redman Ltd, 1954), pp. 246–9.

76. MAX PLANCK, *The Origin and Development of the Quantum Theory*, translated by H. T. Clarke and L. Silberstein (Oxford, 1922), pp. 3–4, 11–13.

77. NIELS BOHR, 'Discussion with Einstein on Epistemological Problems in Atomic Physics', from *Albert Einstein, Philosopher-Scientist*, ed. P. A. Schilpp (Tudor Publishing Co., New York, fourth printing 1957), pp. 222–3.

78. WERNER HEISENBERG, 'Recent Changes in the Foundations of Exact Science', translated by F. C. Hayes, from Werner Heisenberg, *Philosophic Problems of Nuclear Science* (Faber and Faber, 1952), pp. 16–18.

79. EDWIN HUBBLE, *The Observational Approach to Cosmology* (Oxford, 1937), pp. 29–31.

80. ALBERT EINSTEIN, *The Meaning of Relativity* (Methuen, 4th ed. 1950), pp. 121–2.

81. H. BONDI, *Cosmology* (Cambridge, 1952), pp. 12, 140–1, 143–4.

82. ISAAC NEWTON, *Opticks*, as extract 44, p. 400.

83. ERNST MACH, 'The Economical Nature of Physical Enquiry', from Ernst Mach, *Popular Scientific Lectures* (Open Court Publishing Co., Chicago, 1895), pp. 192–3, 194, 198–9, 206–7.

84. HENRI POINCARÉ, *La Science et l'hypothèse*, translated by W.J.G. (Walter Scott Publishing Co. Ltd, 1905), pp. 160–1.

85. ALBERT EINSTEIN, *On the Method of Theoretical Physics*, as extract 75, pp. 271–4.

86. Sir ARTHUR EDDINGTON, *The Philosophy of Physical Science* (Cambridge, 1939), pp. 184–6.

87. RAPHAEL DEMOS, *Doubts about Empiricism*, from *Philosophy of Science* **14**, no. 3 (July 1947), pp. 203–4, 205–6, 207–8, 210, 215–16.

88. V. I. LENIN, *Materialism and Empirio-Criticism* (Lawrence and Wishart, London, 1948), pp. 95, 97, 99–100.

89. FREDERICK ENGELS, *Dialectics of Nature*, translated from the German by Clemens Dutt (Lawrence and Wishart, London, 1940), pp. 26–7.

90. M. POLANYI, letter on 'The Cultural Significance of Science', from *Nature, Lond.*, **147**, no. 3717, p. 119.

91. C. D. DARLINGTON, 'Freedom and Responsibility in Academic Life', *Bulletin of the Atomic Scientists*, **13**, no. 4 (April 1957), p. 133.

CLASSIFICATION OF THE
NATURAL SCIENCES

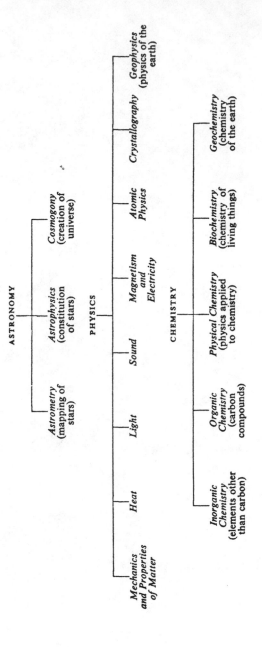

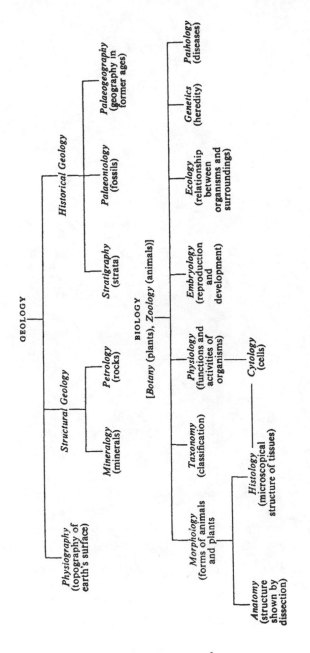

INDEX

Italic numerals indicate selections from the literature.

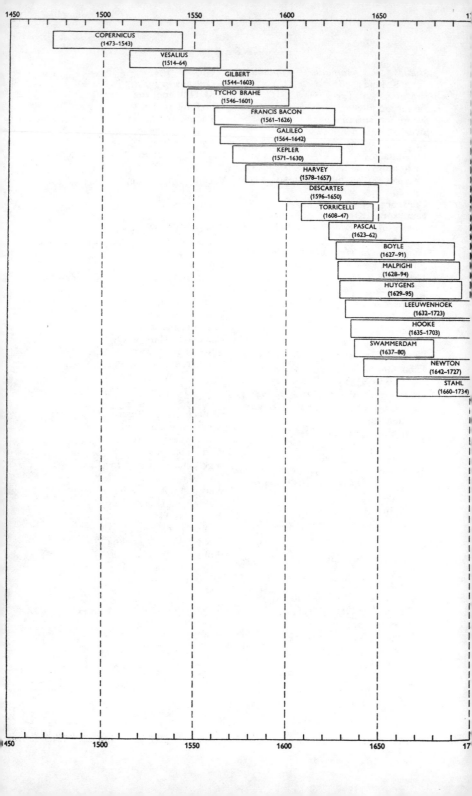

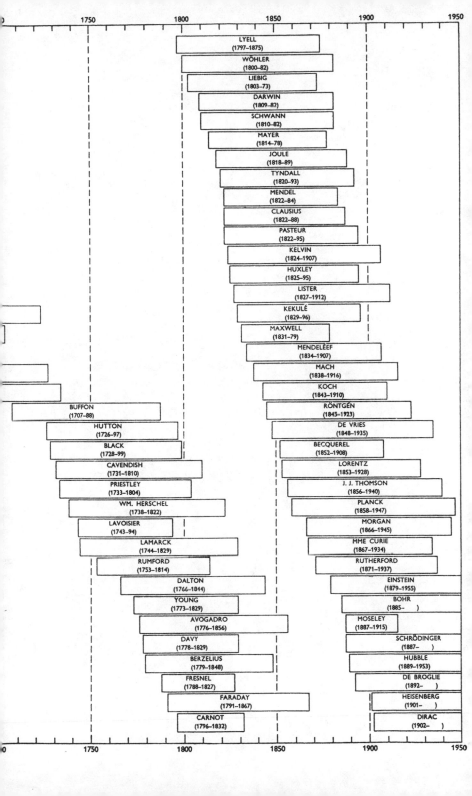